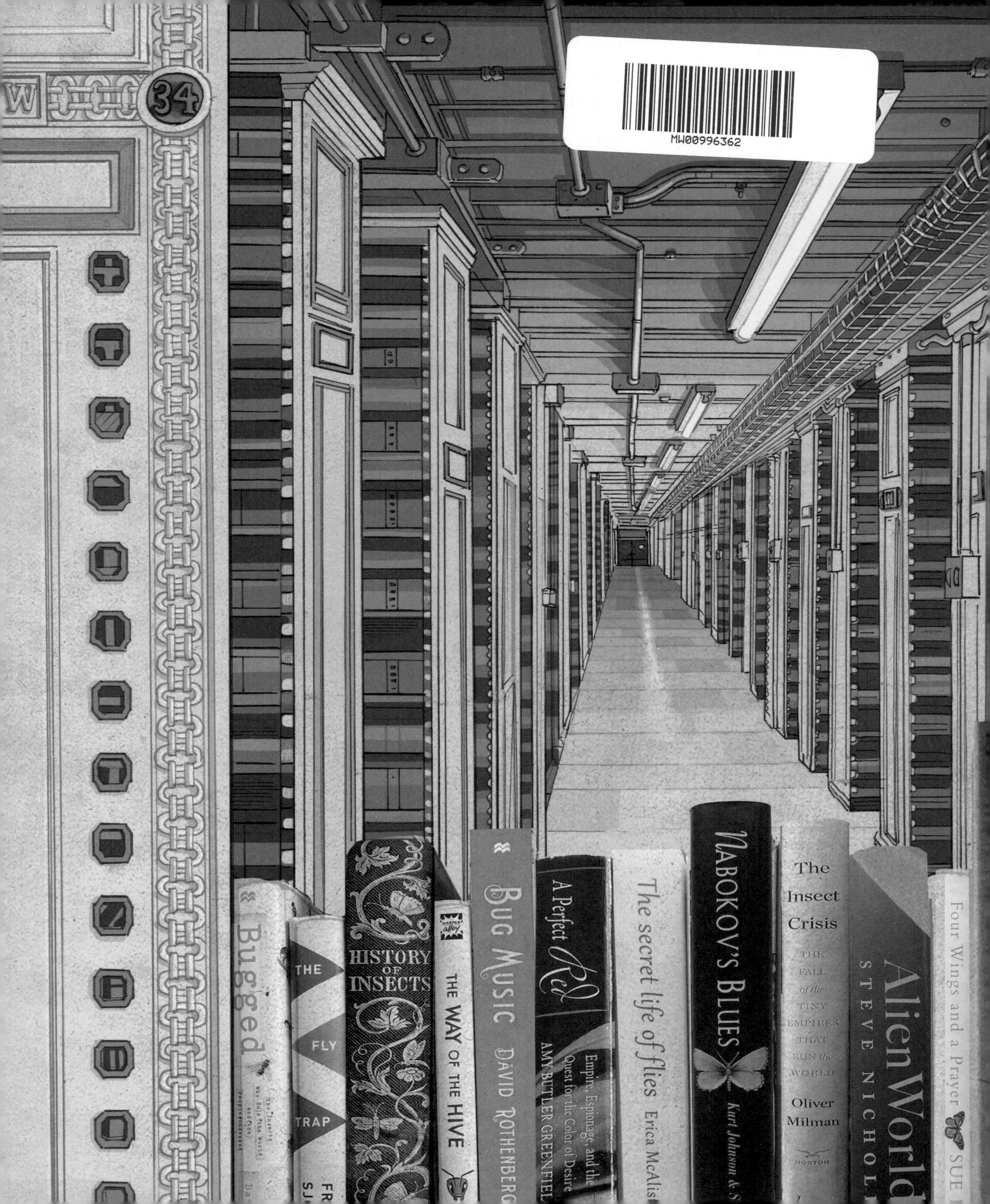
W
34
Bugged
THE
FLY
TRAP
HISTORY OF INSECTS
alley
THE WAY OF THE HIVE
BUG MUSIC
DAVID ROTHENBERG
A Perfect Red
Empire, Espionage, and the Quest for the Color of Desire
The secret life of flies
NABOKOV'S BLUES
The Insect Crisis
THE FALL of the TINY EMPIRES THAT RUN the WORLD
Oliver Milman
NORTON
Alien World
Four Wings and a Prayer

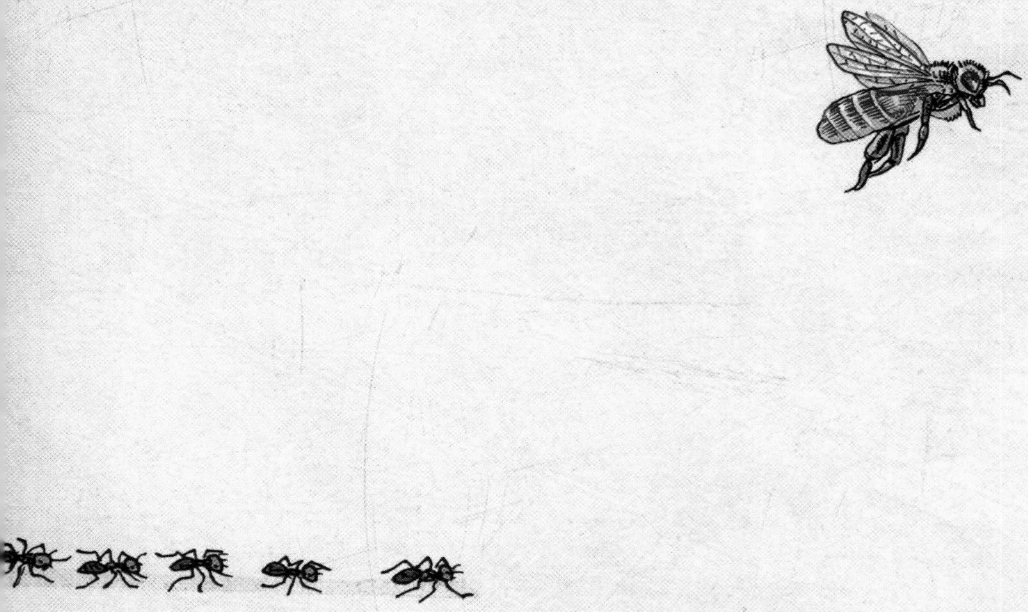

INSECT

A Natura

XV
Wunder= rau=
pen.

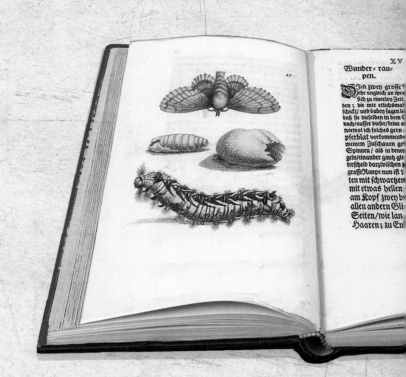

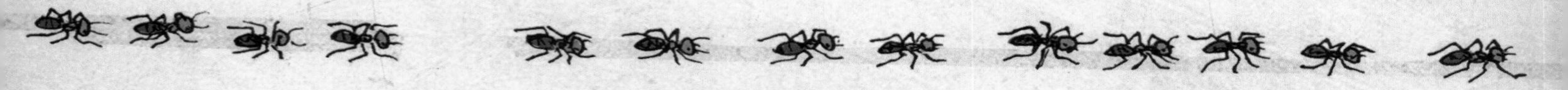

INSECTOPOI
A Natural History

Peter Kuper

INSECTOPOLIS

A Natural History

XVII.
35
Wunder-raupen.
Vermes miraculosi.
Diese zwey grosse Raupen / welche einander sehr ungleich an ihrer Gestalt und Farbe / haben sich zu einerley Zeit / und zwar im May / befunden; die mir etlichsmal einige hohe Liebhaber überschickt / und dabey sagen lassen / auch selbst zu mir gesagt / daß sie dieselben in dem Gras angetroffen: Weiß demnach / ausser dieser / keine andere Speise für sie zu melden / wiewol ich solches gern glaube / weil die am 32. Kupferblat vorkommende Raupe ebenmässig Gras bey meinem Zuschauen gessen; und sie alle drey so wol im Spinnen / als in denen Dattelkernen und Motten-vögeln / einander gantz gleich worden / also daß gar kein Unterscheid darzwischen zu verspüren war. Diese unterste grosse Raupe nun ist weiß / von oben zu beeden Seiten mit schwartzen / grossen Flecken / von unten mit etwas hellen / kleinern geziert / hat vornen am Kopf zwey braune Büschel Haare / und an allen andern Gliedern hat sie / unten an beeden Seiten / wie lange / runde Wartzen / mit braunen Haaren; zu Ende auf dem Rucken findt sich ein

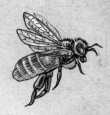

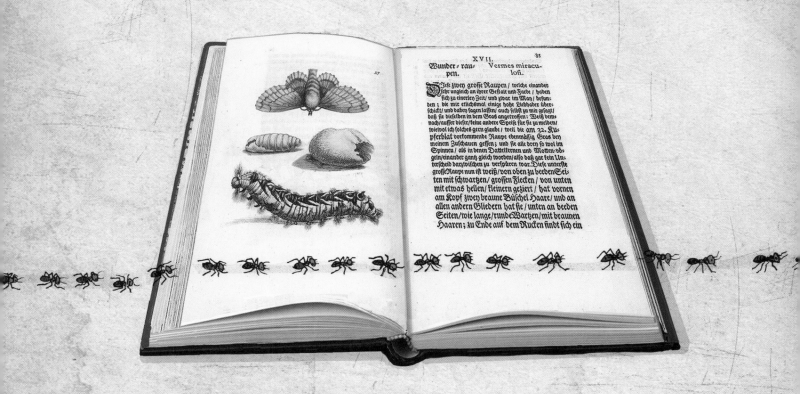

W. W. NORTON & COMPANY
Independent Publishers Since 1923

Dedicated to
entomologists
and arthropods
the world over.

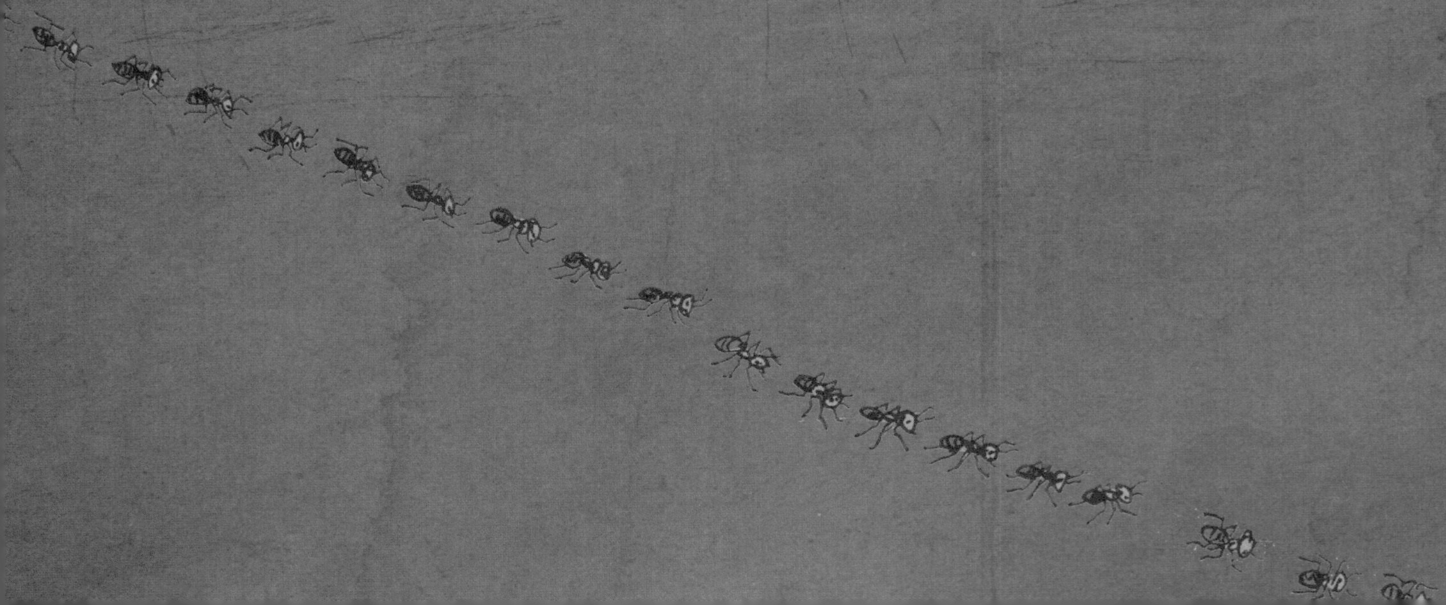

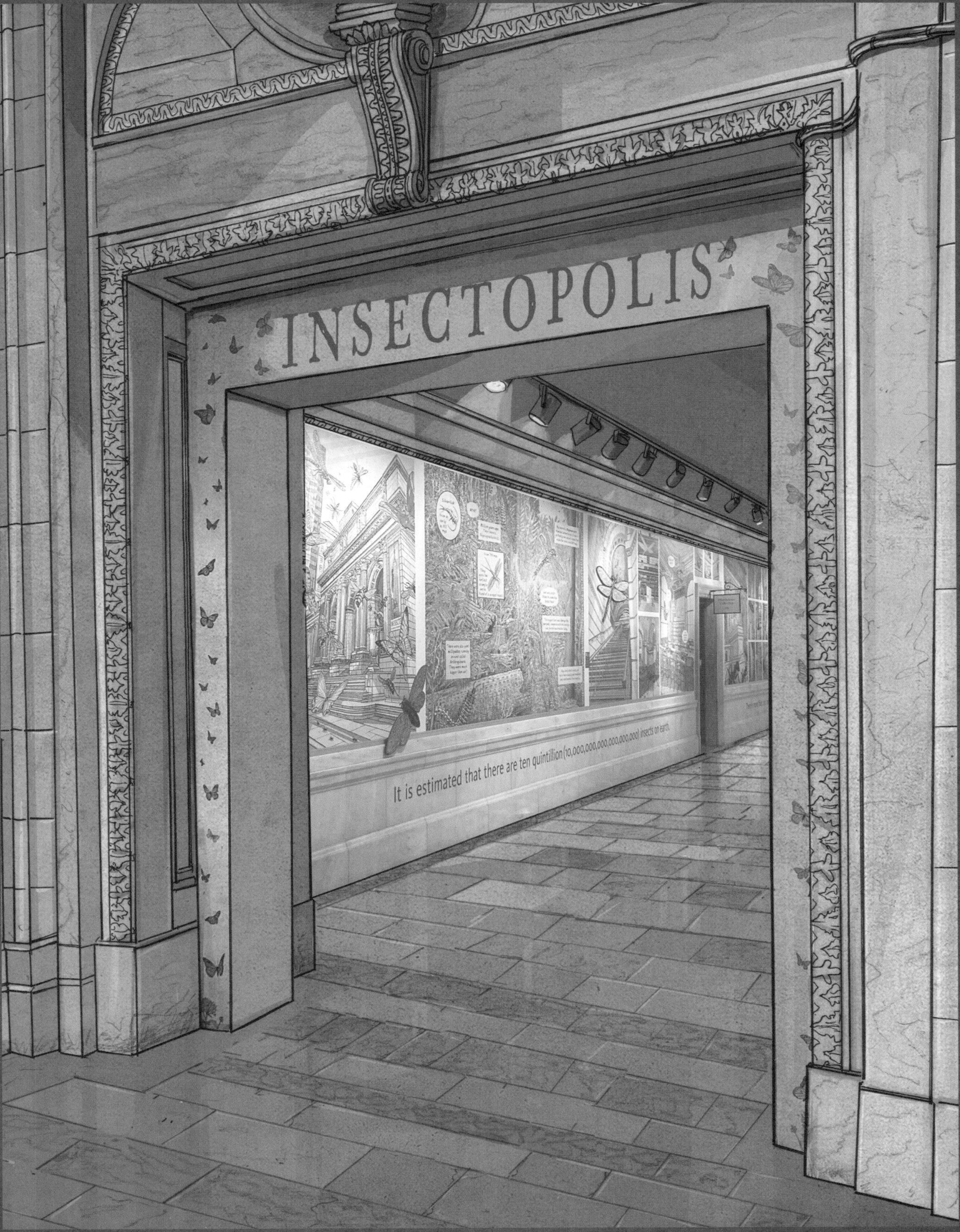
INSECTOPOLIS
It is estimated that there are ten quintillion (10,000,000,000,000,000,000) insects on earth.

If all mankind were to disappear, the world would regenerate back to the rich state of equilibrium that existed ten thousand years ago.

If insects were to vanish...

the environment would
collapse into chaos.

—E. O. Wilson
(1929–2021)

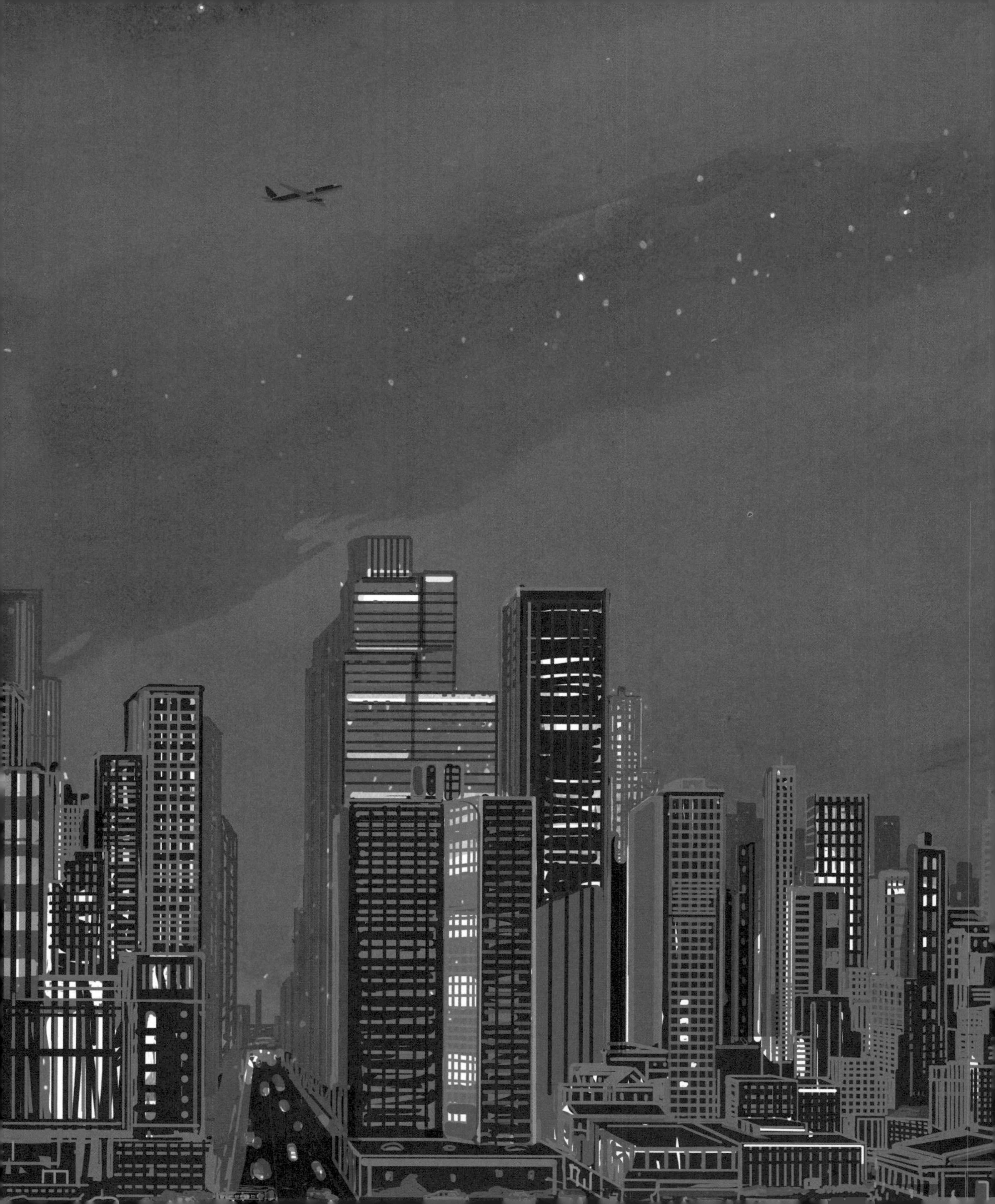

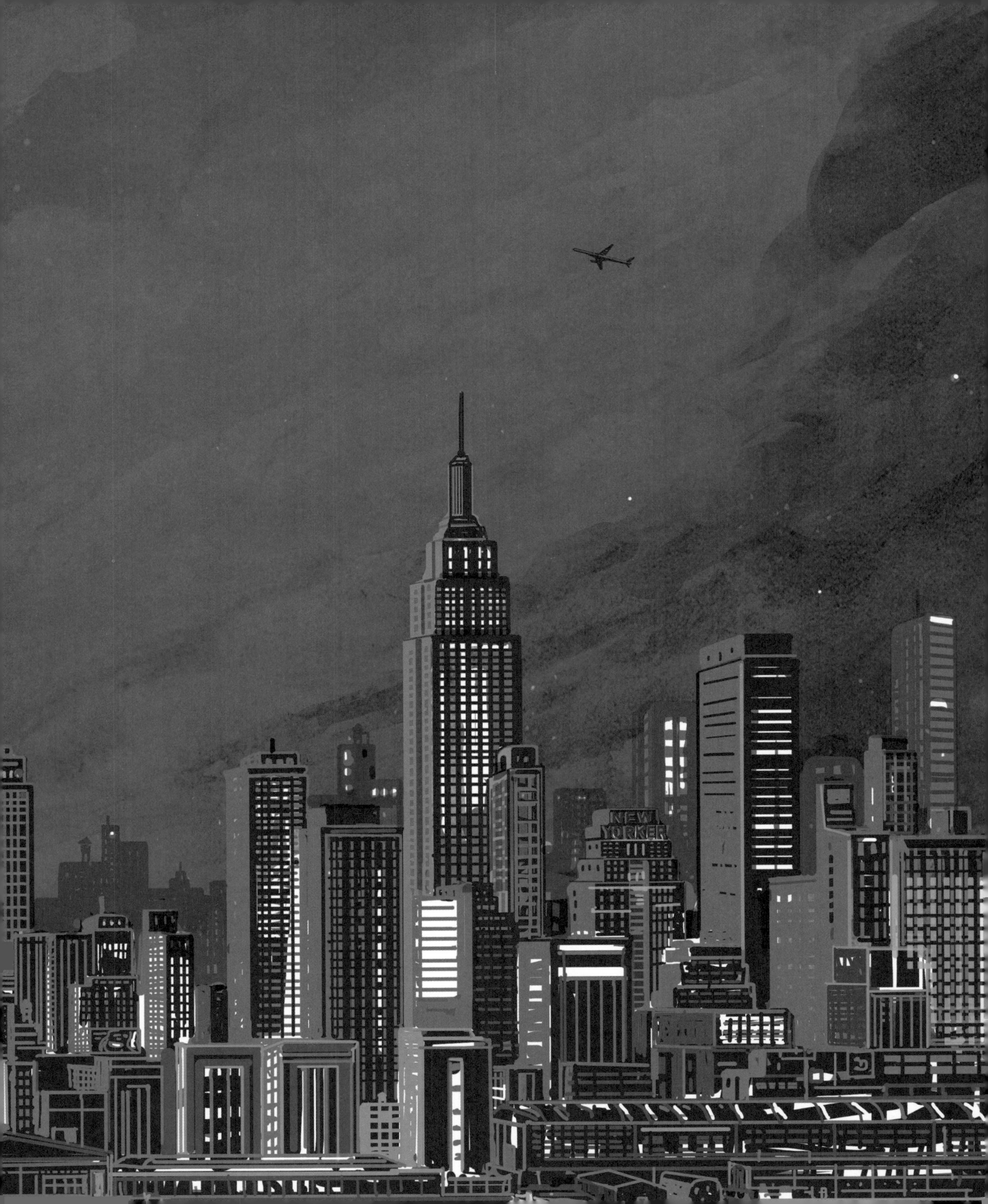
NEW
YORKER

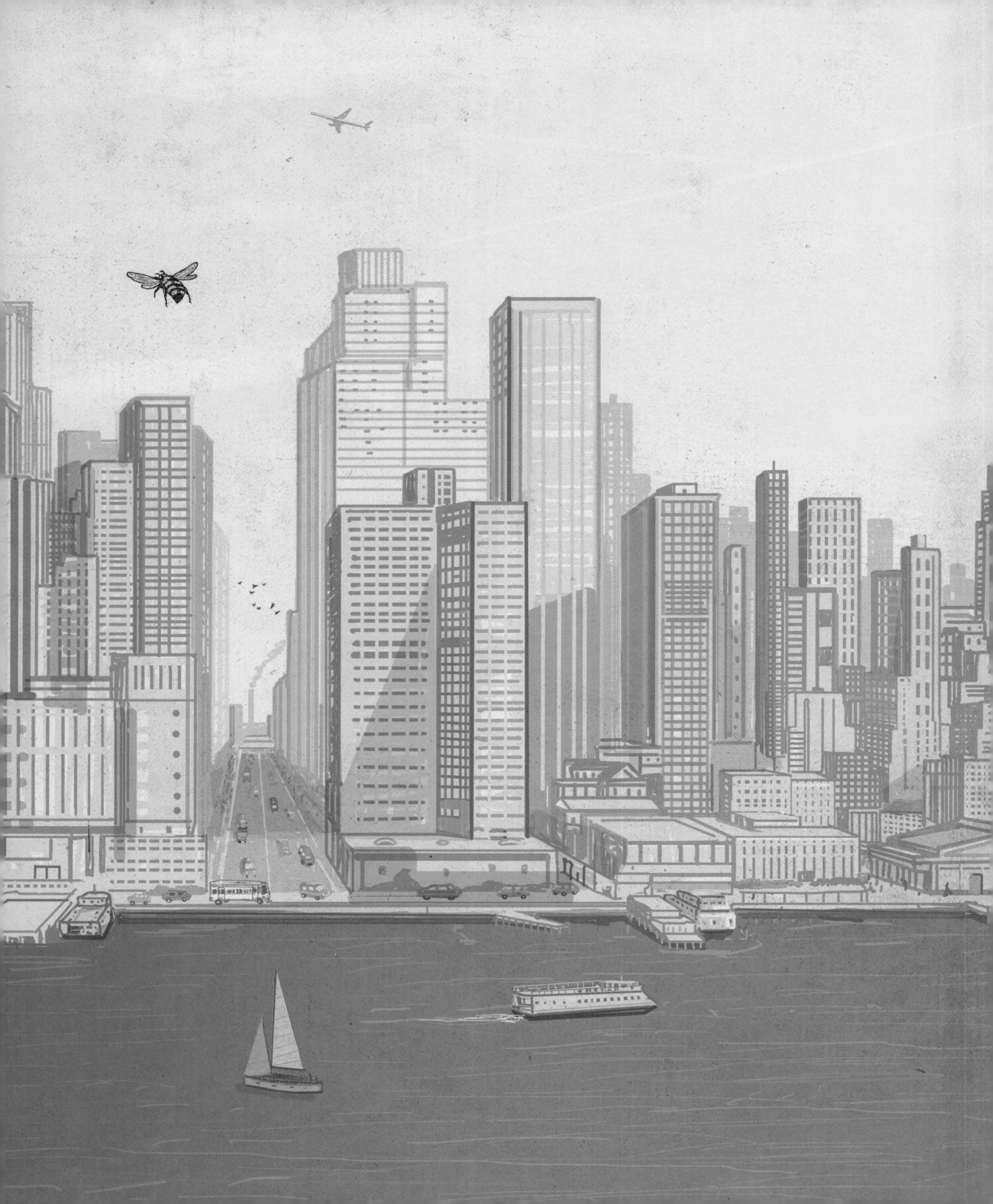

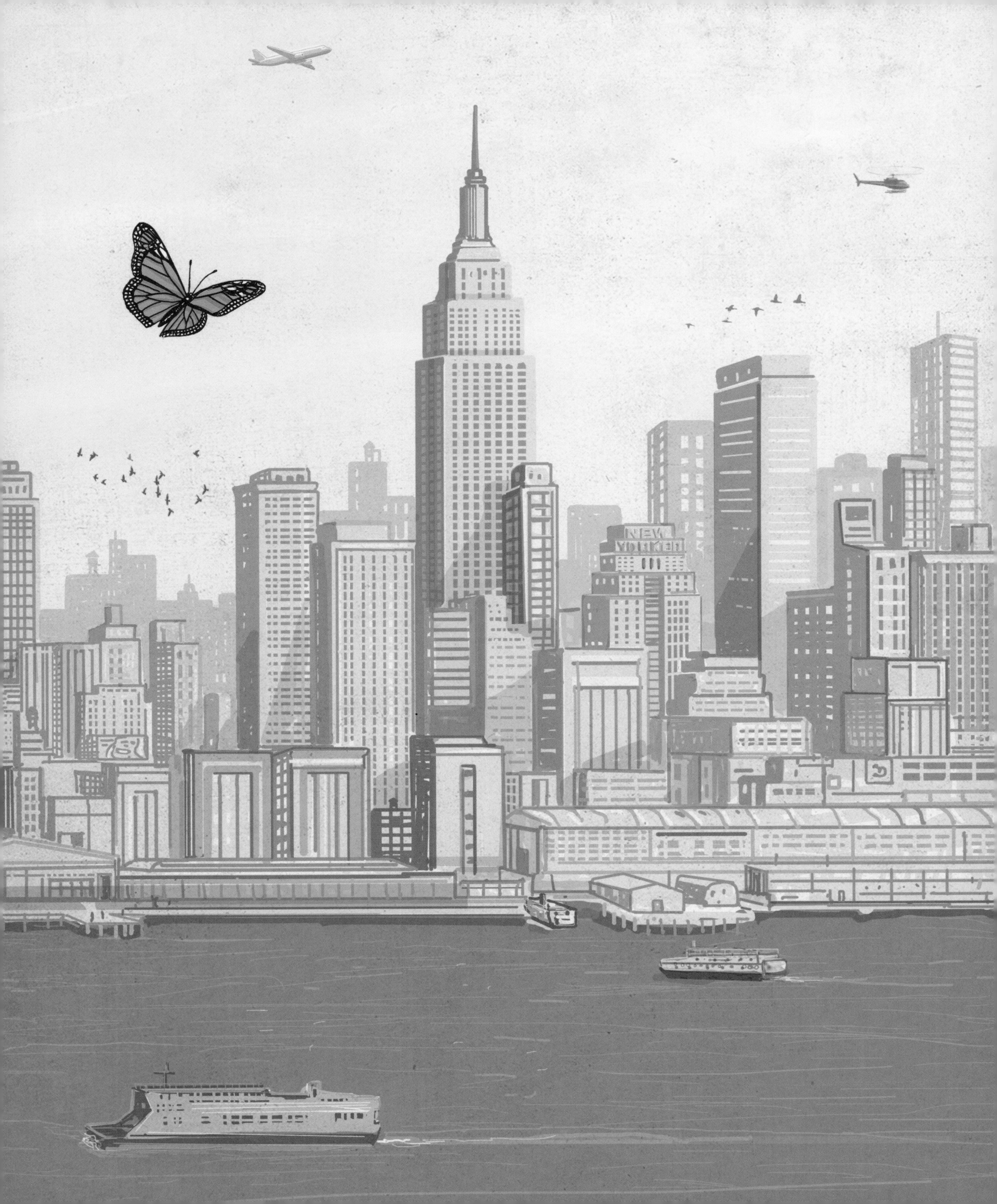
NEW
YORKER

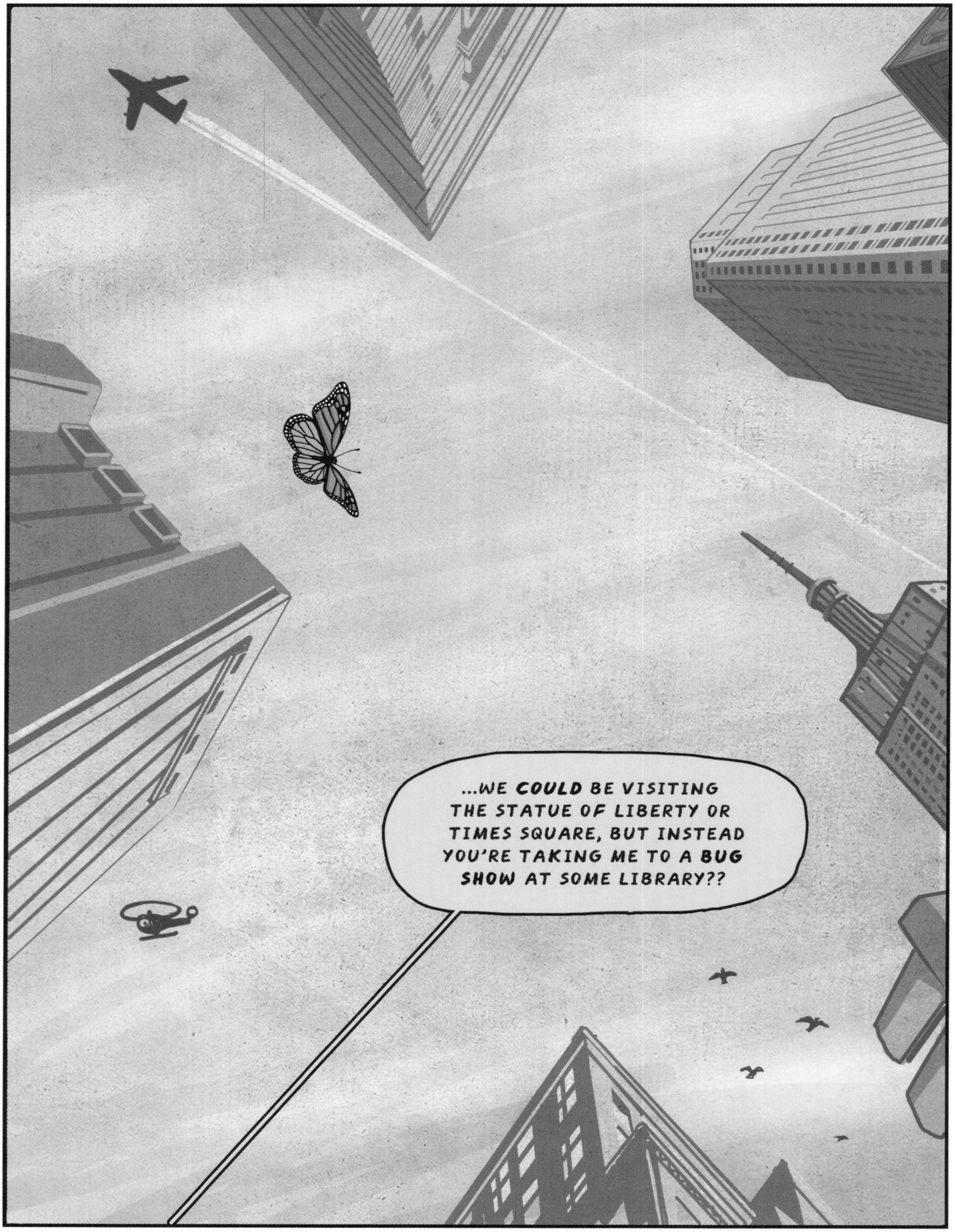
...WE COULD BE VISITING THE STATUE OF LIBERTY OR TIMES SQUARE, BUT INSTEAD YOU'RE TAKING ME TO A BUG SHOW AT SOME LIBRARY??

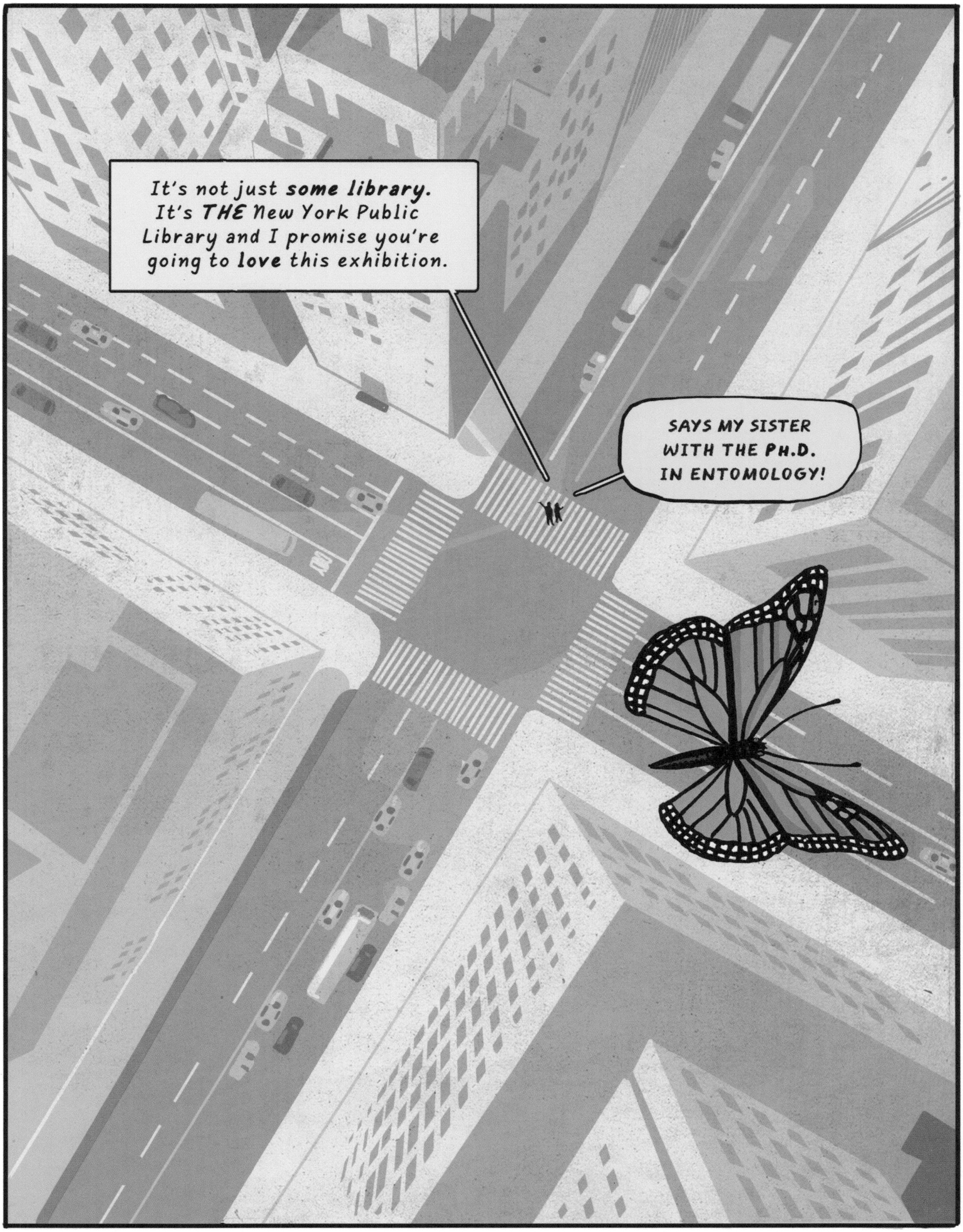
It's not just some library. It's THE New York Public Library and I promise you're going to love this exhibition.
SAYS MY SISTER WITH THE PH.D. IN ENTOMOLOGY!

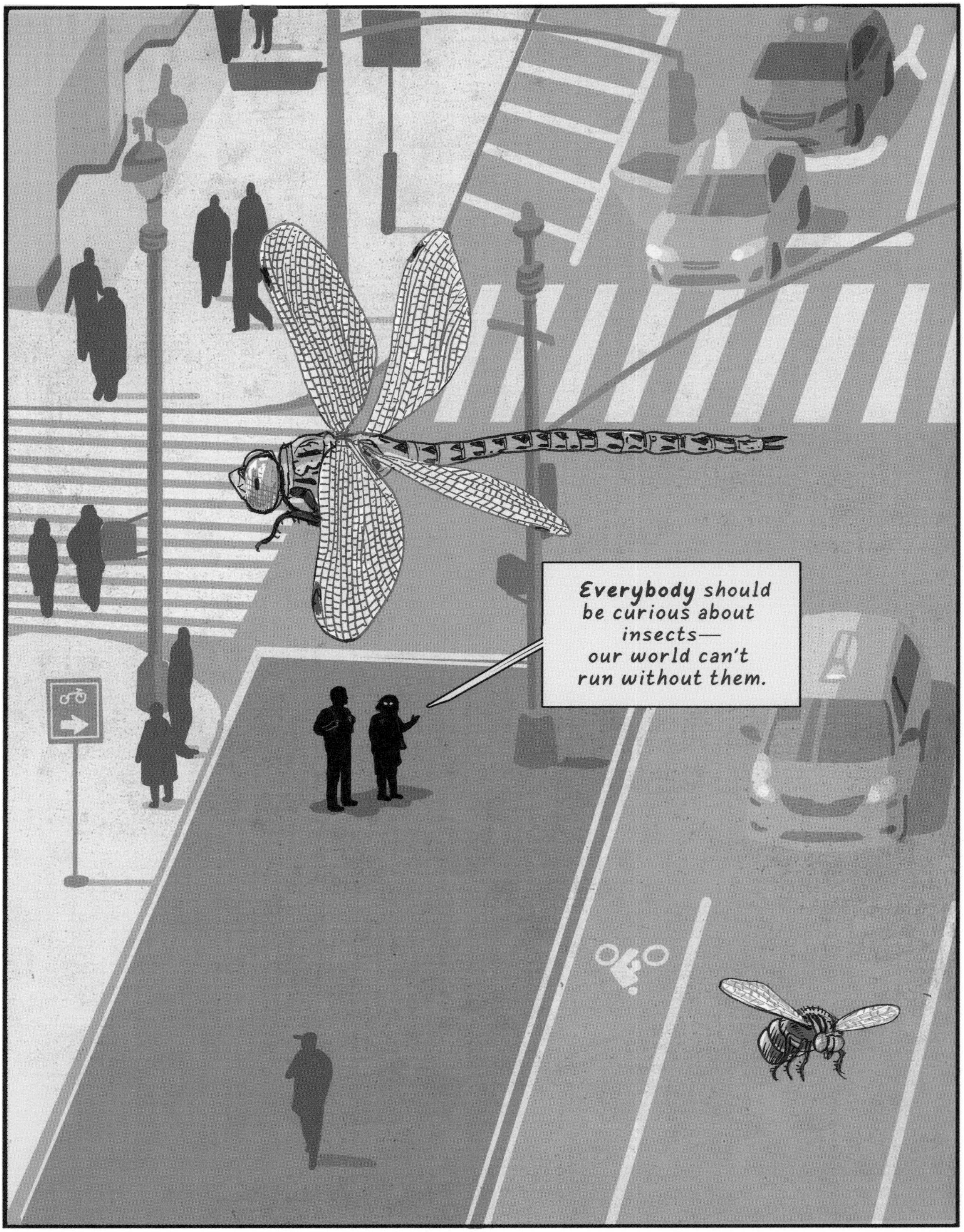
Everybody should be curious about insects— our world can't run without them.

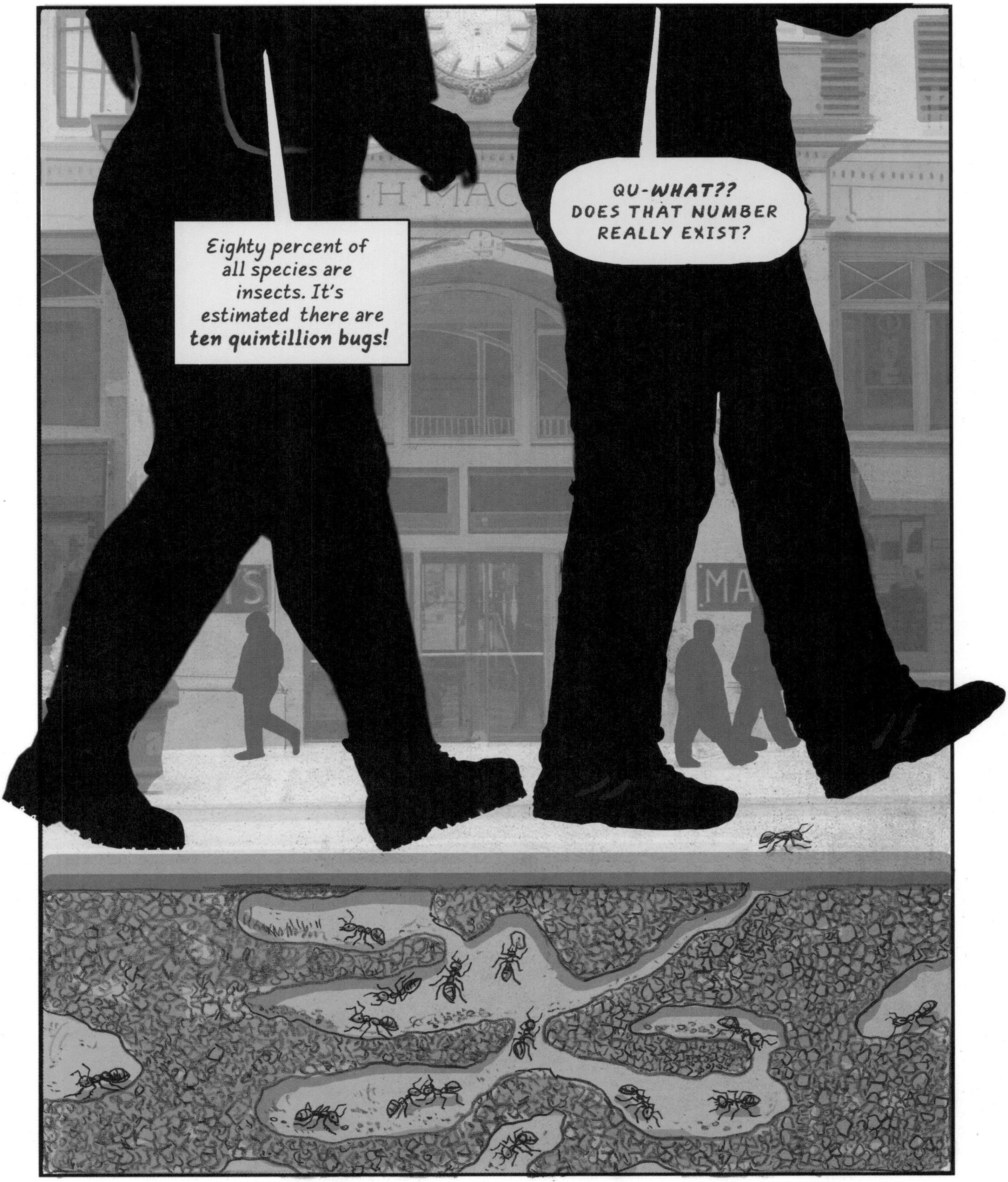
Eighty percent of all species are insects. It's estimated there are **ten quintillion bugs!**
QU-*WHAT??* DOES THAT NUMBER REALLY EXIST?

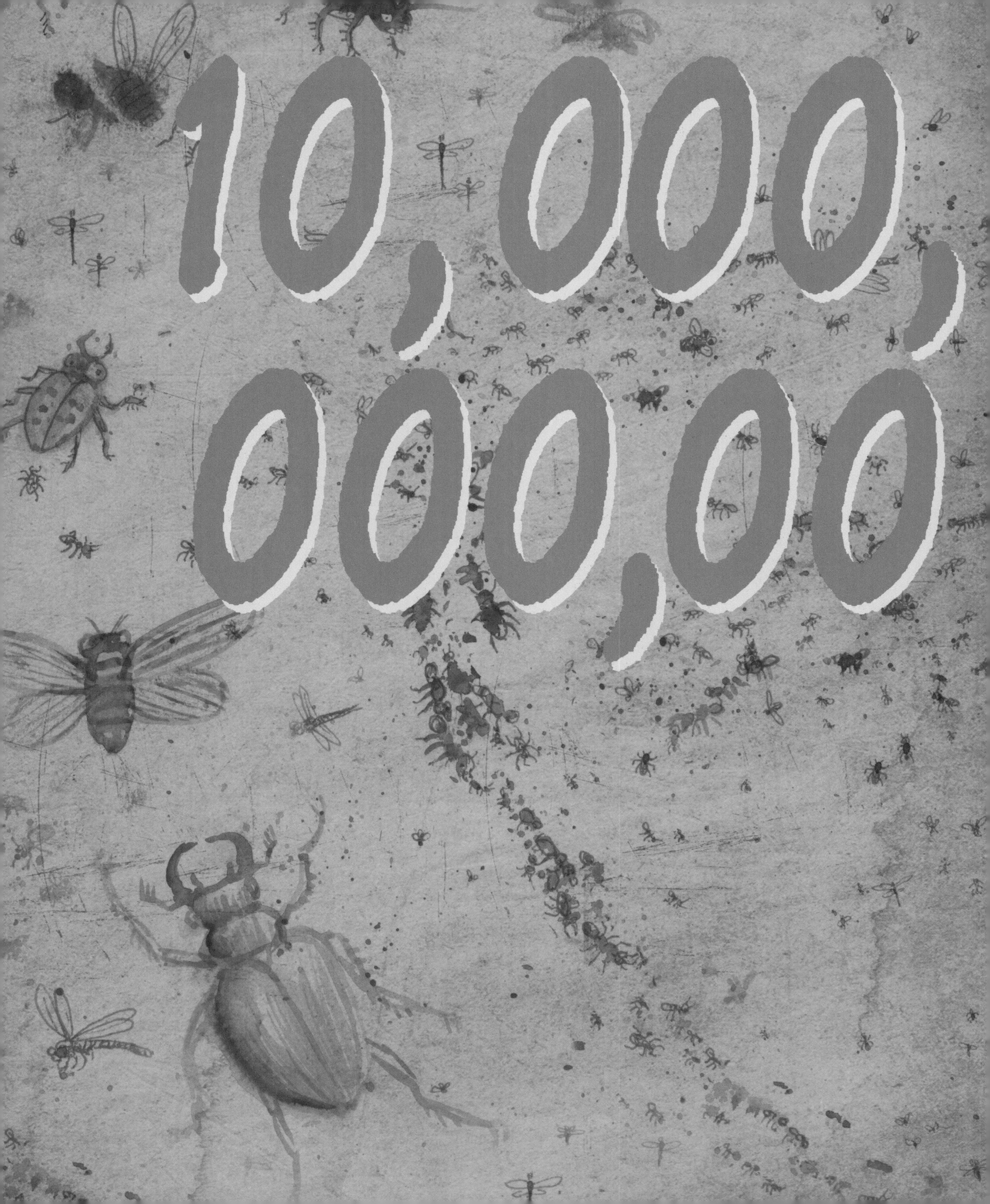
10,000,
000,00

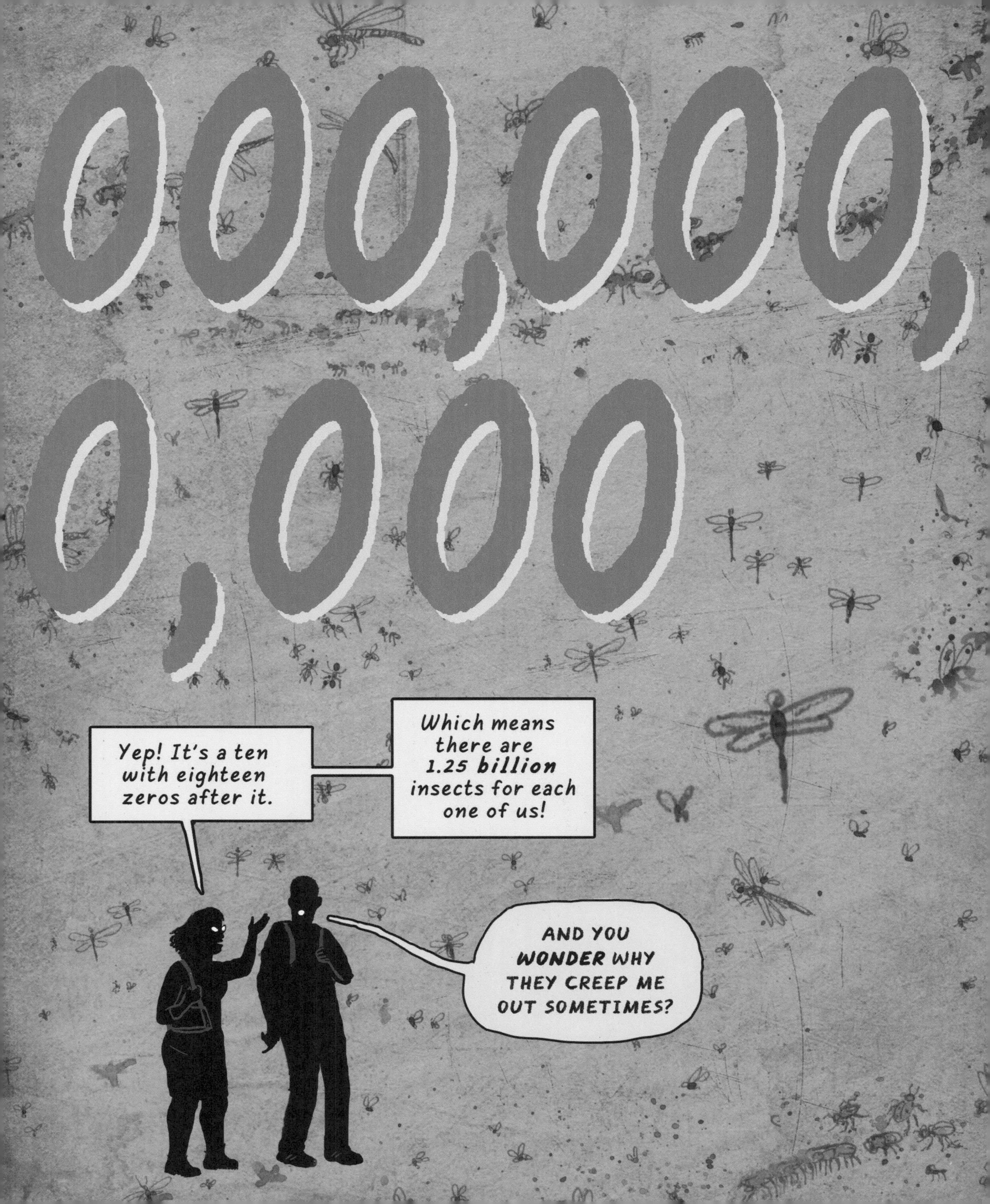
000,000,
0,000
Yep! It's a ten with eighteen zeros after it.
Which means there are 1.25 billion insects for each one of us!
AND YOU WONDER WHY THEY CREEP ME OUT SOMETIMES?

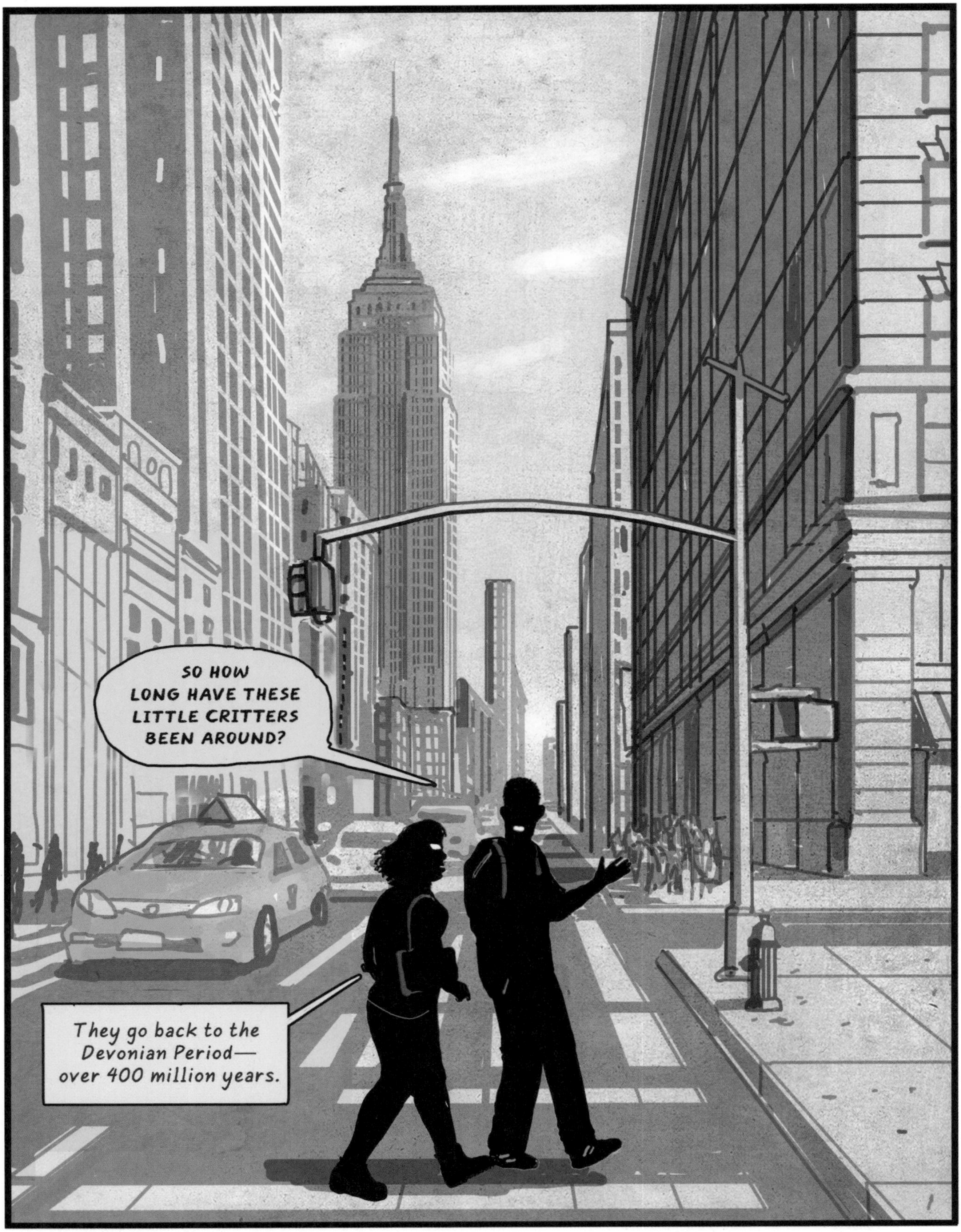
SO HOW LONG HAVE THESE LITTLE CRITTERS BEEN AROUND?
They go back to the Devonian Period— over 400 million years.

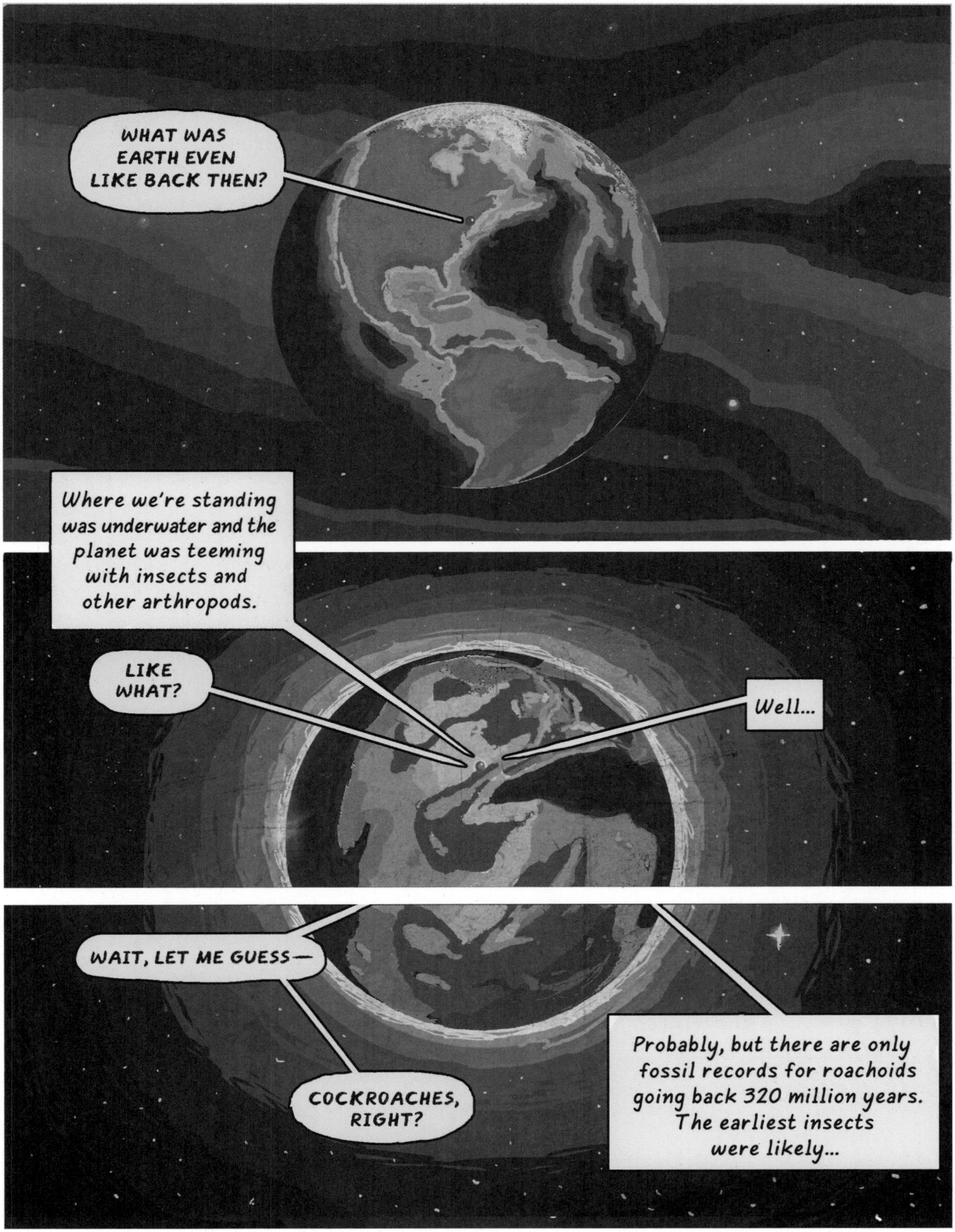
WHAT WAS EARTH EVEN LIKE BACK THEN?
Where we're standing was underwater and the planet was teeming with insects and other arthropods.
LIKE WHAT?
Well...
WAIT, LET ME GUESS—
COCKROACHES, RIGHT?
Probably, but there are only fossil records for roachoids going back 320 million years. The earliest insects were likely...

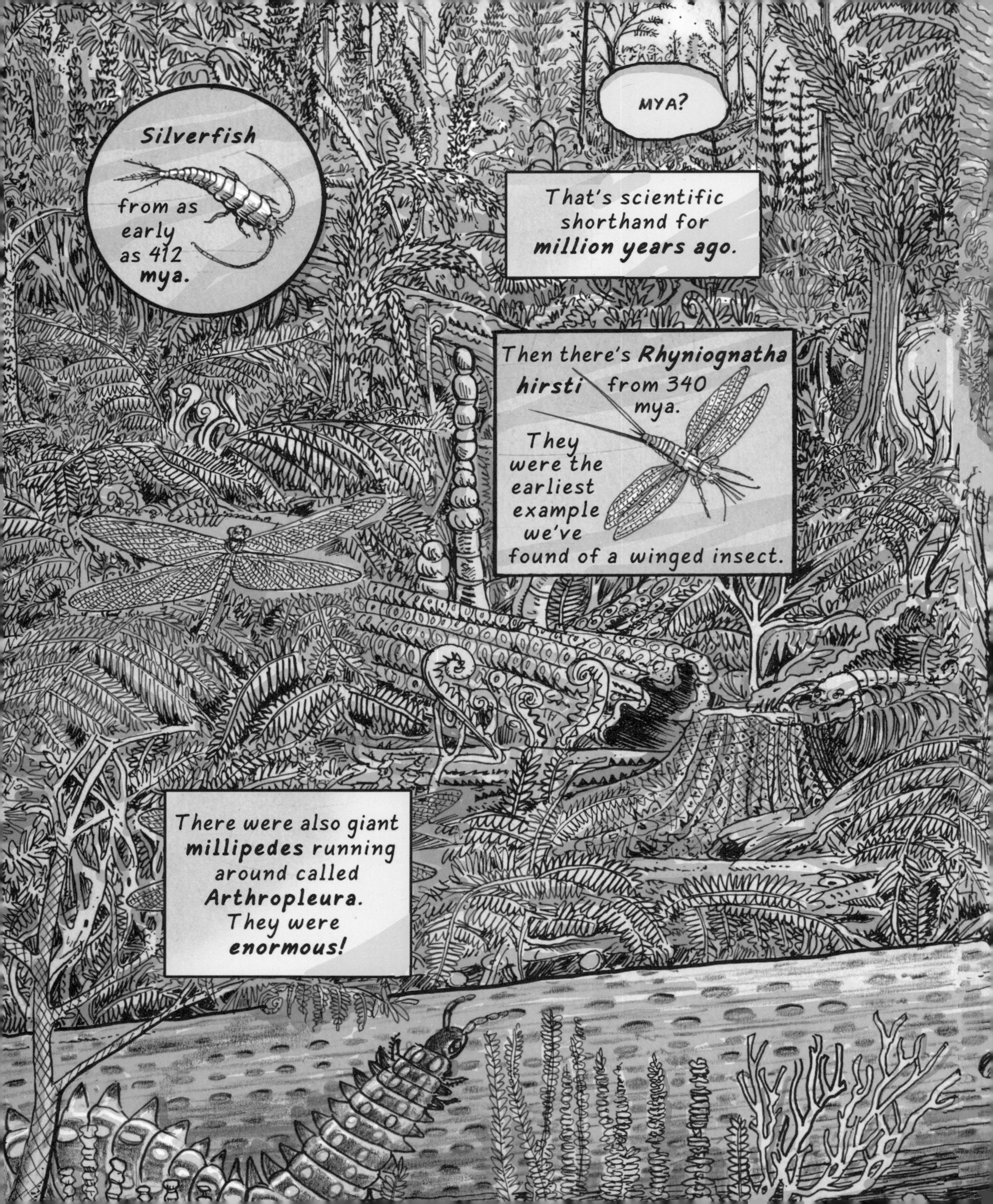
Silverfish
from as early as 412 mya.
MYA?
That's scientific shorthand for million years ago.
Then there's Rhyniognatha hirsti from 340 mya.
They were the earliest example we've found of a winged insect.
There were also giant millipedes running around called Arthropleura. They were enormous!

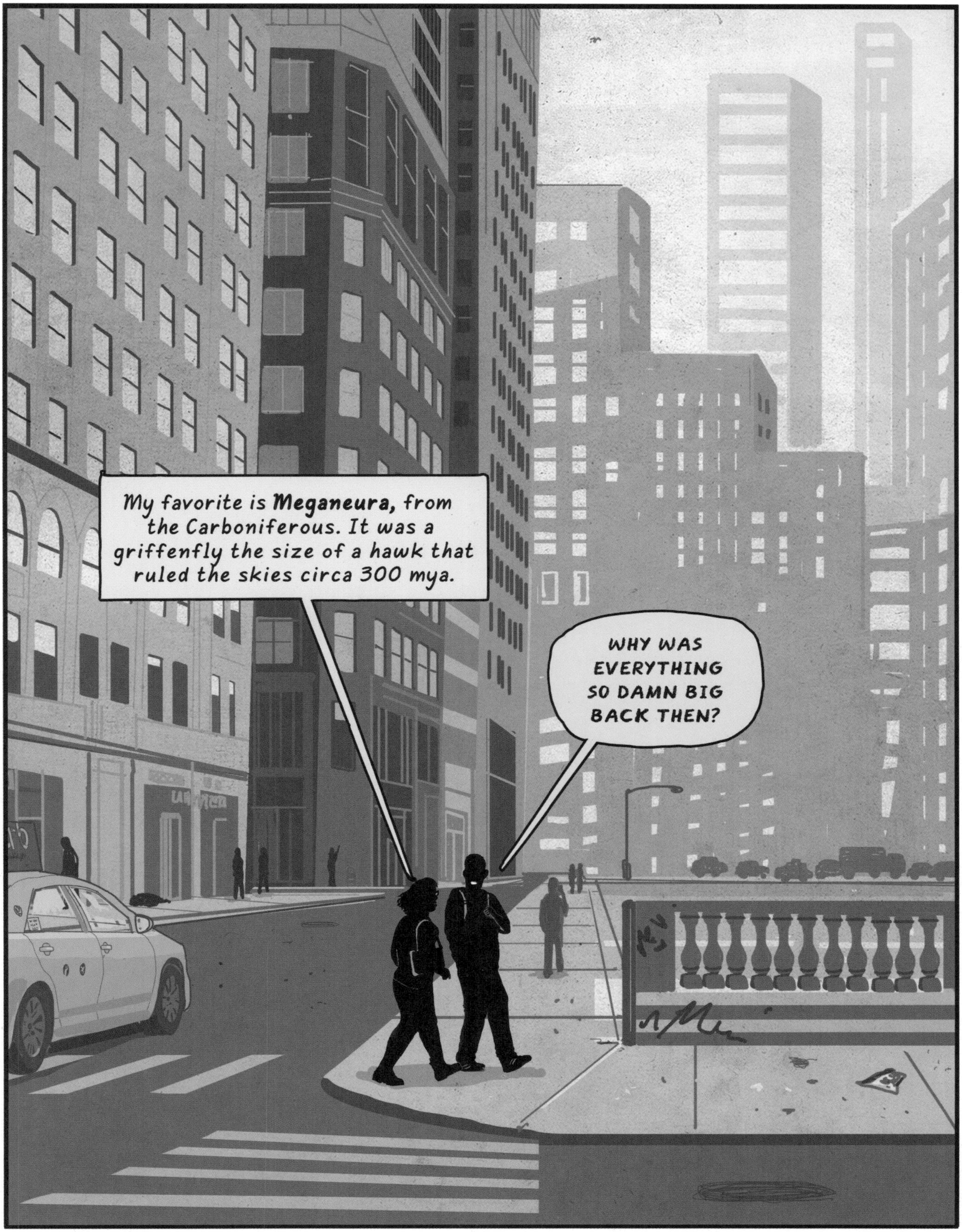
My favorite is **Meganeura,** from the Carboniferous. It was a griffenfly the size of a hawk that ruled the skies circa 300 mya.
WHY WAS EVERYTHING SO DAMN BIG BACK THEN?

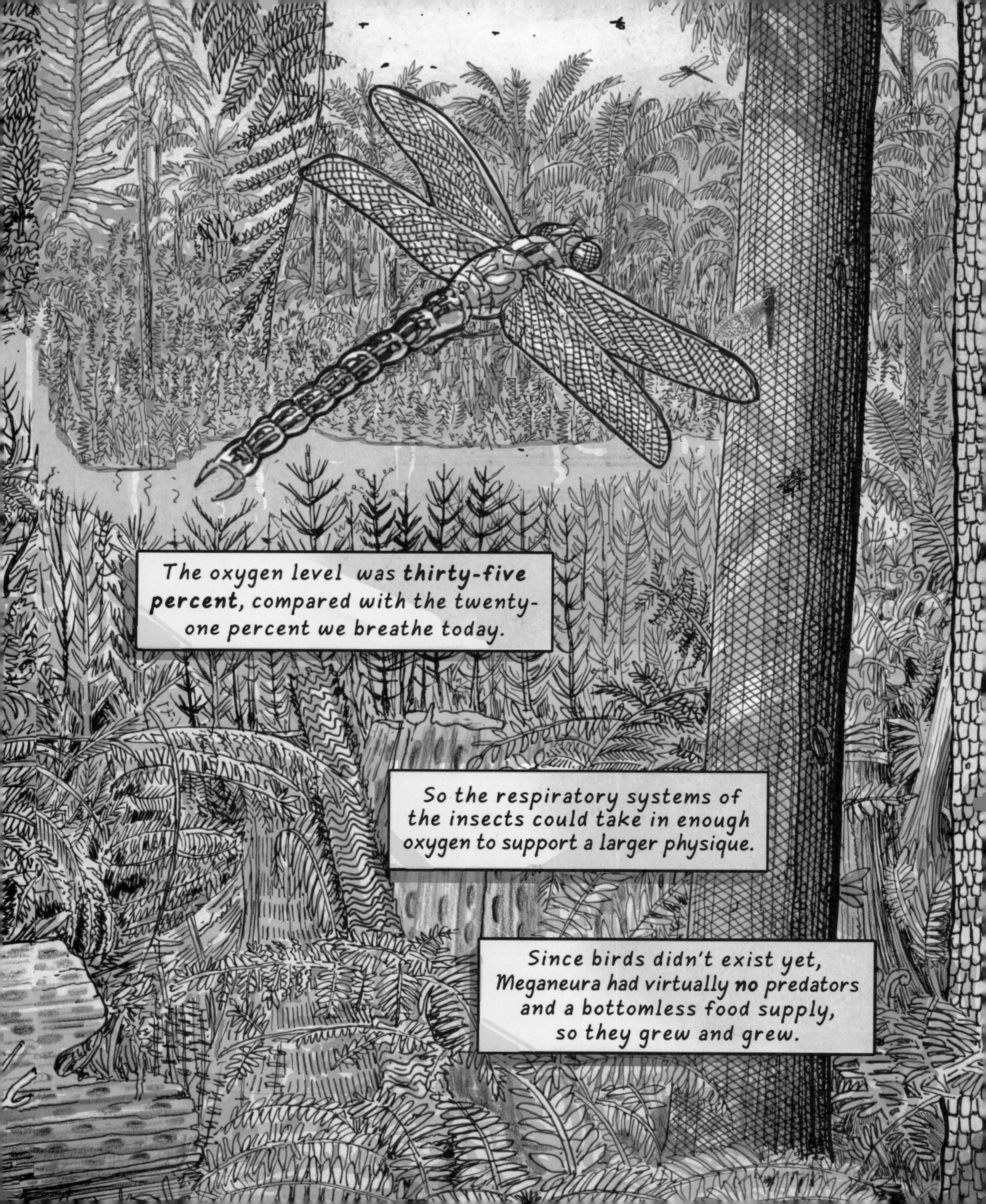
The oxygen level was **thirty-five percent**, compared with the twenty-one percent we breathe today.
So the respiratory systems of the insects could take in enough oxygen to support a larger physique.
Since birds didn't exist yet, Meganeura had virtually **no** predators and a bottomless food supply, so they grew and grew.

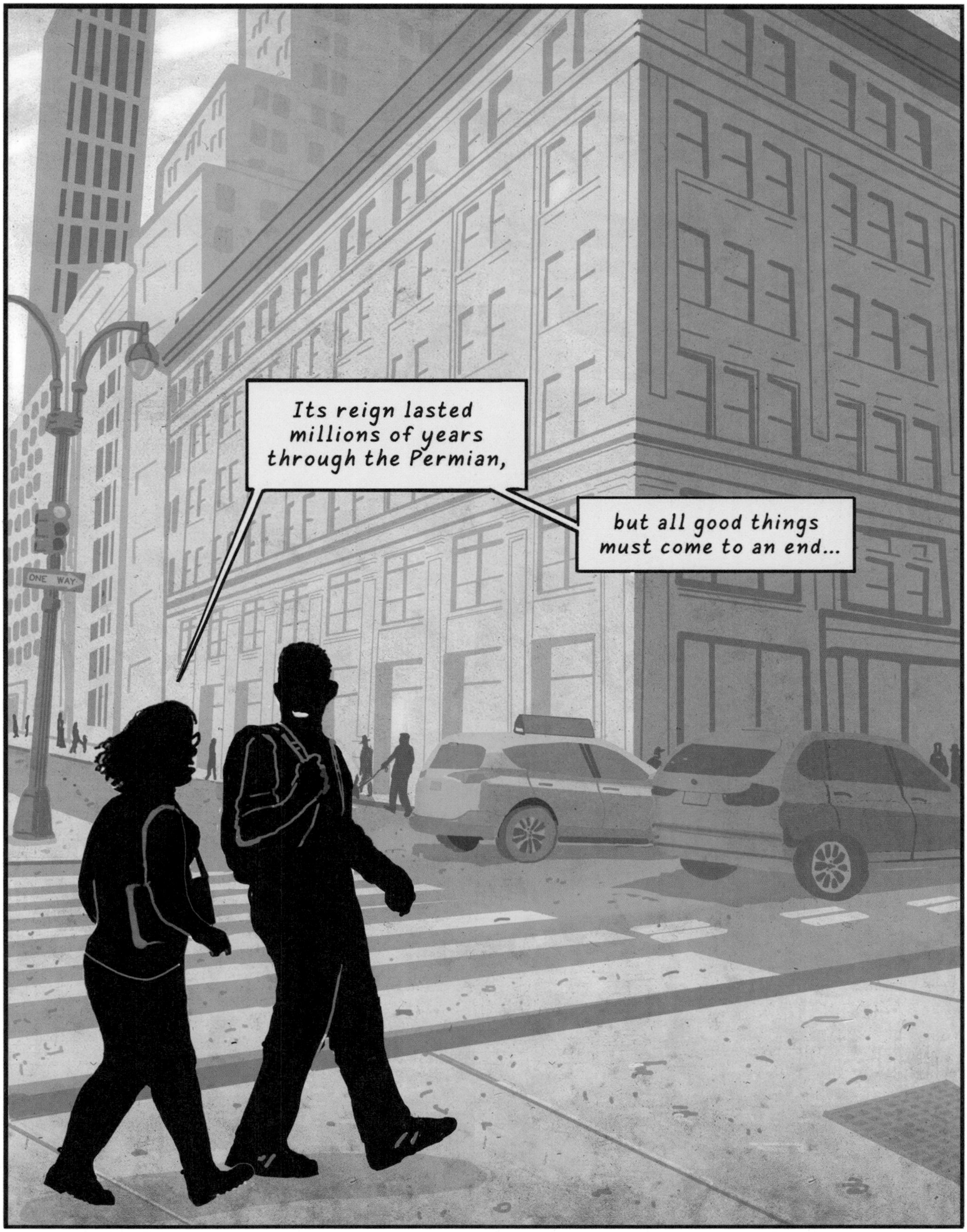
Its reign lasted millions of years through the Permian,
but all good things must come to an end...

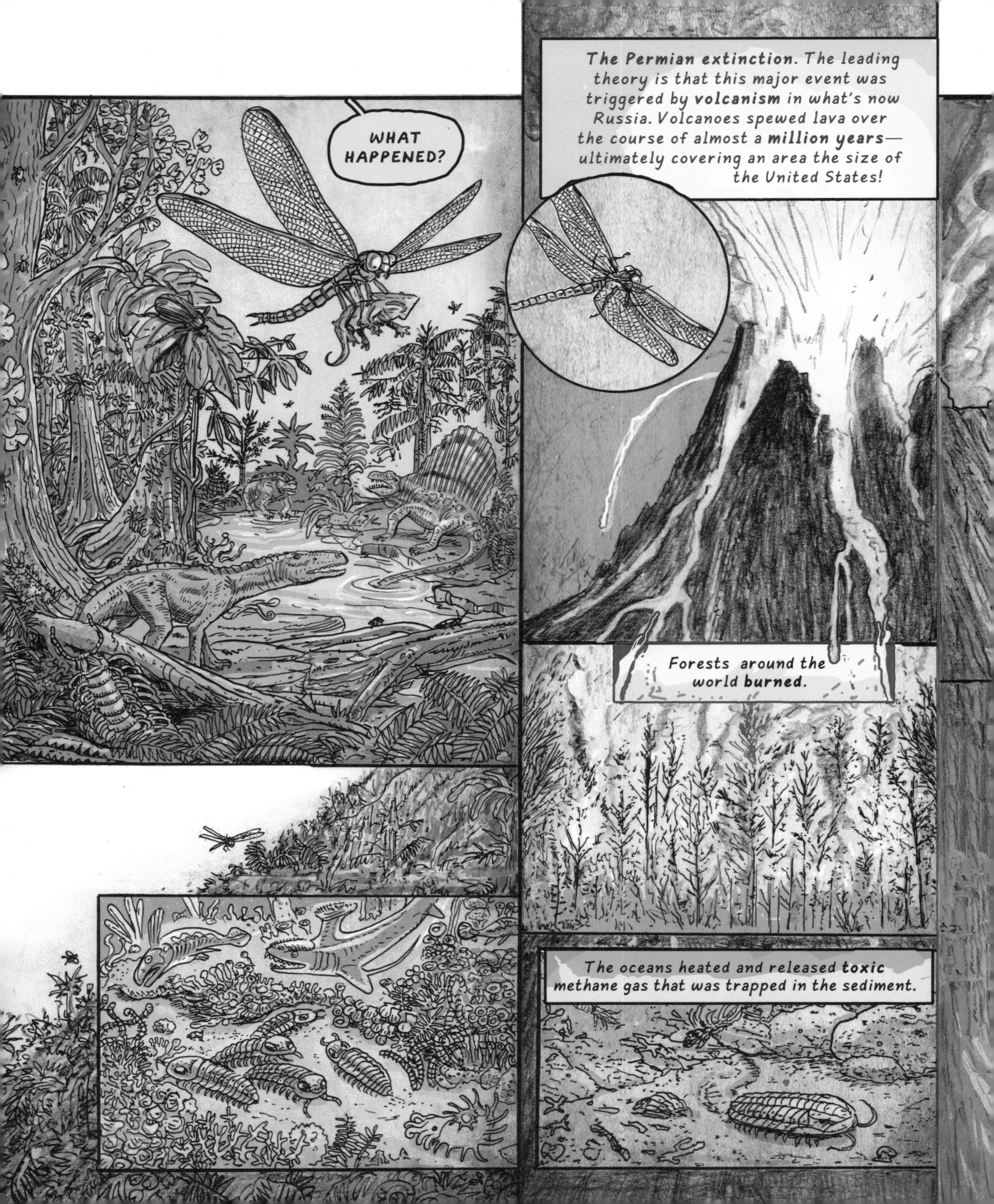
WHAT HAPPENED?
The Permian extinction. The leading theory is that this major event was triggered by volcanism in what's now Russia. Volcanoes spewed lava over the course of almost a million years—ultimately covering an area the size of the United States!
Forests around the world burned.
The oceans heated and released toxic methane gas that was trapped in the sediment.

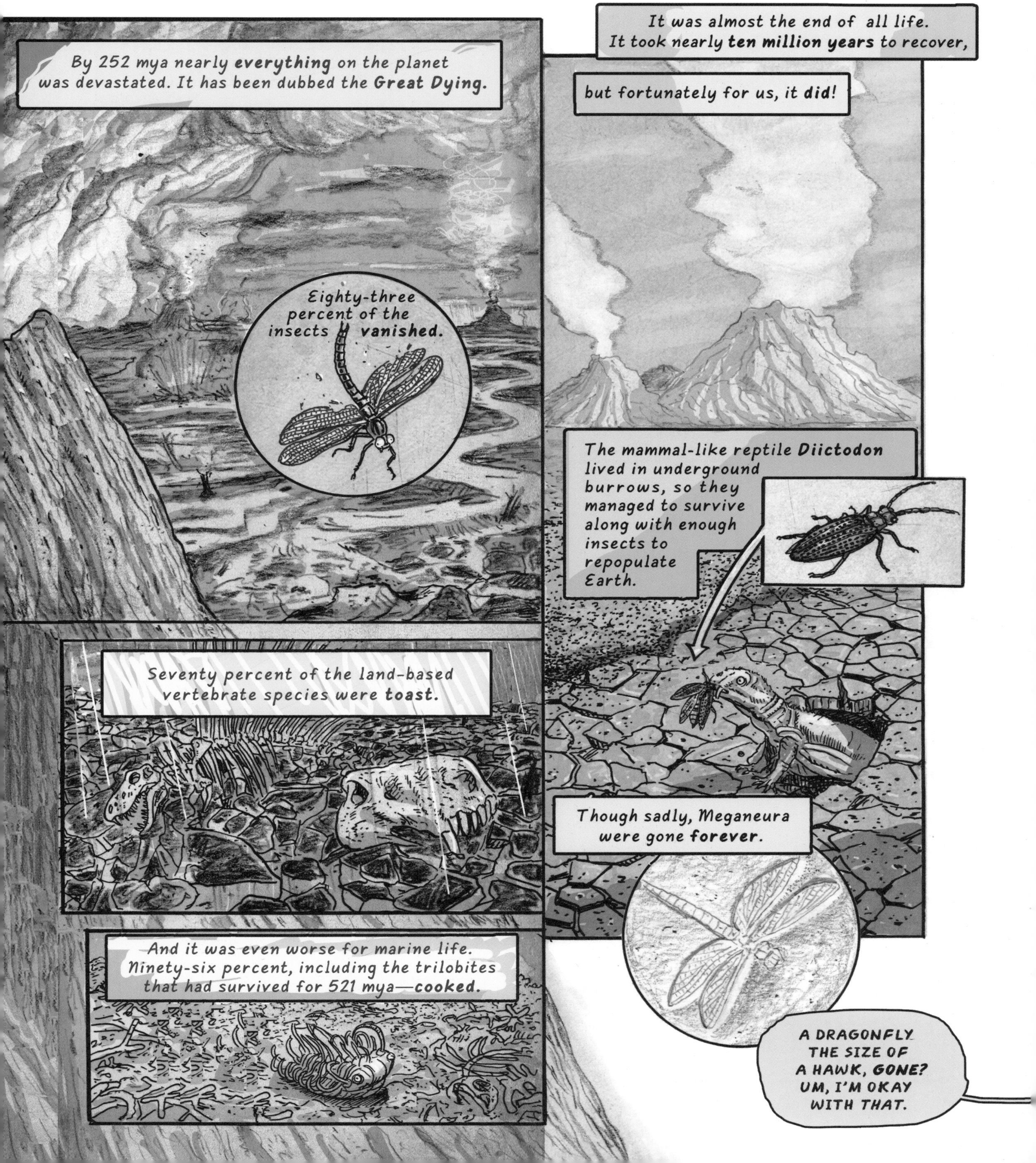

By 252 mya nearly **everything** on the planet was devastated. It has been dubbed the **Great Dying.**
Eighty-three percent of the insects **vanished.**
Seventy percent of the land-based vertebrate species were **toast.**
And it was even worse for marine life. Ninety-six percent, including the trilobites that had survived for 521 mya—**cooked.**
It was almost the end of all life. It took nearly **ten million years** to recover,
but fortunately for us, it **did**!
The mammal-like reptile ***Diictodon*** lived in underground burrows, so they managed to survive along with enough insects to repopulate Earth.
Though sadly, *Meganeura* were gone **forever.**
A DRAGONFLY THE SIZE OF A HAWK, **GONE?** UM, I'M OKAY WITH THAT.

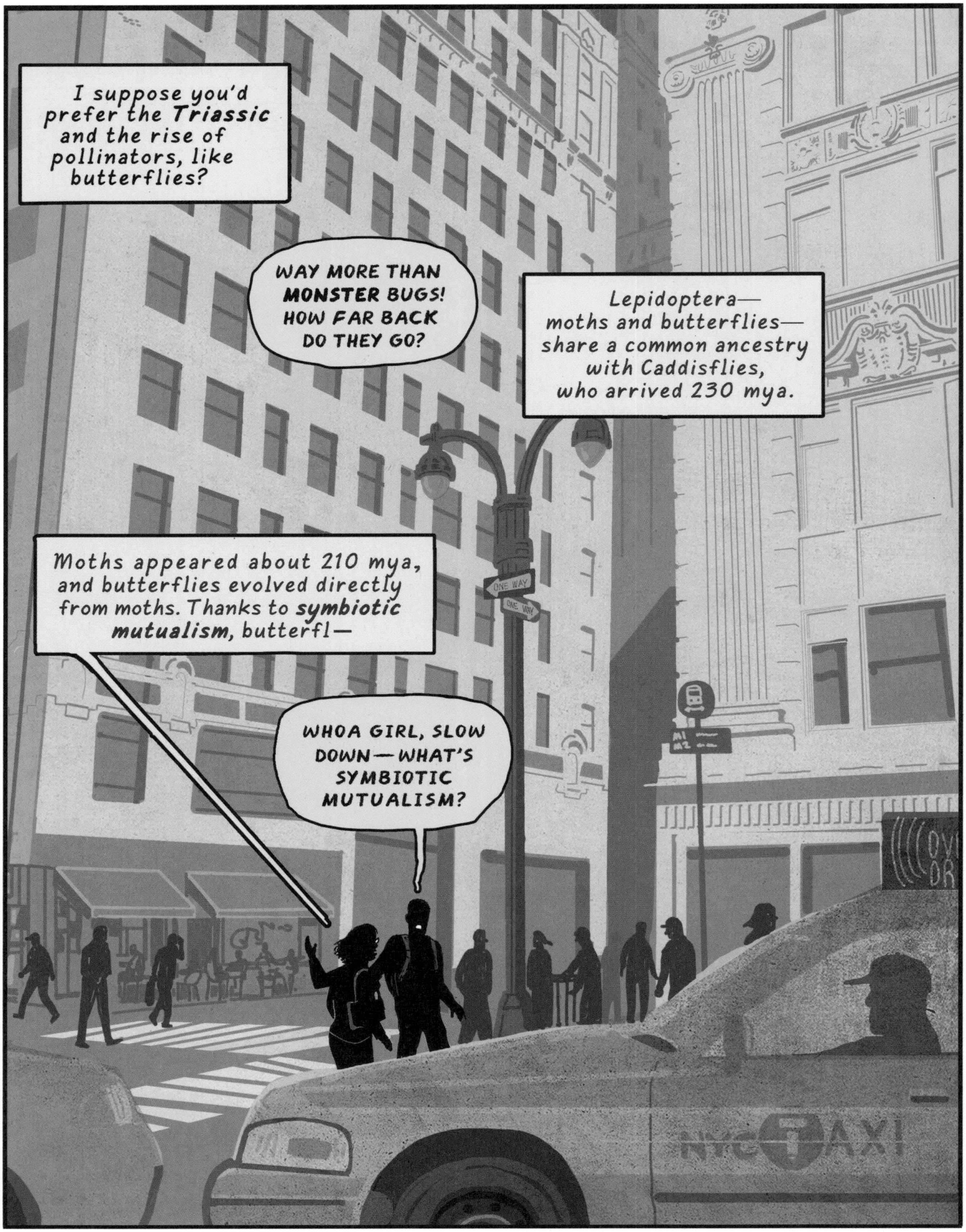
I suppose you'd prefer the *Triassic* and the rise of pollinators, like butterflies?
WAY MORE THAN **MONSTER** BUGS! HOW FAR BACK DO THEY GO?
Lepidoptera— moths and butterflies— share a common ancestry with Caddisflies, who arrived 230 mya.
Moths appeared about 210 mya, and butterflies evolved directly from moths. Thanks to ***symbiotic mutualism***, butterfl—
WHOA GIRL, SLOW DOWN—WHAT'S SYMBIOTIC MUTUALISM?
ONE WAY
ONE WAY
NYC TAXI

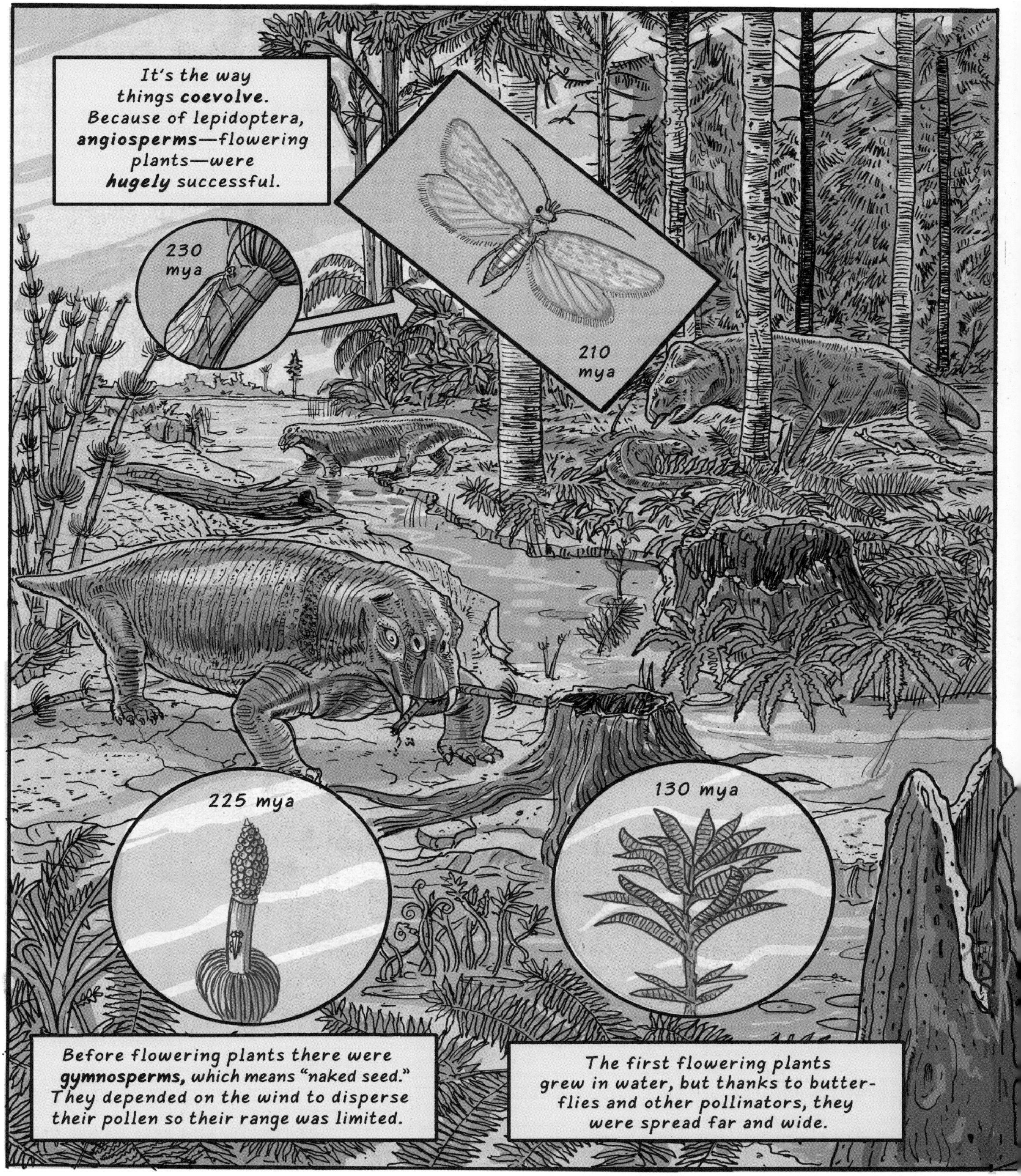

It's the way things **coevolve.** Because of lepidoptera, **angiosperms**—flowering plants—were ***hugely*** successful.
230 mya
210 mya
225 mya
130 mya
Before flowering plants there were **gymnosperms,** which means "naked seed." They depended on the wind to disperse their pollen so their range was limited.
The first flowering plants grew in water, but thanks to butter-flies and other pollinators, they were spread far and wide.

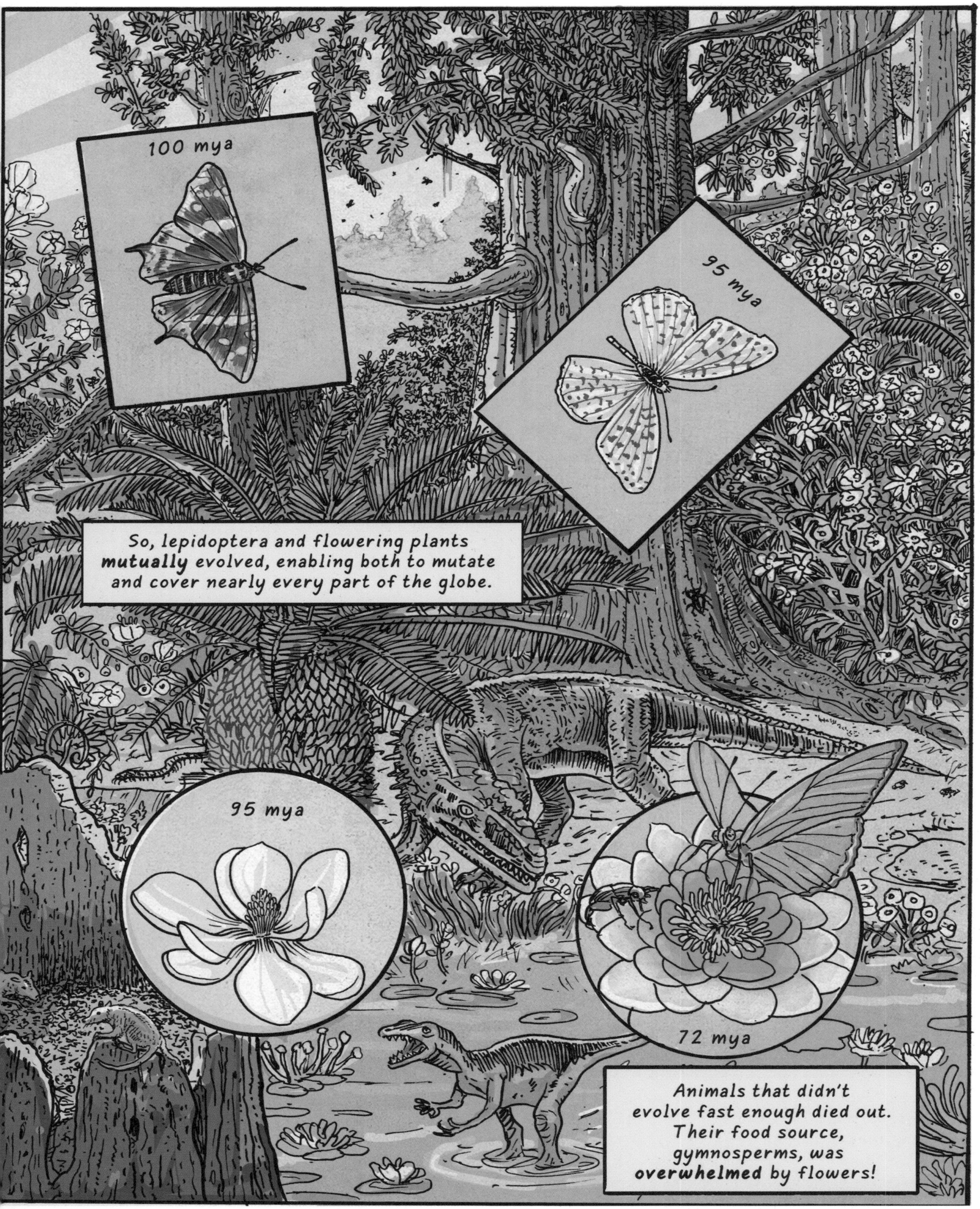
100 mya
95 mya
So, lepidoptera and flowering plants **mutually** evolved, enabling both to mutate and cover nearly every part of the globe.
95 mya
72 mya
Animals that didn't evolve fast enough died out. Their food source, gymnosperms, was **overwhelmed** by flowers!

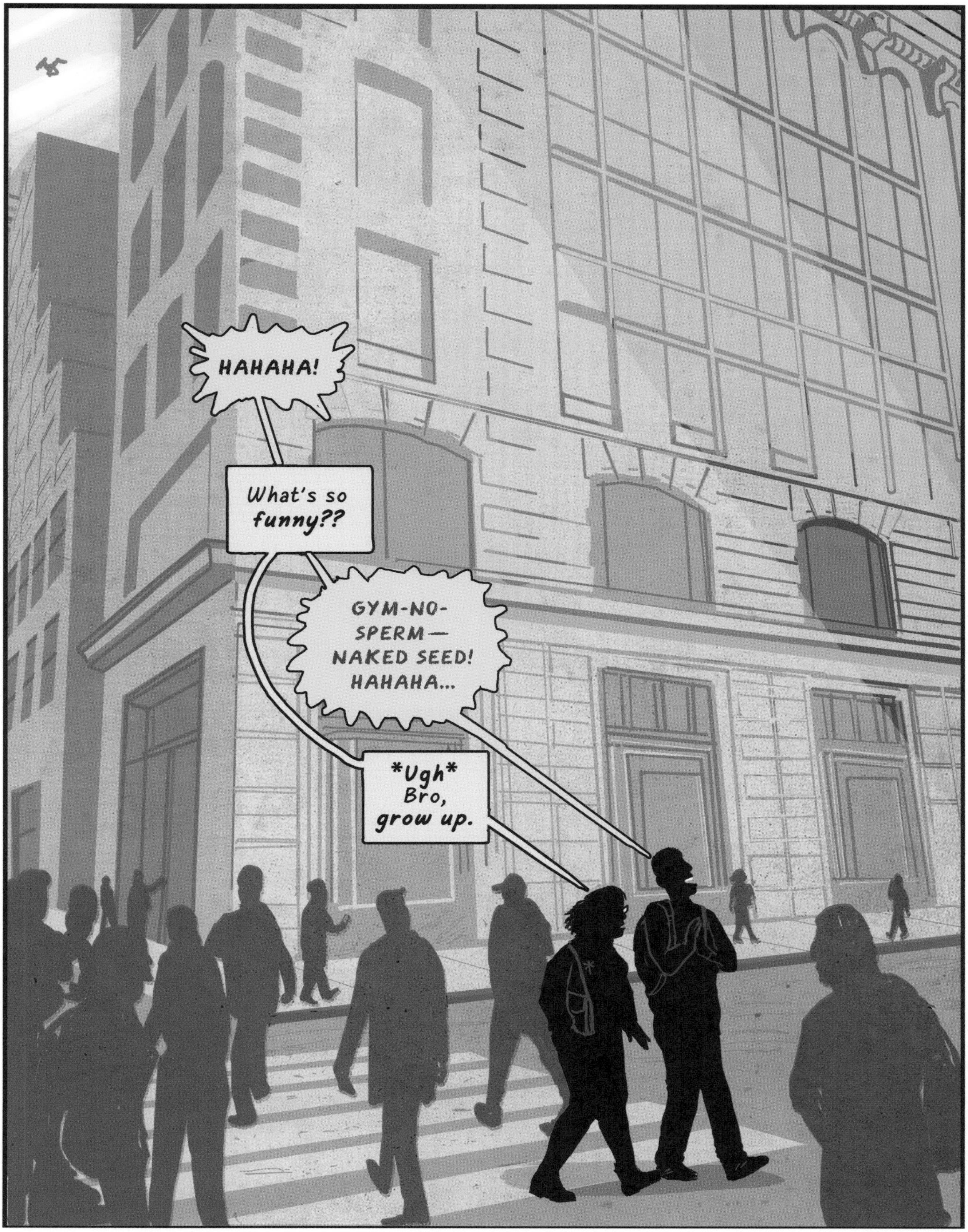
HAHAHA!
What's so funny??
GYM-NO-SPERM—NAKED SEED! HAHAHA...
Ugh Bro, grow up.

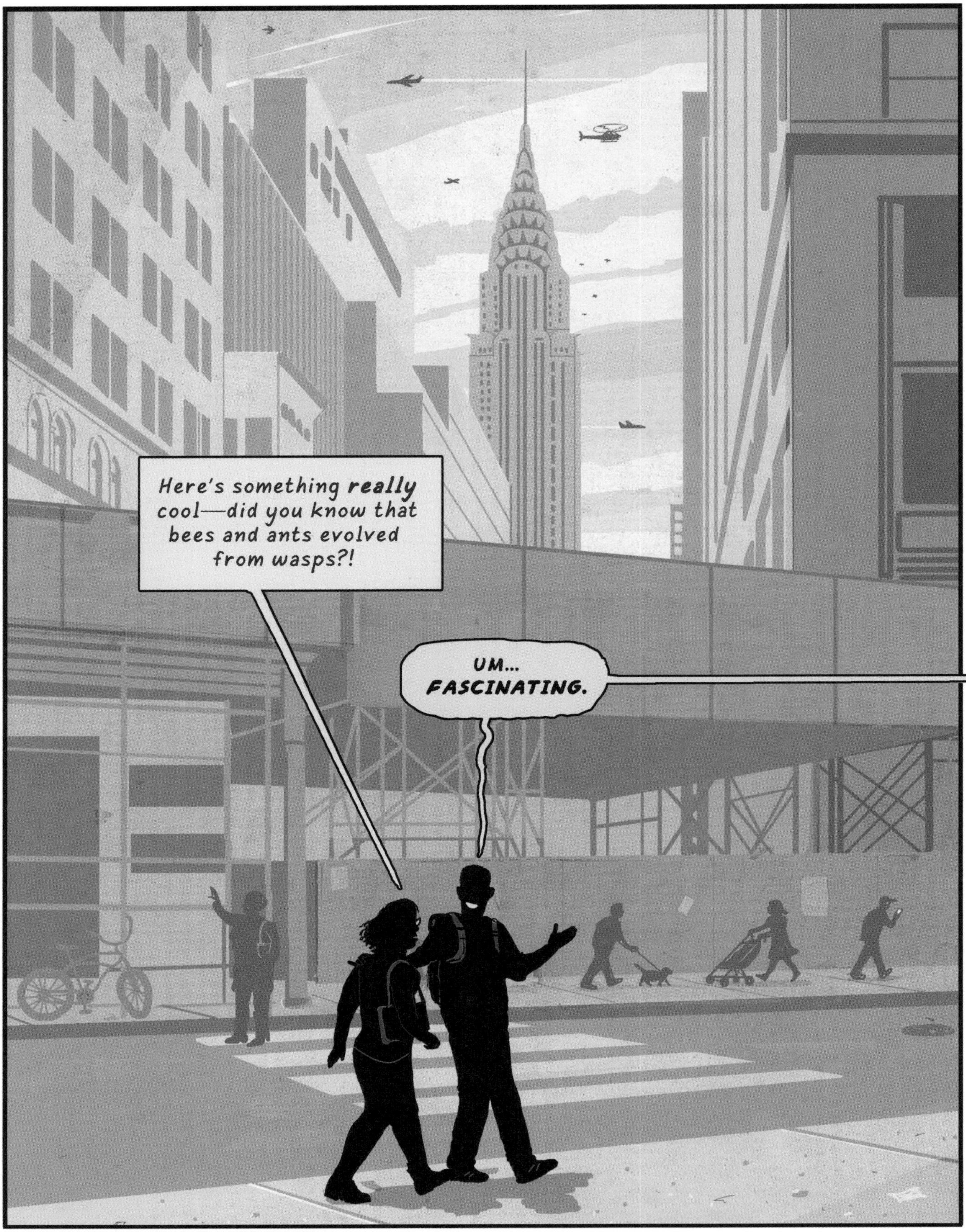
Here's something **really** cool—did you know that bees and ants evolved from wasps?!
UM... **FASCINATING.**

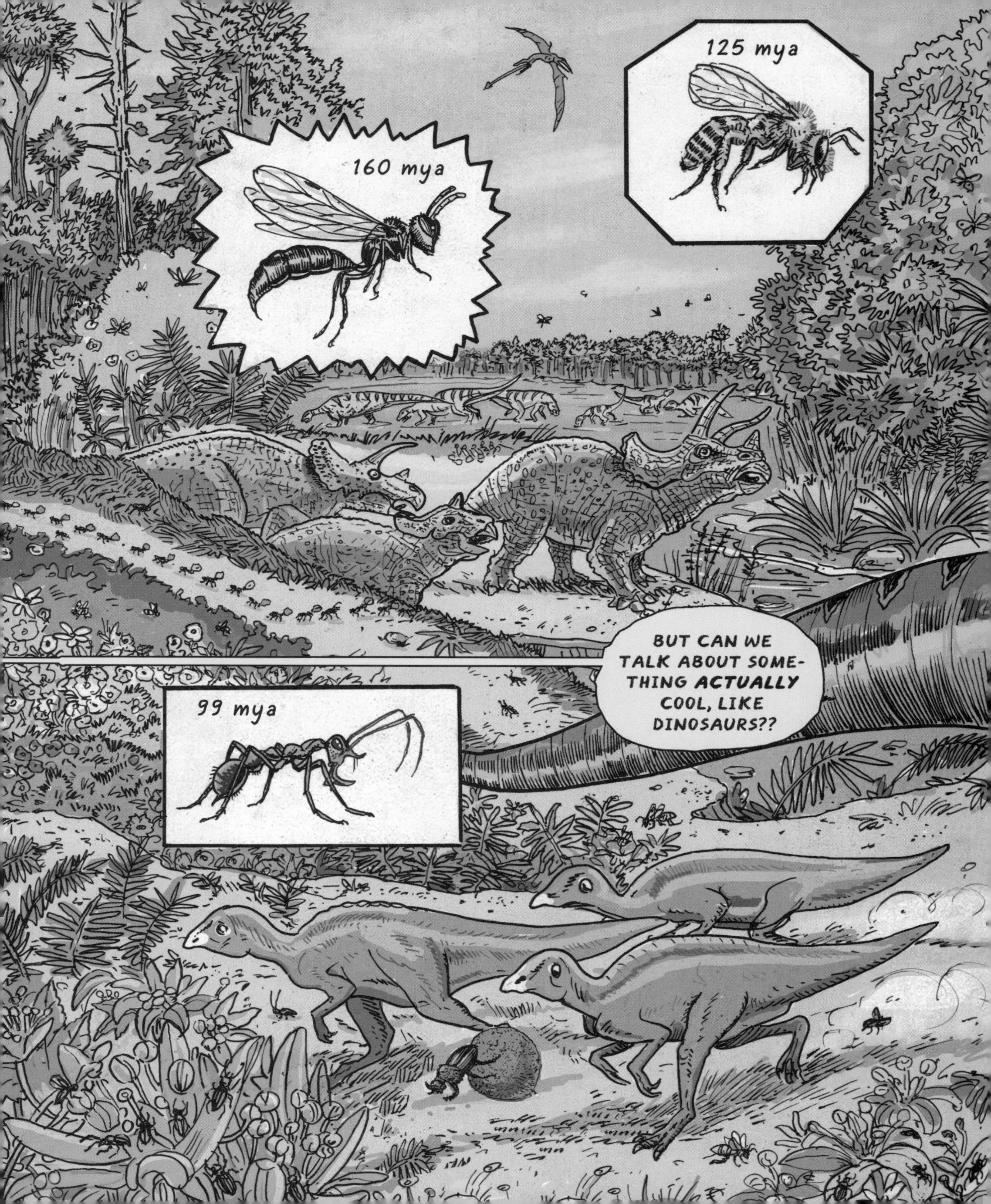
160 mya
125 mya
99 mya
BUT CAN WE TALK ABOUT SOMETHING ACTUALLY COOL, LIKE DINOSAURS??

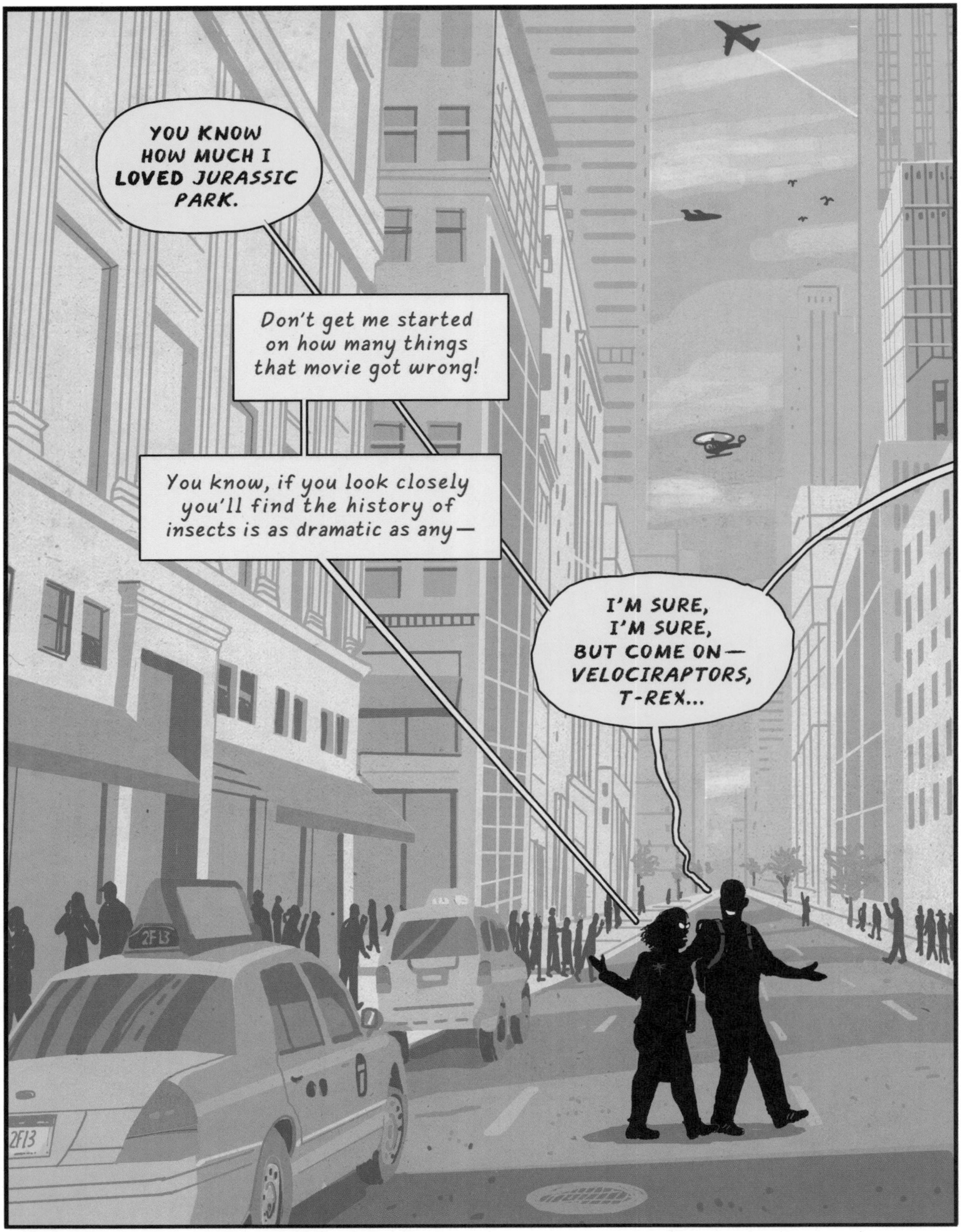
YOU KNOW HOW MUCH I LOVED JURASSIC PARK.
Don't get me started on how many things that movie got wrong!
You know, if you look closely you'll find the history of insects is as dramatic as any—
I'M SURE, I'M SURE, BUT COME ON— VELOCIRAPTORS, T-REX...
2F13
2F13

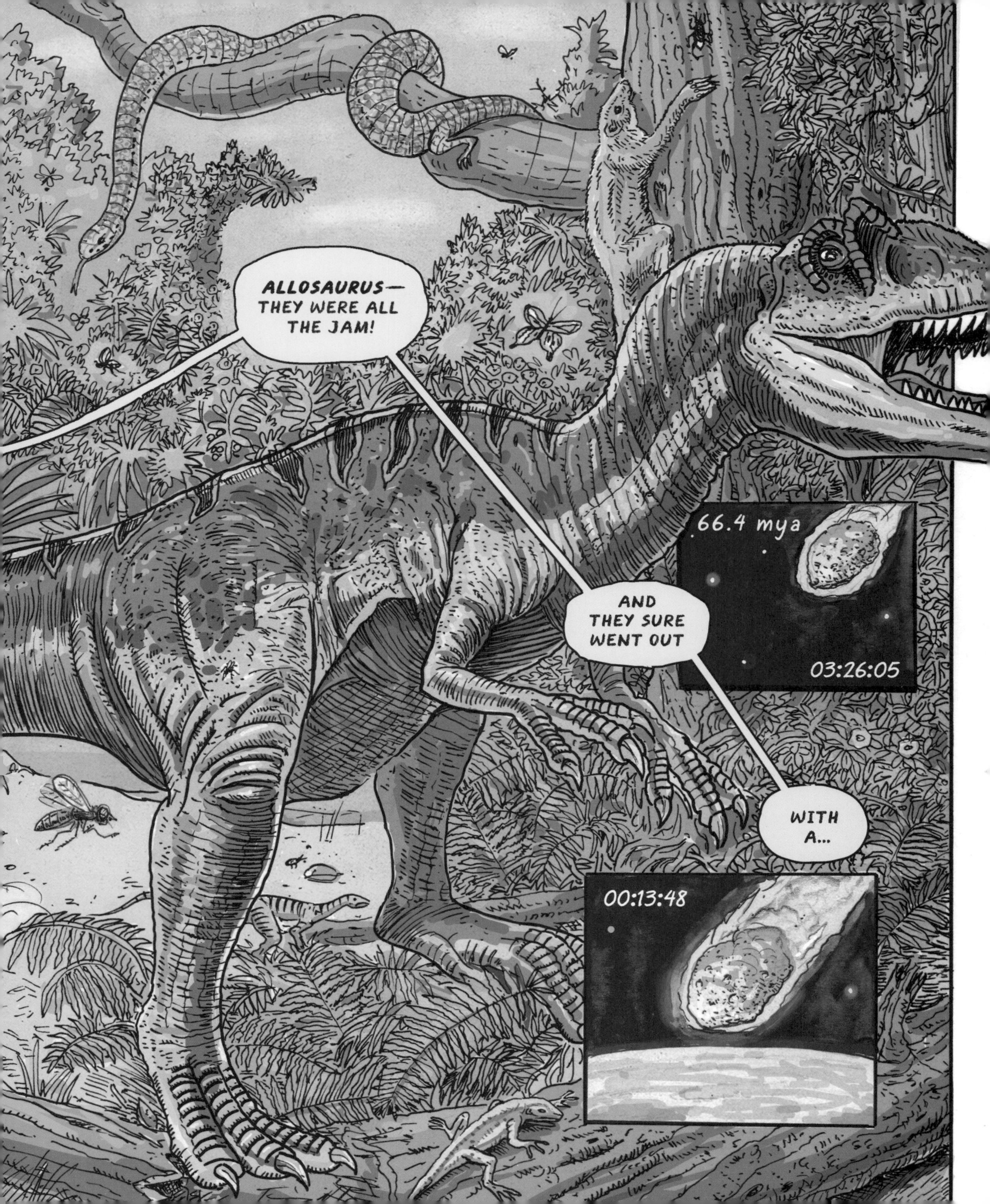
ALLOSAURUS—
THEY WERE ALL
THE JAM!
66.4 mya
03:26:05
AND
THEY SURE
WENT OUT
WITH
A...
00:13:48

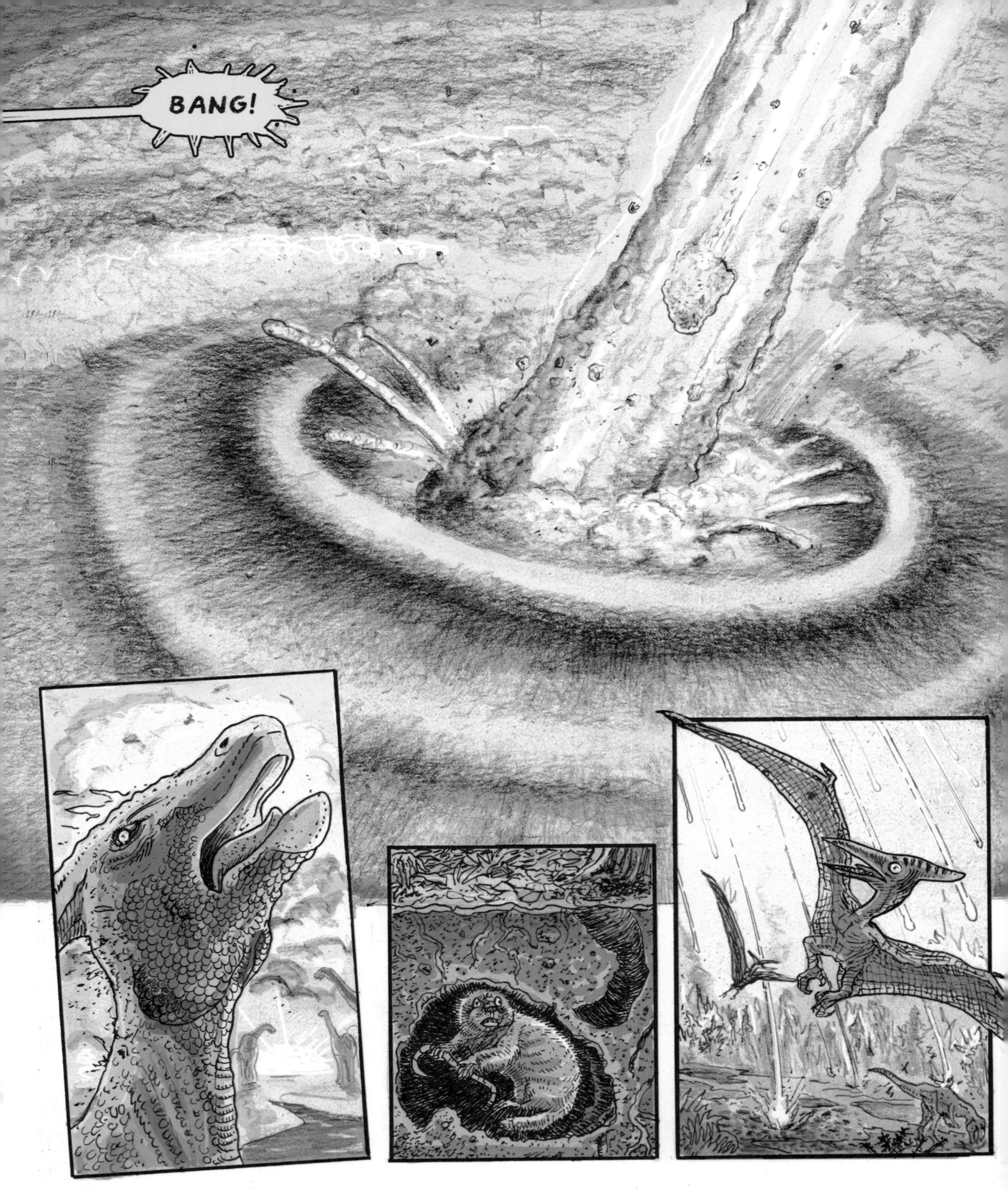
BANG!

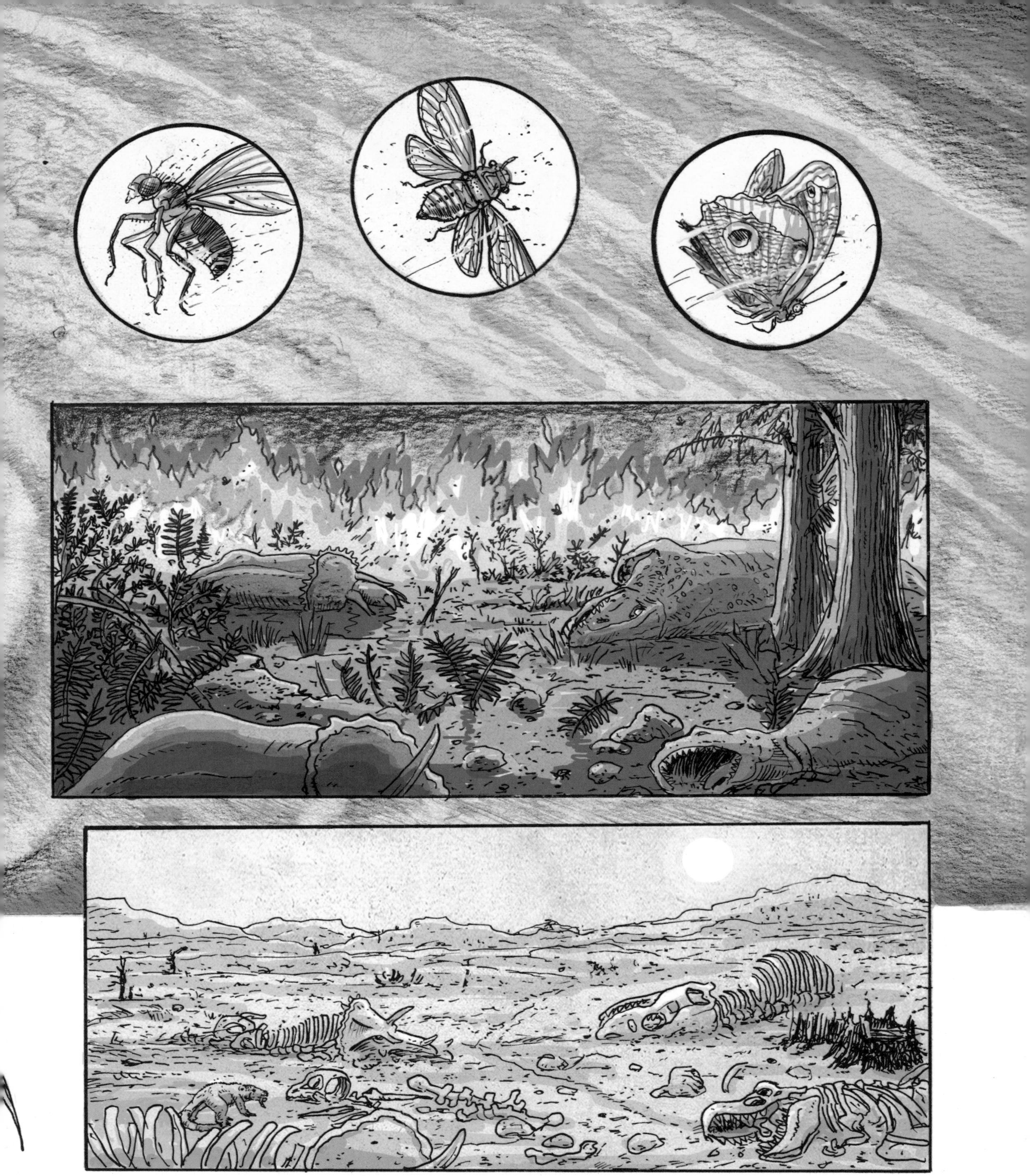

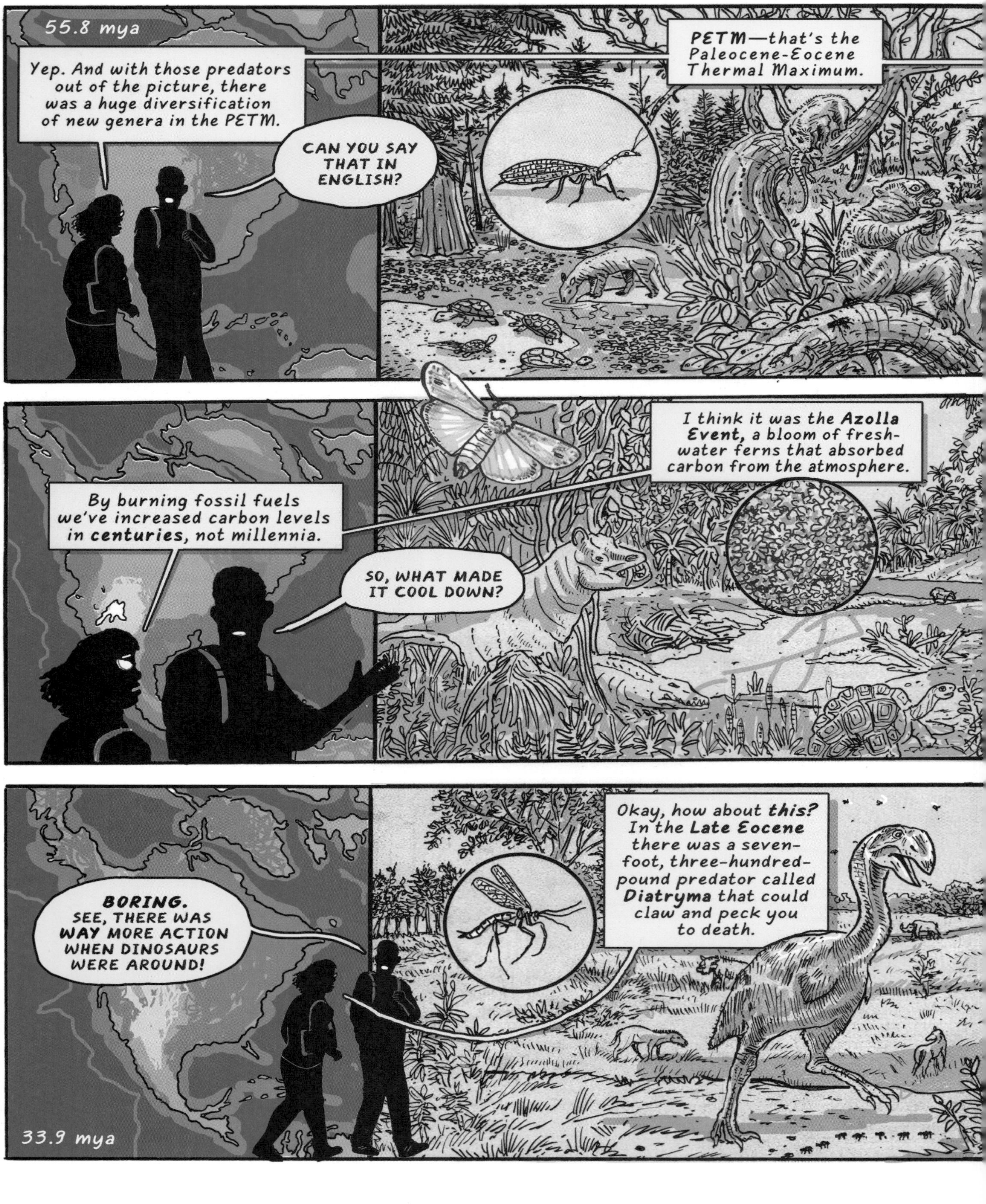
55.8 mya
Yep. And with those predators out of the picture, there was a huge diversification of new genera in the PETM.
CAN YOU SAY THAT IN ENGLISH?
PETM—that's the Paleocene-Eocene Thermal Maximum.
By burning fossil fuels we've increased carbon levels in centuries, not millennia.
SO, WHAT MADE IT COOL DOWN?
I think it was the Azolla Event, a bloom of fresh-water ferns that absorbed carbon from the atmosphere.
BORING. SEE, THERE WAS WAY MORE ACTION WHEN DINOSAURS WERE AROUND!
Okay, how about this? In the Late Eocene there was a seven-foot, three-hundred-pound predator called Diatryma that could claw and peck you to death.
33.9 mya

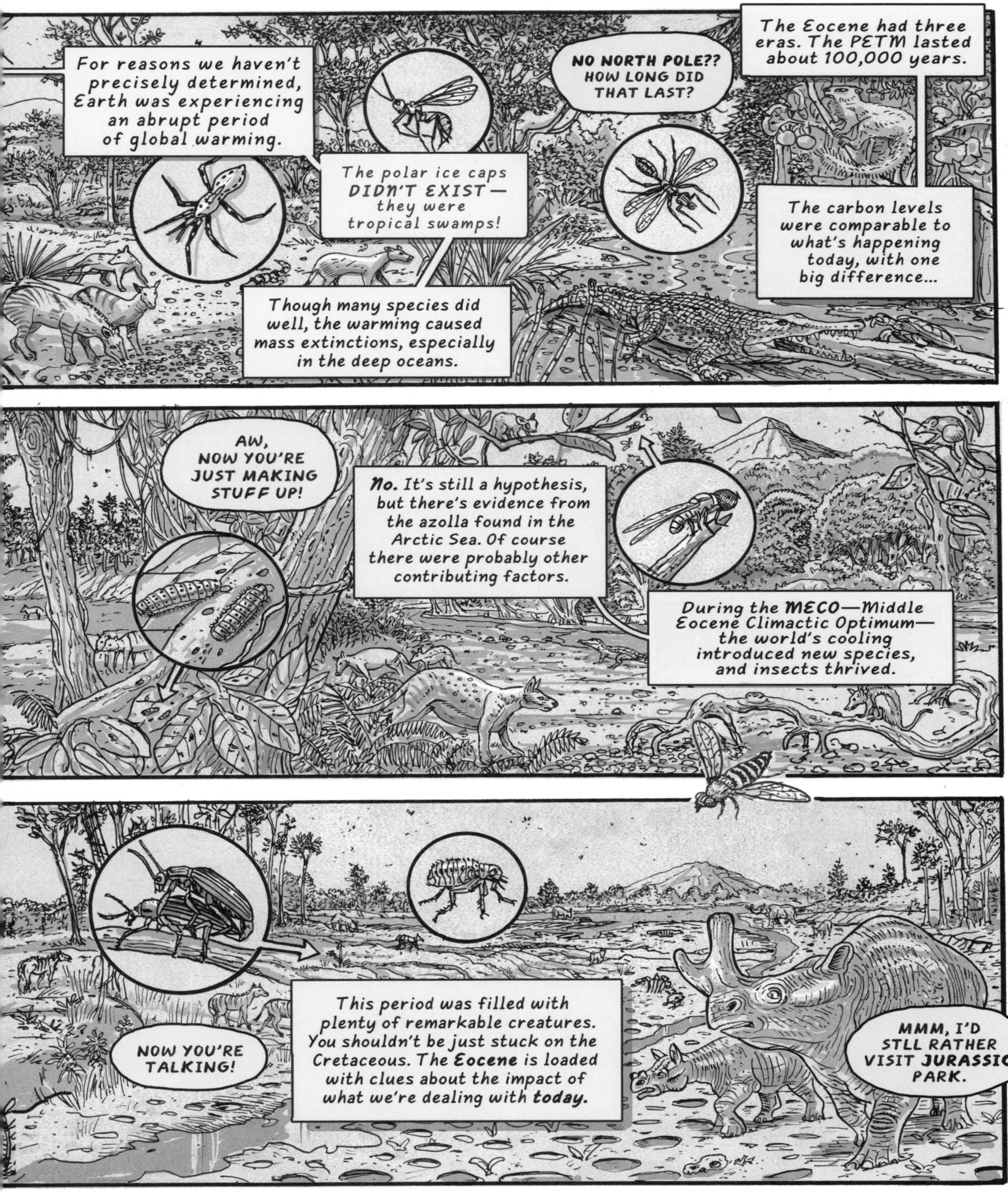

For reasons we haven't precisely determined, Earth was experiencing an abrupt period of global warming.
The polar ice caps DIDN'T EXIST— they were tropical swamps!
Though many species did well, the warming caused mass extinctions, especially in the deep oceans.
NO NORTH POLE?? HOW LONG DID THAT LAST?
The Eocene had three eras. The PETM lasted about 100,000 years.
The carbon levels were comparable to what's happening today, with one big difference...
AW, NOW YOU'RE JUST MAKING STUFF UP!
No. It's still a hypothesis, but there's evidence from the azolla found in the Arctic Sea. Of course there were probably other contributing factors.
During the MECO—Middle Eocene Climactic Optimum— the world's cooling introduced new species, and insects thrived.
NOW YOU'RE TALKING!
This period was filled with plenty of remarkable creatures. You shouldn't be just stuck on the Cretaceous. The Eocene is loaded with clues about the impact of what we're dealing with today.
MMM, I'D STLL RATHER VISIT JURASSIC PARK.

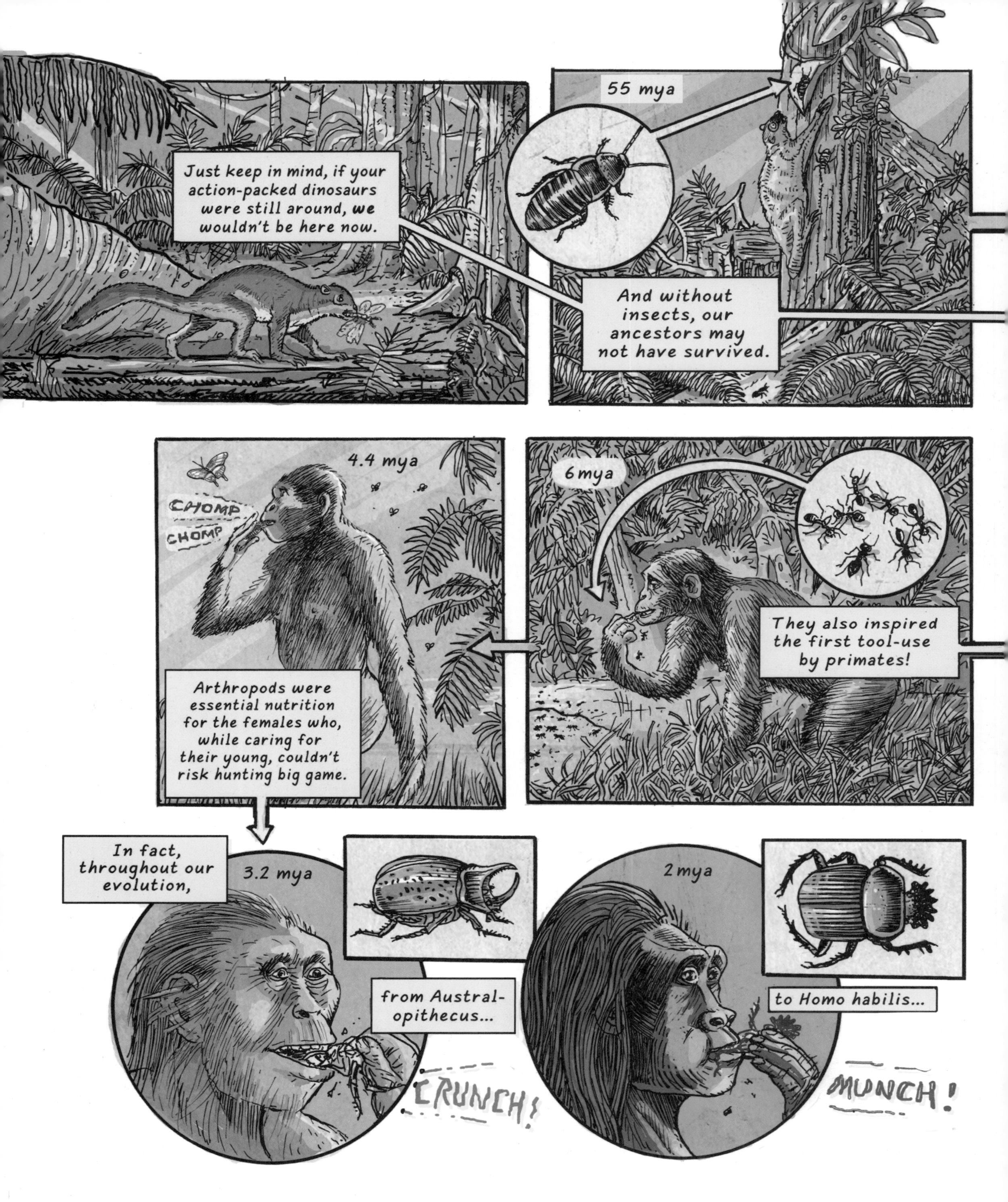
Just keep in mind, if your action-packed dinosaurs were still around, we wouldn't be here now.
55 mya
And without insects, our ancestors may not have survived.
6 mya
They also inspired the first tool-use by primates!
4.4 mya
CHOMP
CHOMP
Arthropods were essential nutrition for the females who, while caring for their young, couldn't risk hunting big game.
In fact, throughout our evolution,
3.2 mya
from Austral-opithecus...
CRUNCH!
2 mya
to Homo habilis...
MUNCH!

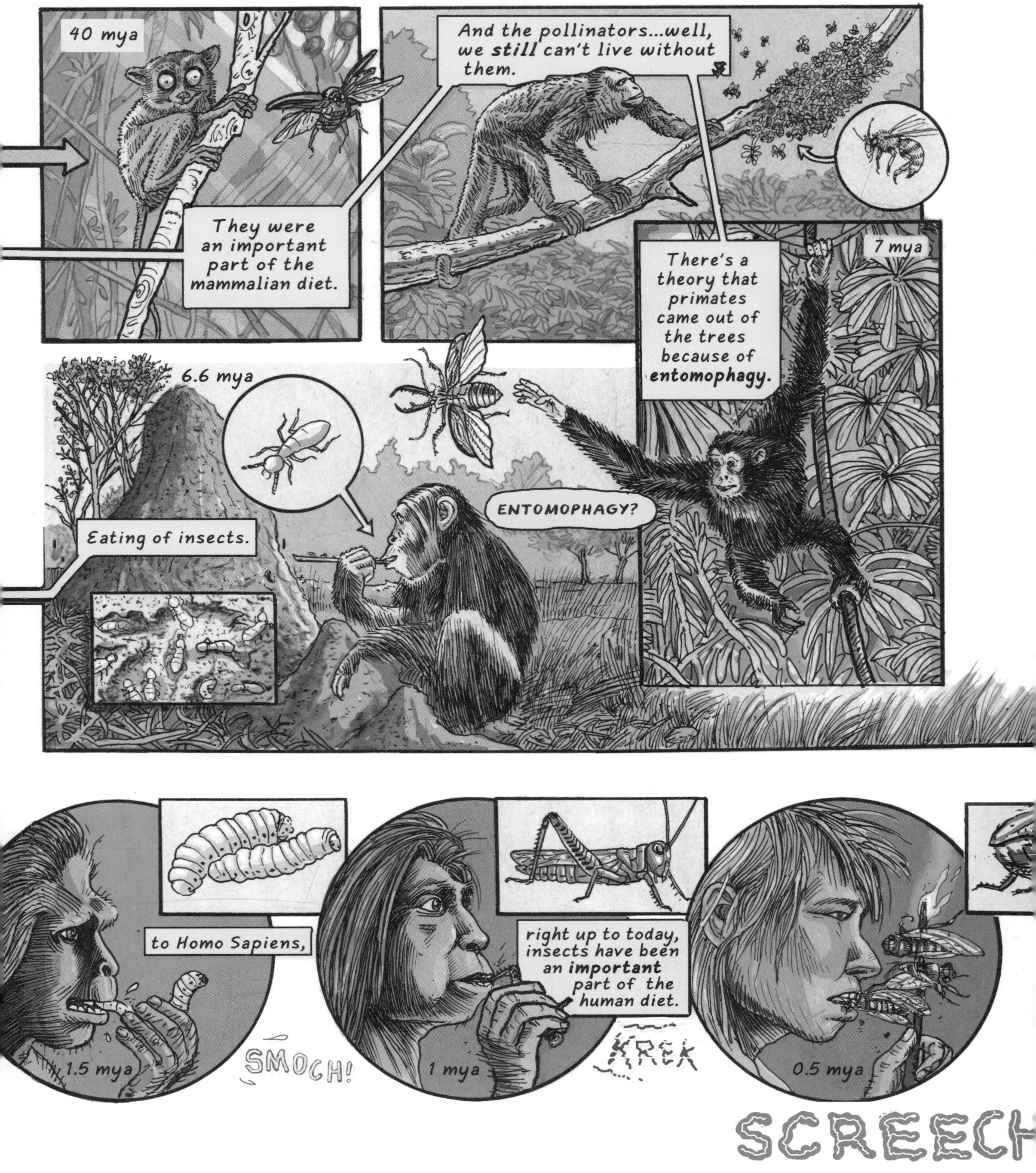

40 mya
They were an important part of the mammalian diet.
And the pollinators...well, we still can't live without them.
7 mya
There's a theory that primates came out of the trees because of entomophagy.
6.6 mya
ENTOMOPHAGY?
Eating of insects.
to Homo Sapiens,
1.5 mya
SMOCH!
right up to today, insects have been an important part of the human diet.
1 mya
KREK
0.5 mya
SCREECH

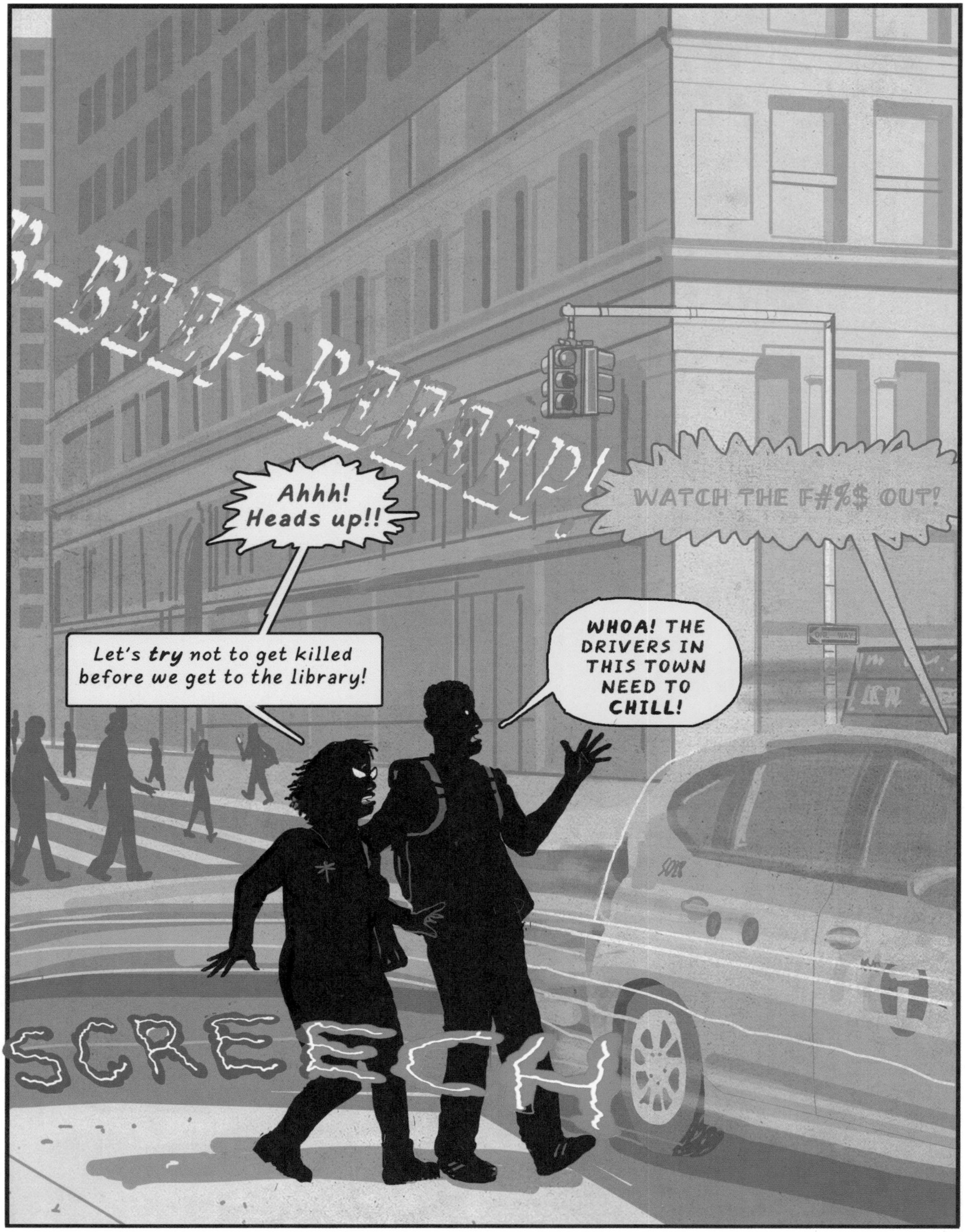

B-BEEP-BEEEEP!
Ahhh! Heads up!!
WATCH THE F#%$ OUT!
Let's *try* not to get killed before we get to the library!
WHOA! THE DRIVERS IN THIS TOWN NEED TO CHILL!
SCREECH

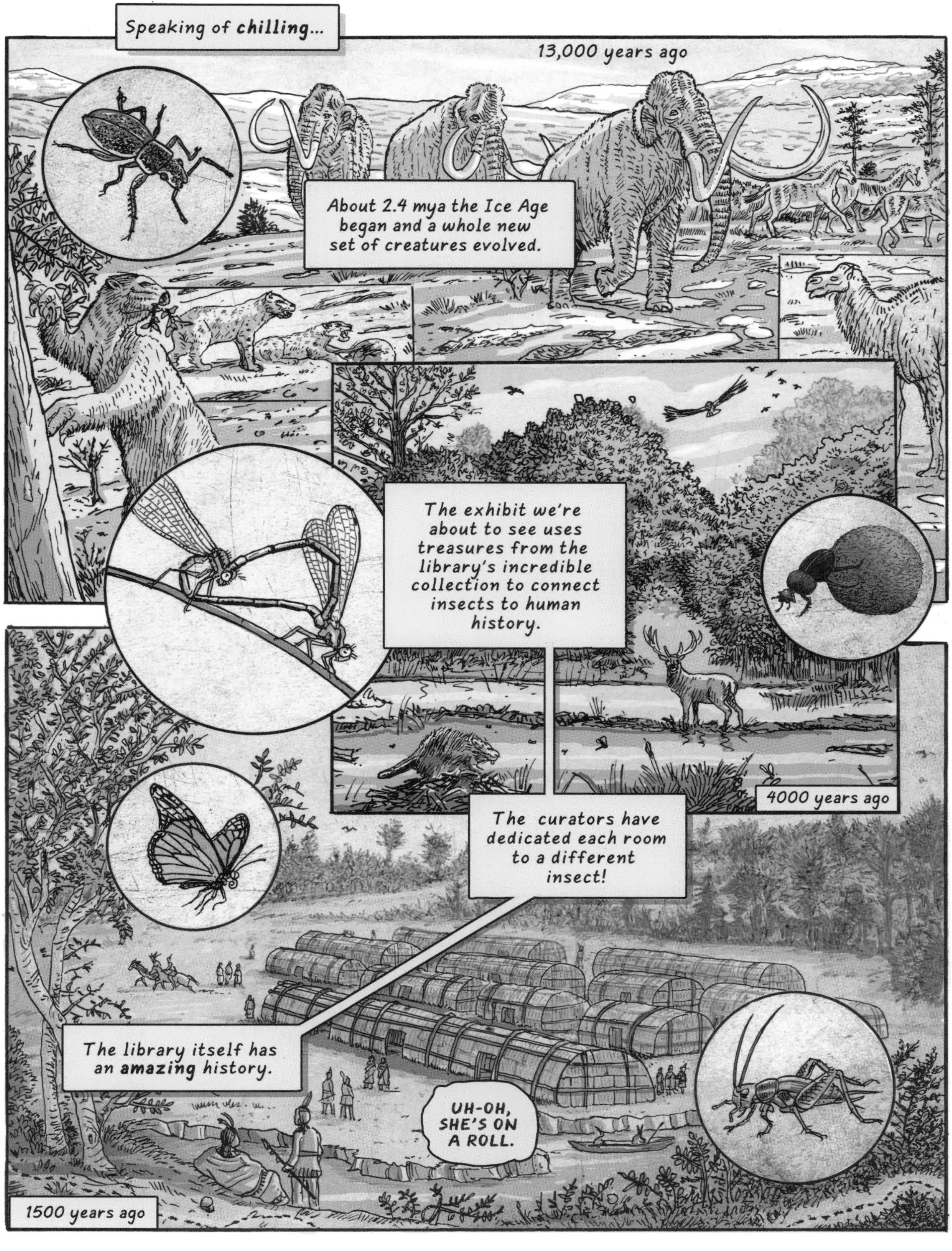
Speaking of **chilling**...
13,000 years ago
About 2.4 mya the Ice Age began and a whole new set of creatures evolved.
The exhibit we're about to see uses treasures from the library's incredible collection to connect insects to human history.
4000 years ago
The curators have dedicated each room to a different insect!
The library itself has an **amazing** history.
UH-OH, SHE'S ON A ROLL.
1500 years ago

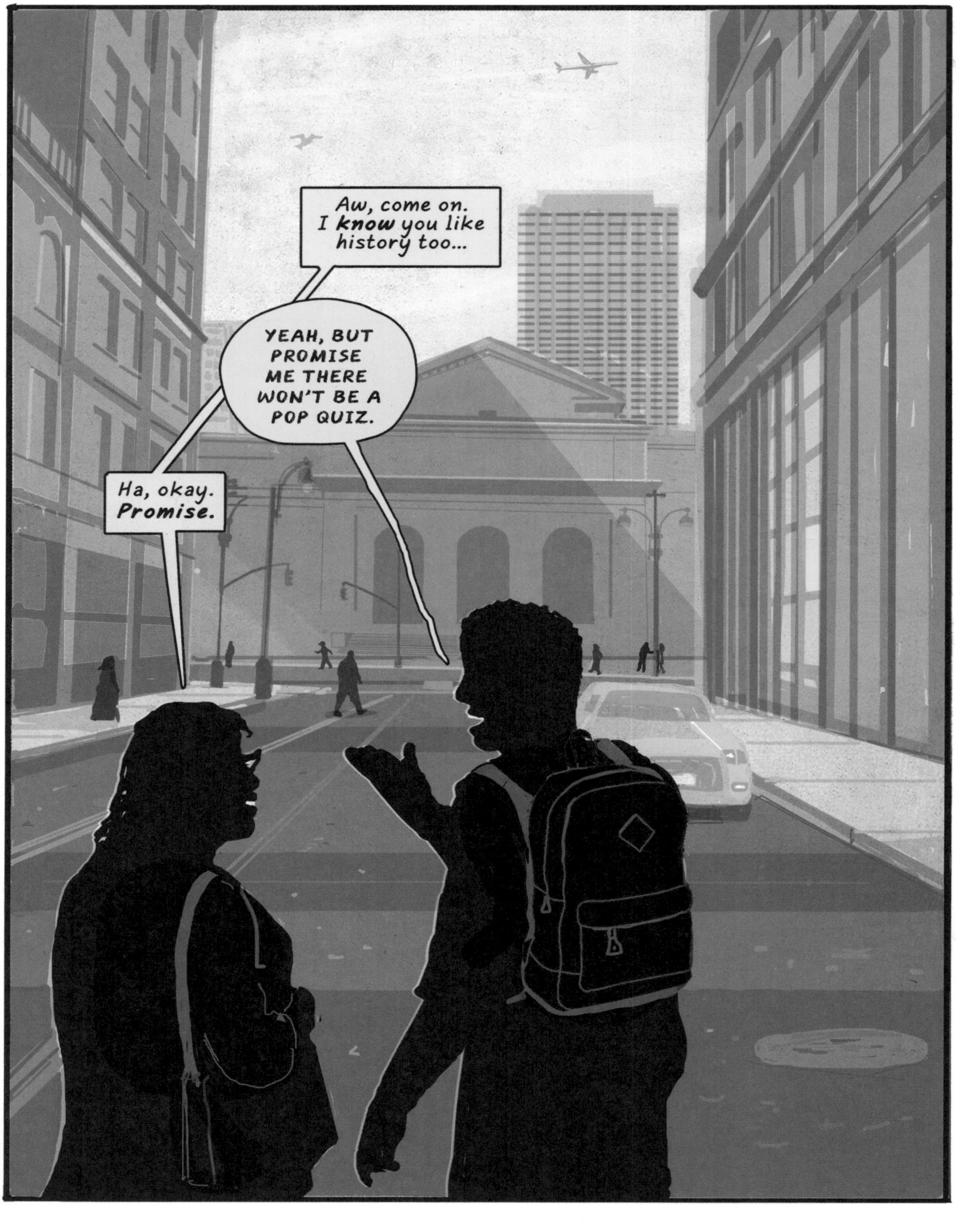
Aw, come on. I know you like history too...
YEAH, BUT PROMISE ME THERE WON'T BE A POP QUIZ.
Ha, okay. Promise.

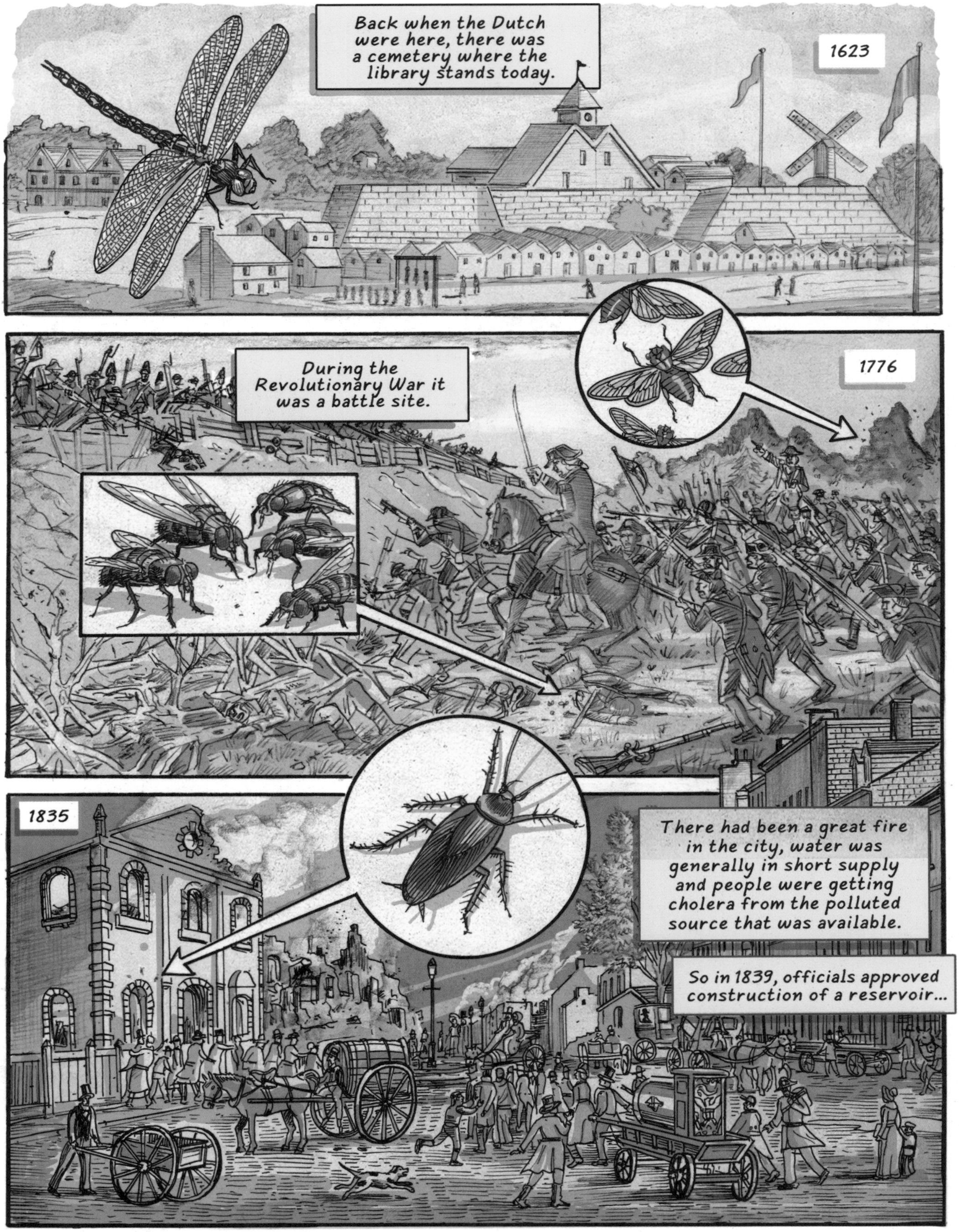
Back when the Dutch were here, there was a cemetery where the library stands today.
1623
During the Revolutionary War it was a battle site.
1776
1835
There had been a great fire in the city, water was generally in short supply and people were getting cholera from the polluted source that was available.
So in 1839, officials approved construction of a reservoir...

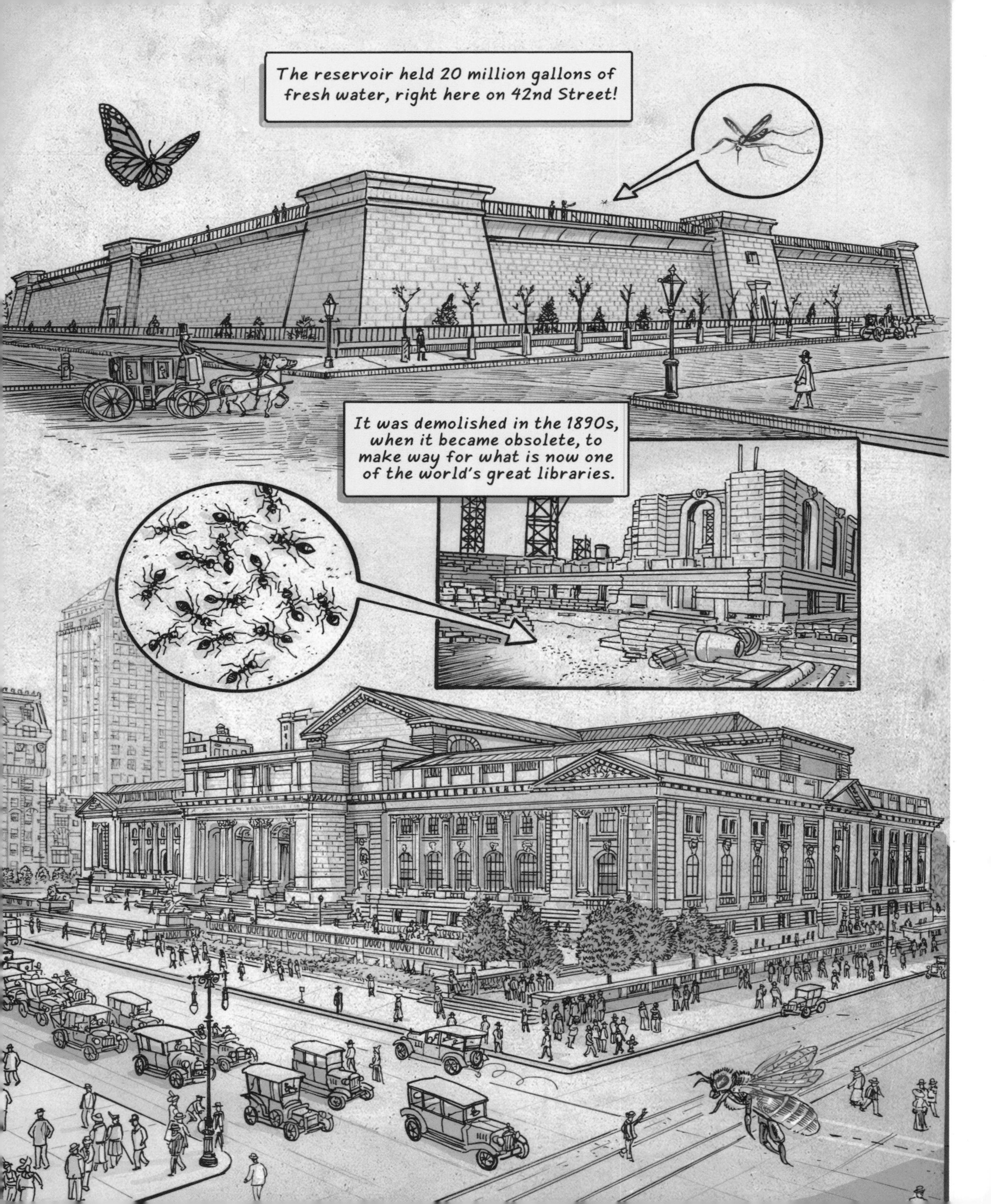
The reservoir held 20 million gallons of fresh water, right here on 42nd Street!
It was demolished in the 1890s, when it became obsolete, to make way for what is now one of the world's great libraries.

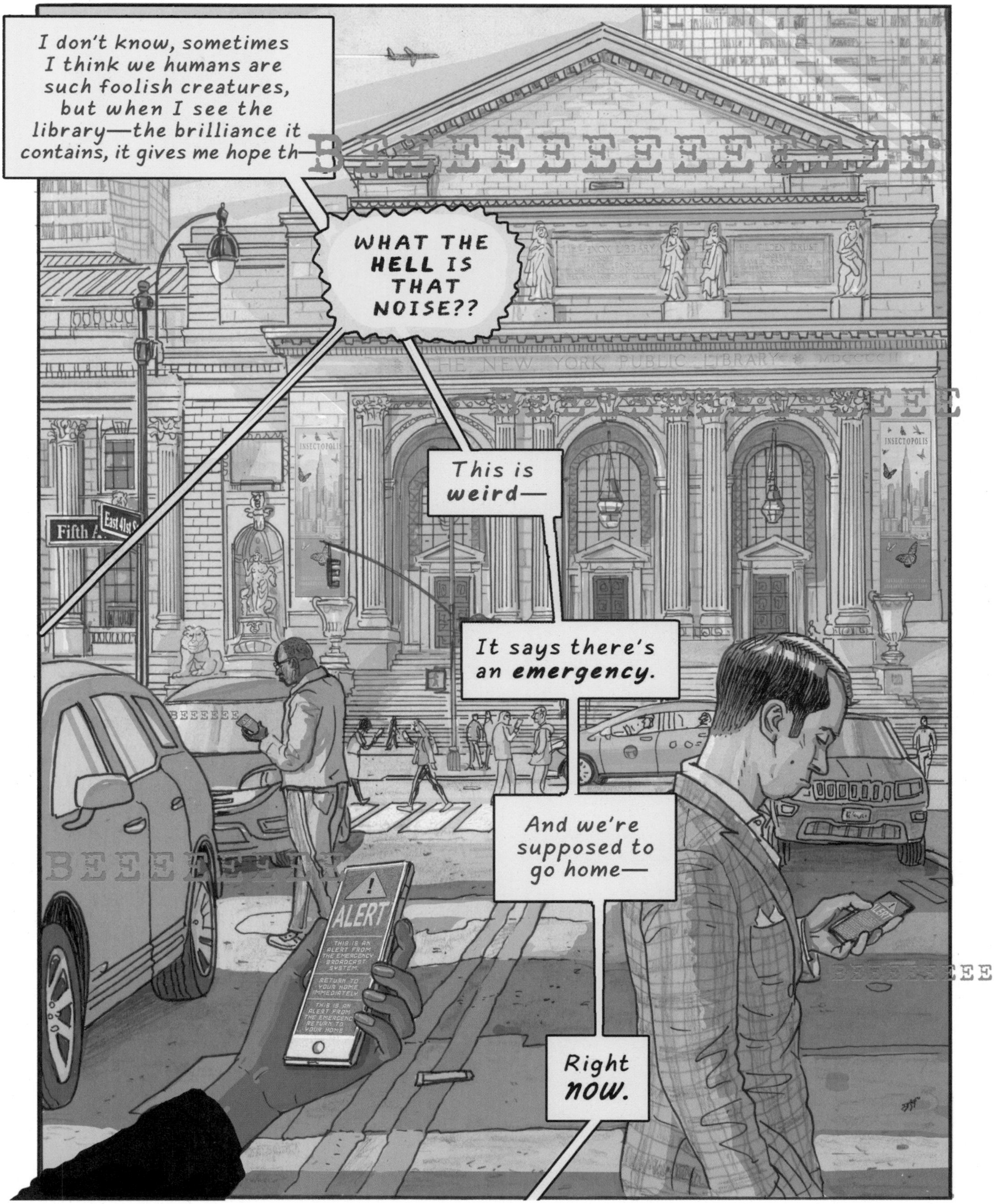

I don't know, sometimes I think we humans are such foolish creatures, but when I see the library—the brilliance it contains, it gives me hope th—
BEEEEEEEEEEEEEEE
WHAT THE HELL IS THAT NOISE??
This is weird—
It says there's an emergency.
And we're supposed to go home—
Right NOW.
THE NEW YORK PUBLIC LIBRARY
INSECTOPOLIS
Fifth A
East 41st St
ALERT
THIS IS AN ALERT FROM THE EMERGENCY BROADCAST SYSTEM.
RETURN TO YOUR HOME IMMEDIATELY
BEEEEEEE

One week later.

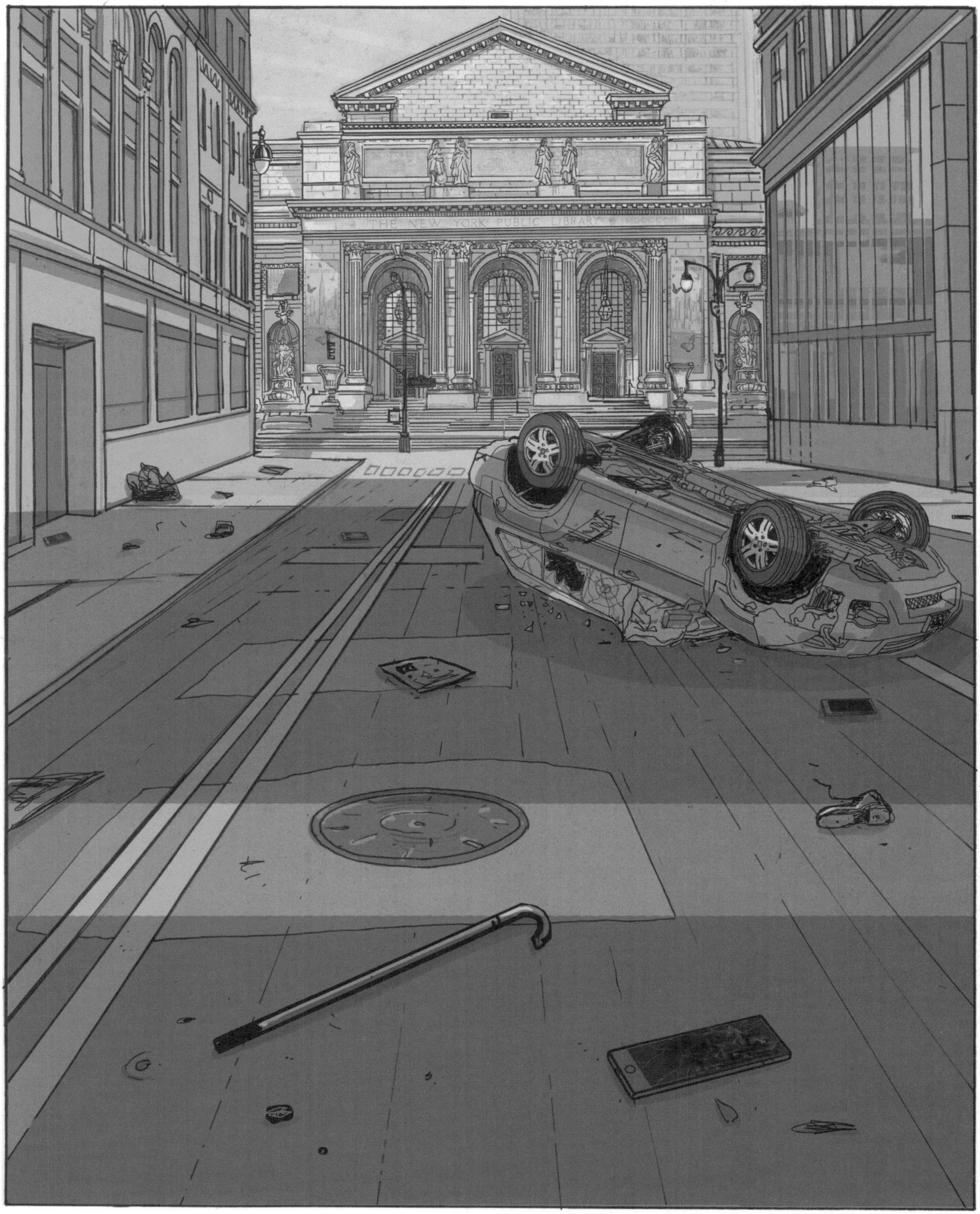
THE NEW YORK PUBLIC LIBRARY

Much (much) later.

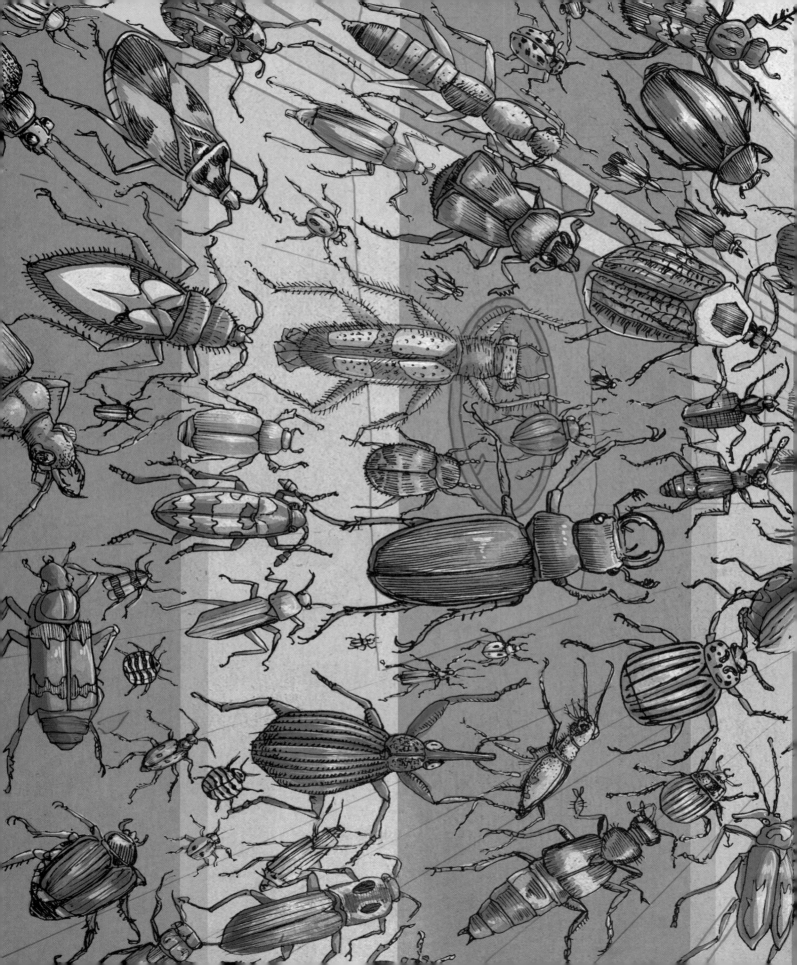

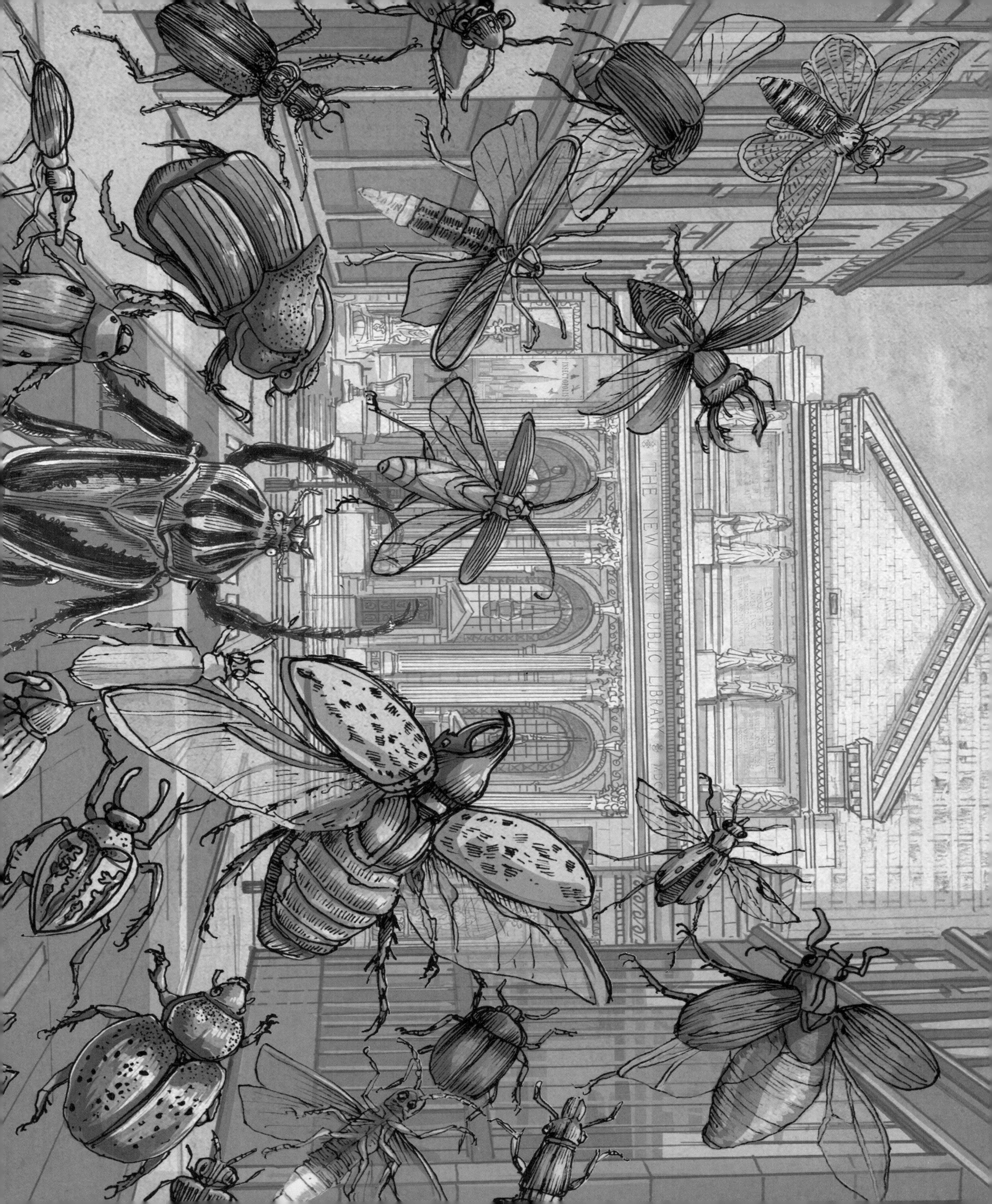
THE NEW YORK PUBLIC LIBRARY
LENOX LIBRARY
INSECTOPOLIS

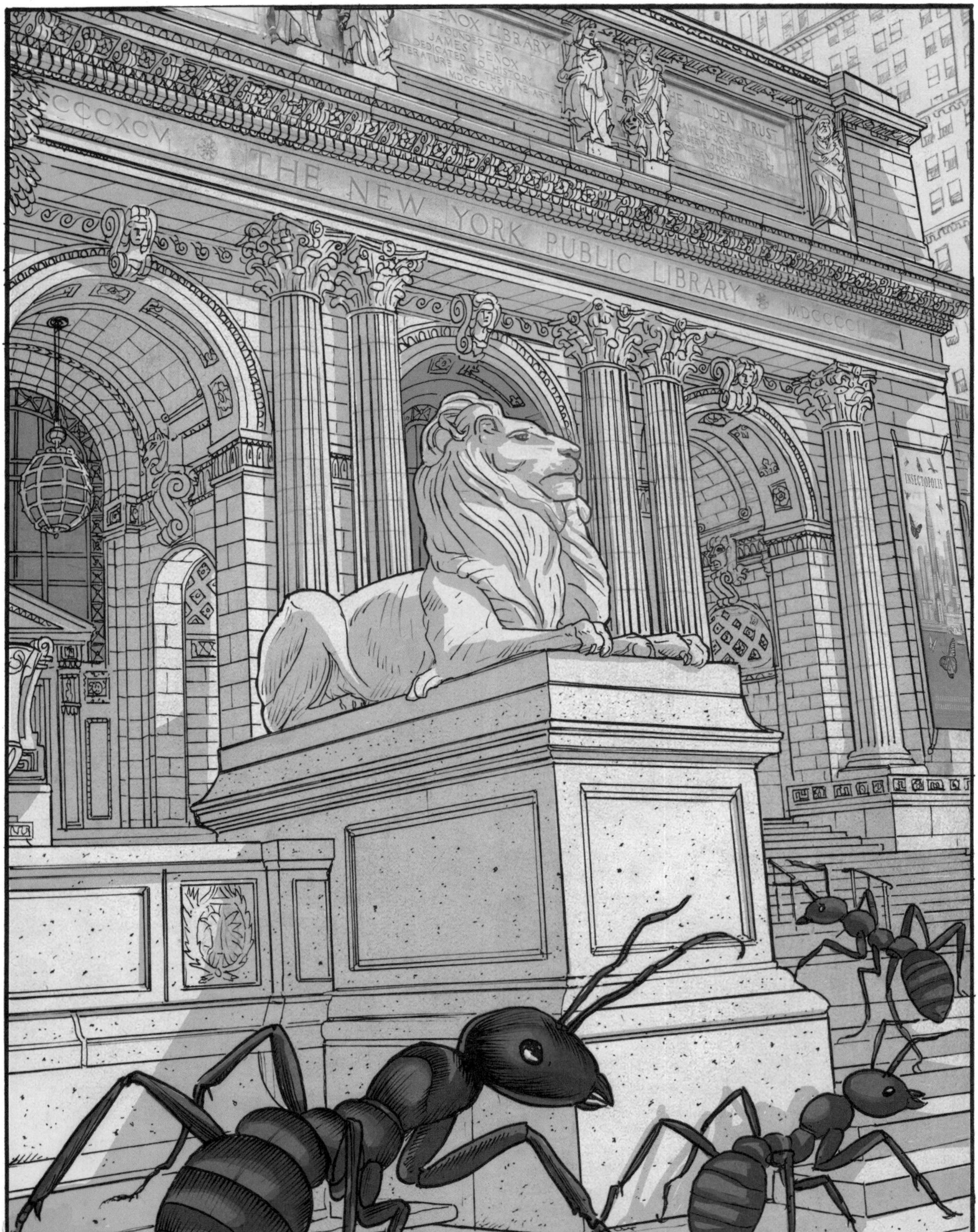
LENOX LIBRARY
FOUNDED BY
JAMES LENOX
DEDICATED TO HISTORY
LITERATURE AND THE FINE ARTS
MDCCCLXX
TILDEN TRUST
THE NEW YORK PUBLIC LIBRARY
MDCCCCII
INSECTOPOLIS

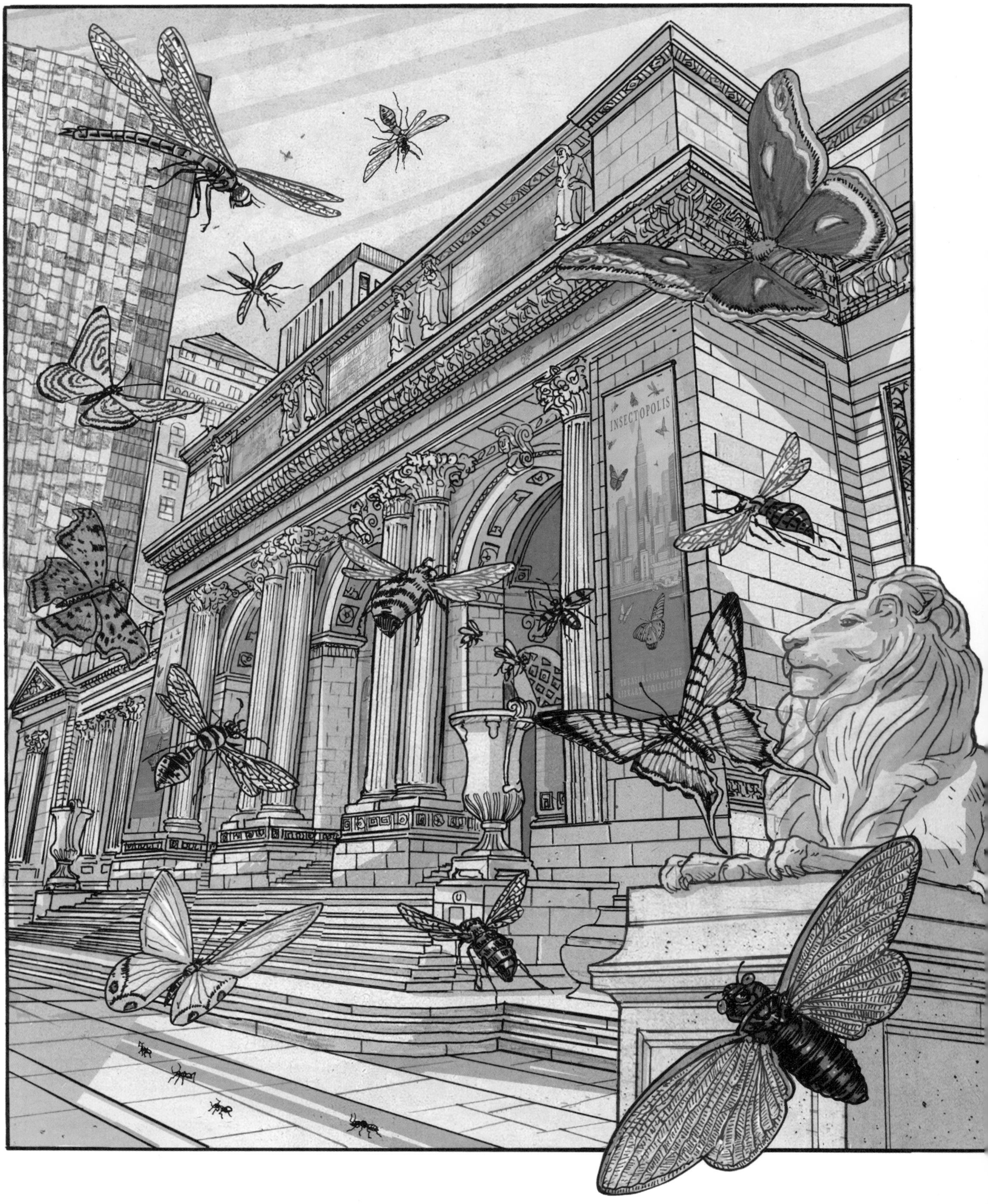

INSECTOPOLIS
THE NEW YORK PUBLIC LIBRARY

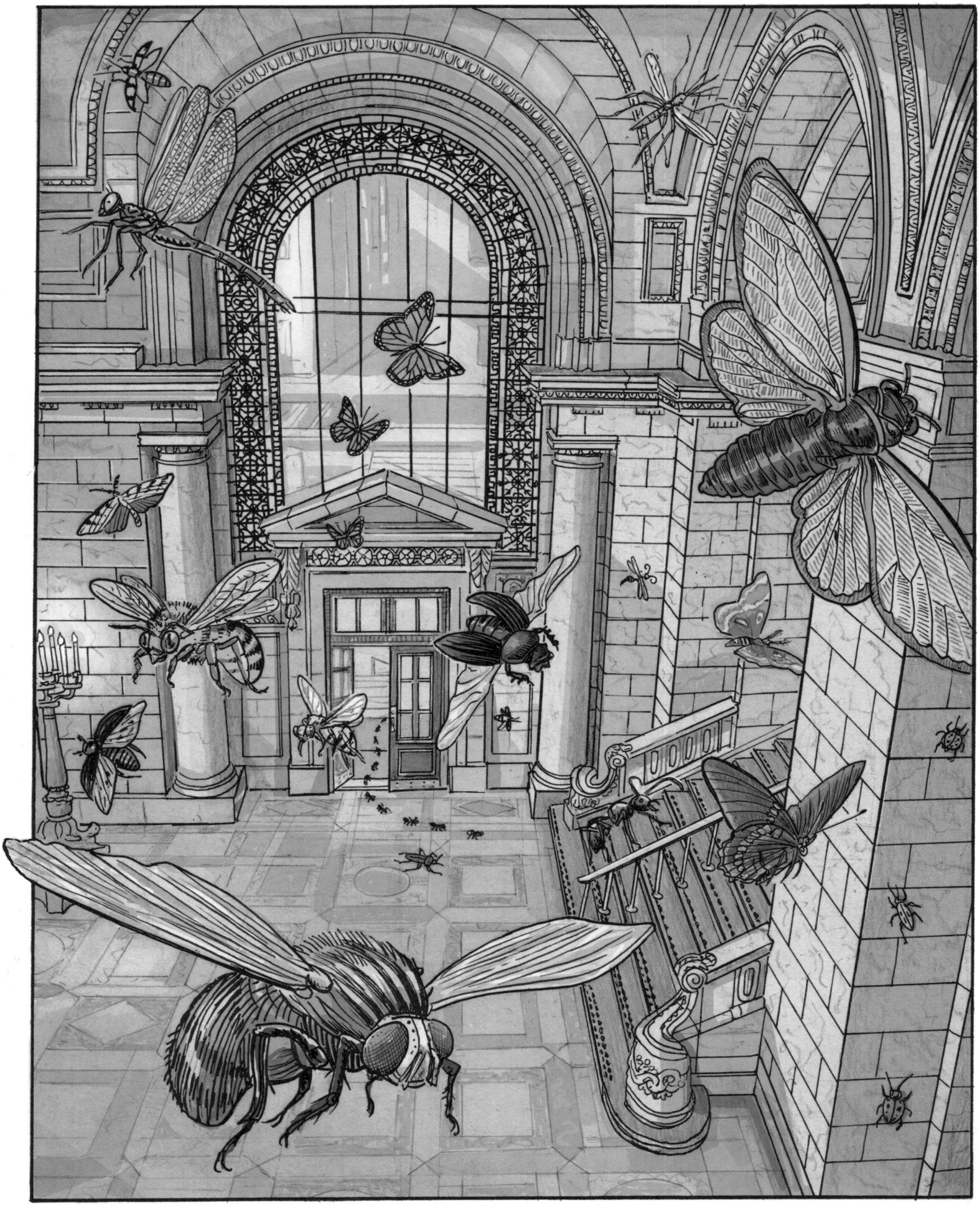

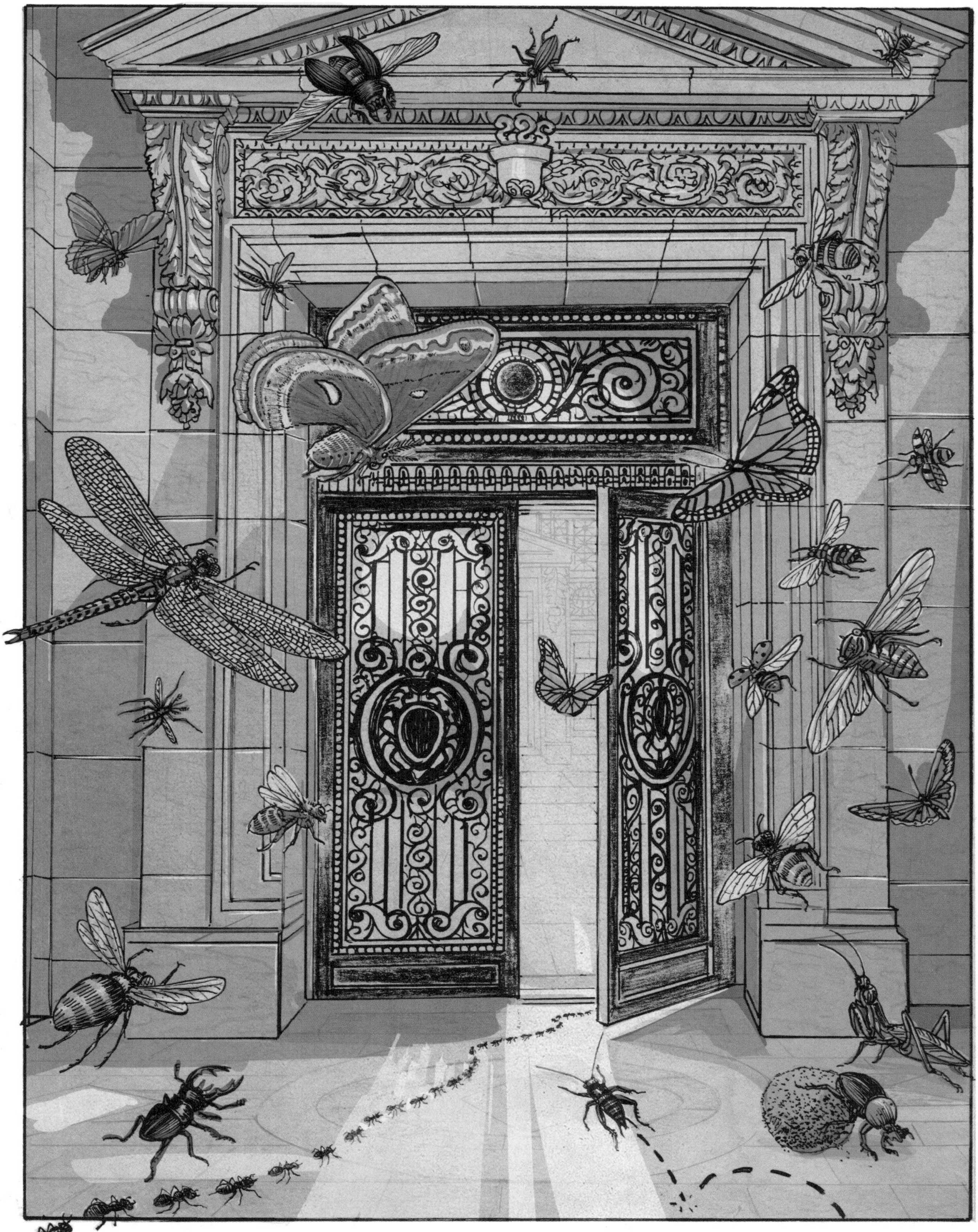

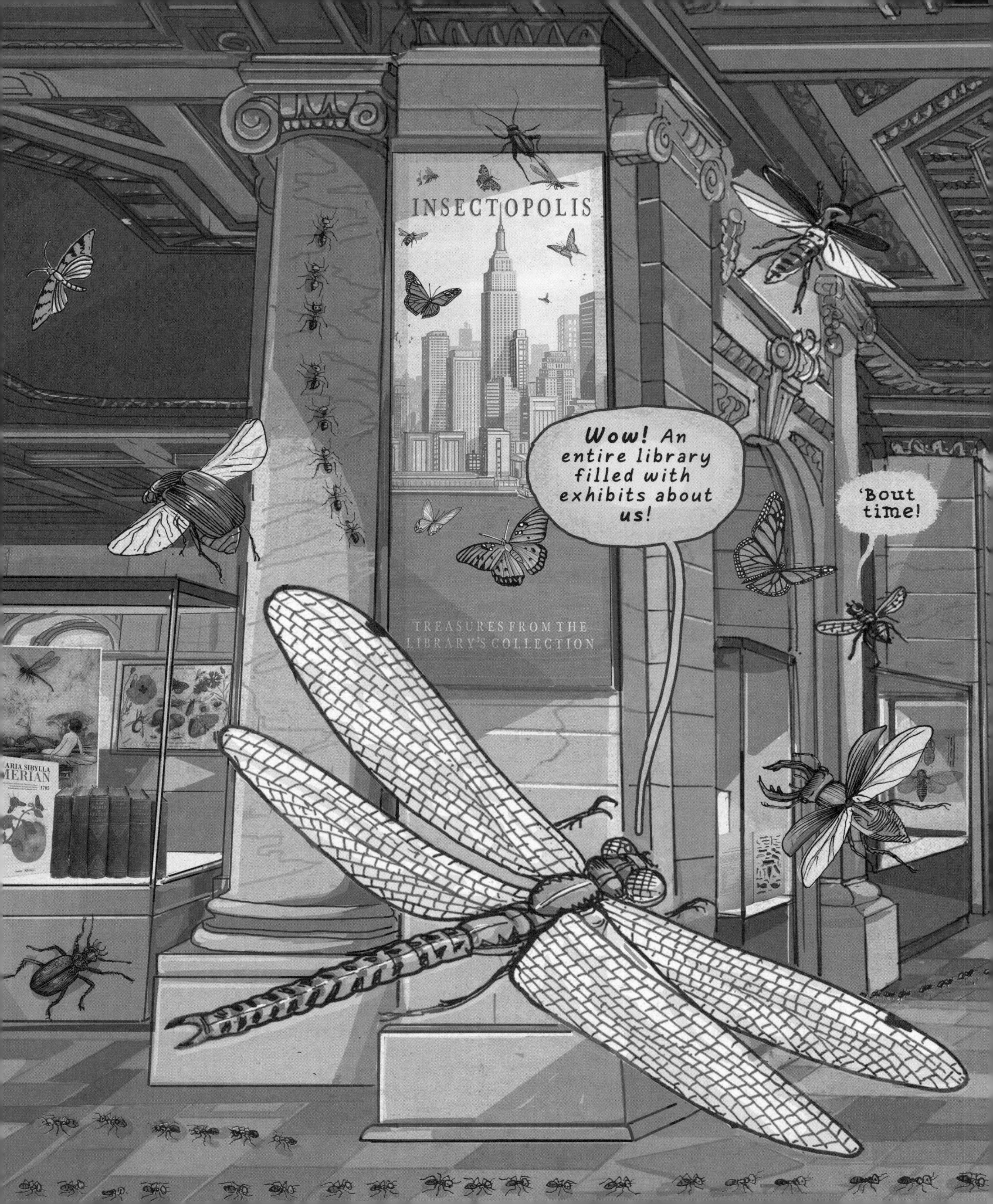
INSECTOPOLIS
TREASURES FROM THE LIBRARY'S COLLECTION
Wow! An entire library filled with exhibits about us!
'Bout time!
MARIA SIBYLLA MERIAN
1705

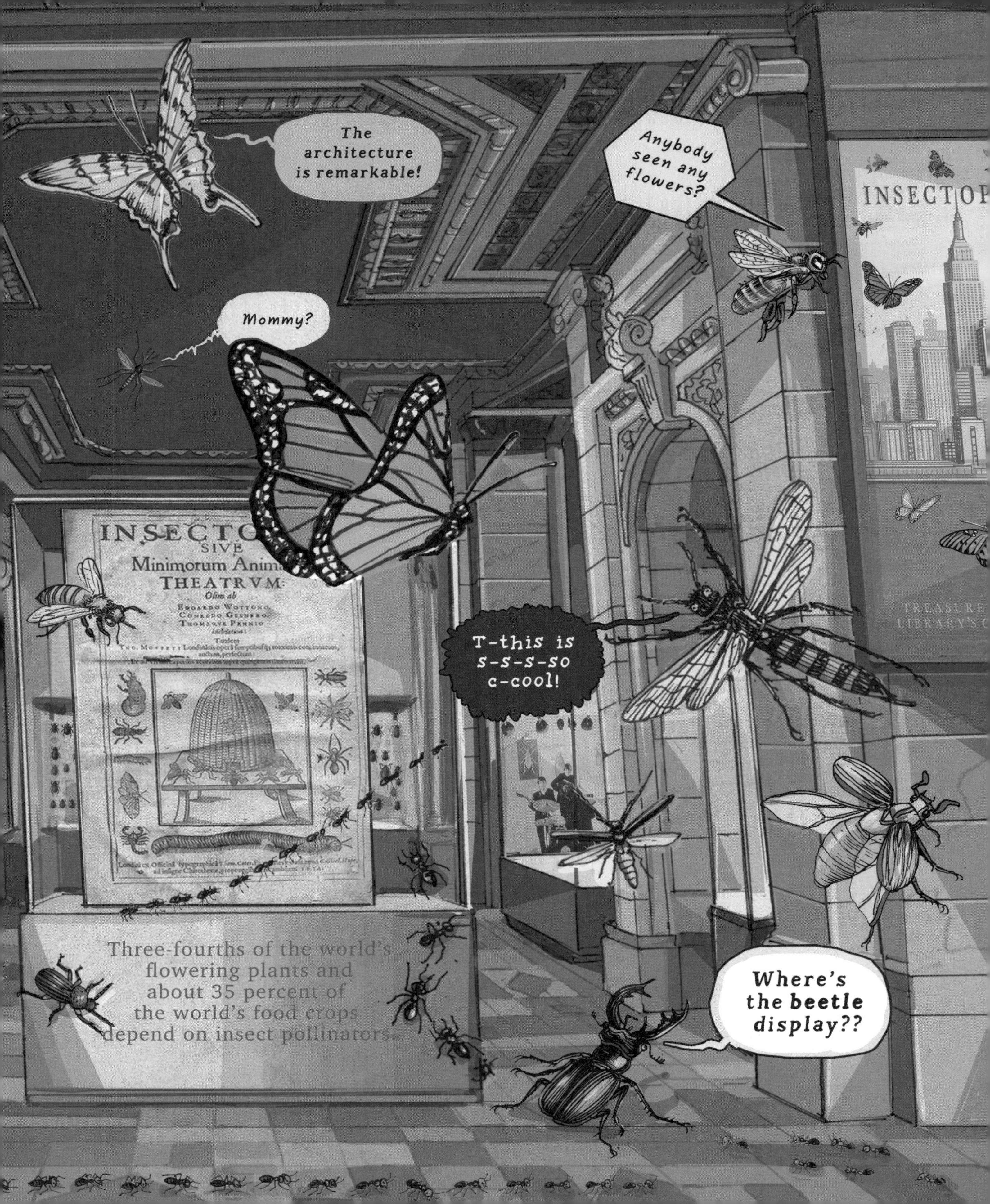
The architecture is remarkable!
Anybody seen any flowers?
INSECTOP
Mommy?
INSECTO
SIVE
Minimorum Anima
THEATRVM
TREASURE
LIBRARY'S C
T-this is s-s-s-so c-cool!
Where's the **beetle** display??

Three-fourths of the world's flowering plants and about 35 percent of the world's food crops depend on insect pollinators.

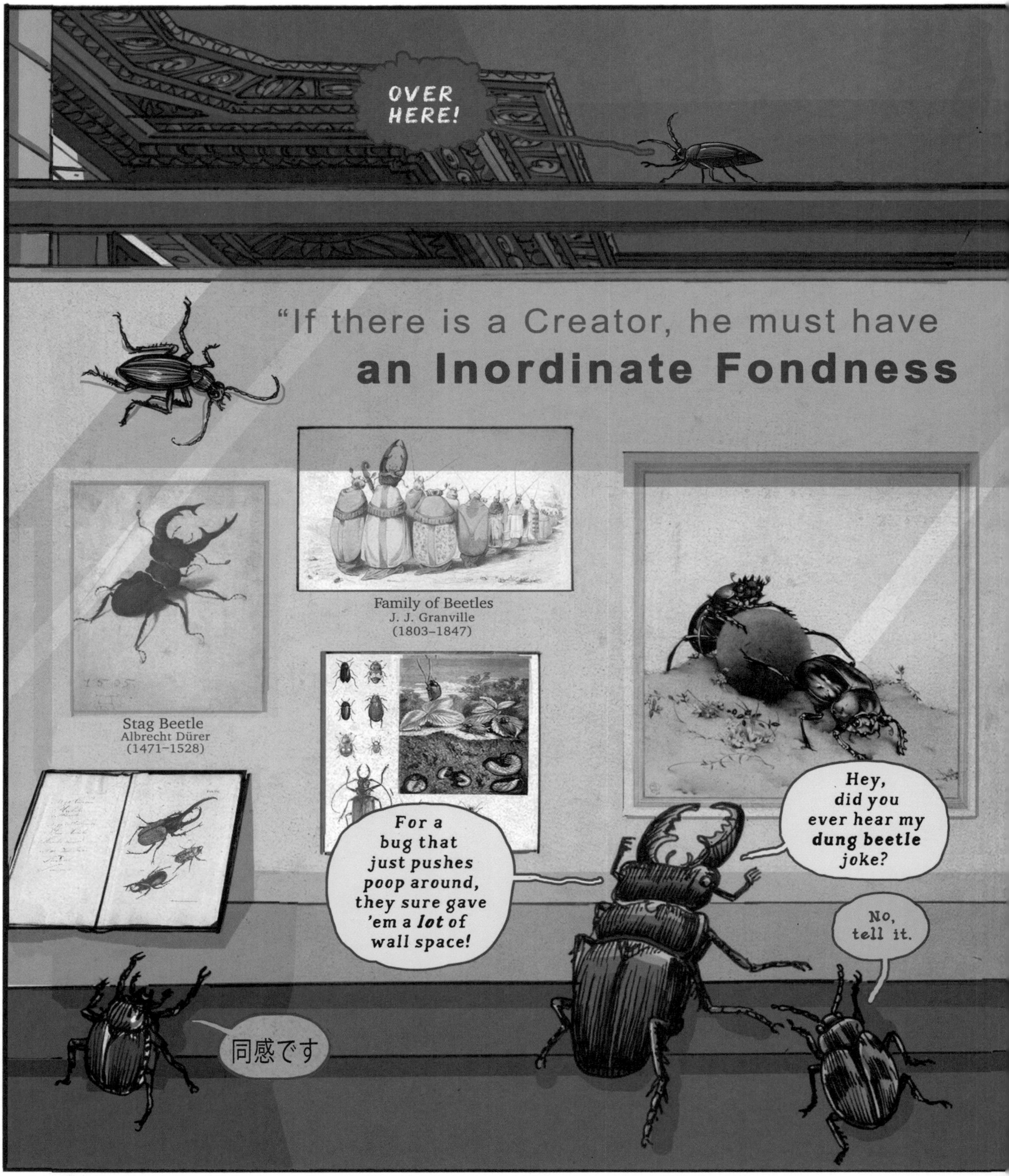
OVER HERE!
"If there is a Creator, he must have
an Inordinate Fondness
Family of Beetles
J. J. Granville
(1803–1847)
Stag Beetle
Albrecht Dürer
(1471–1528)
For a bug that just pushes poop around, they sure gave 'em a *lot* of wall space!
Hey, did you ever hear my **dung beetle** joke?
No, tell it.
同感です

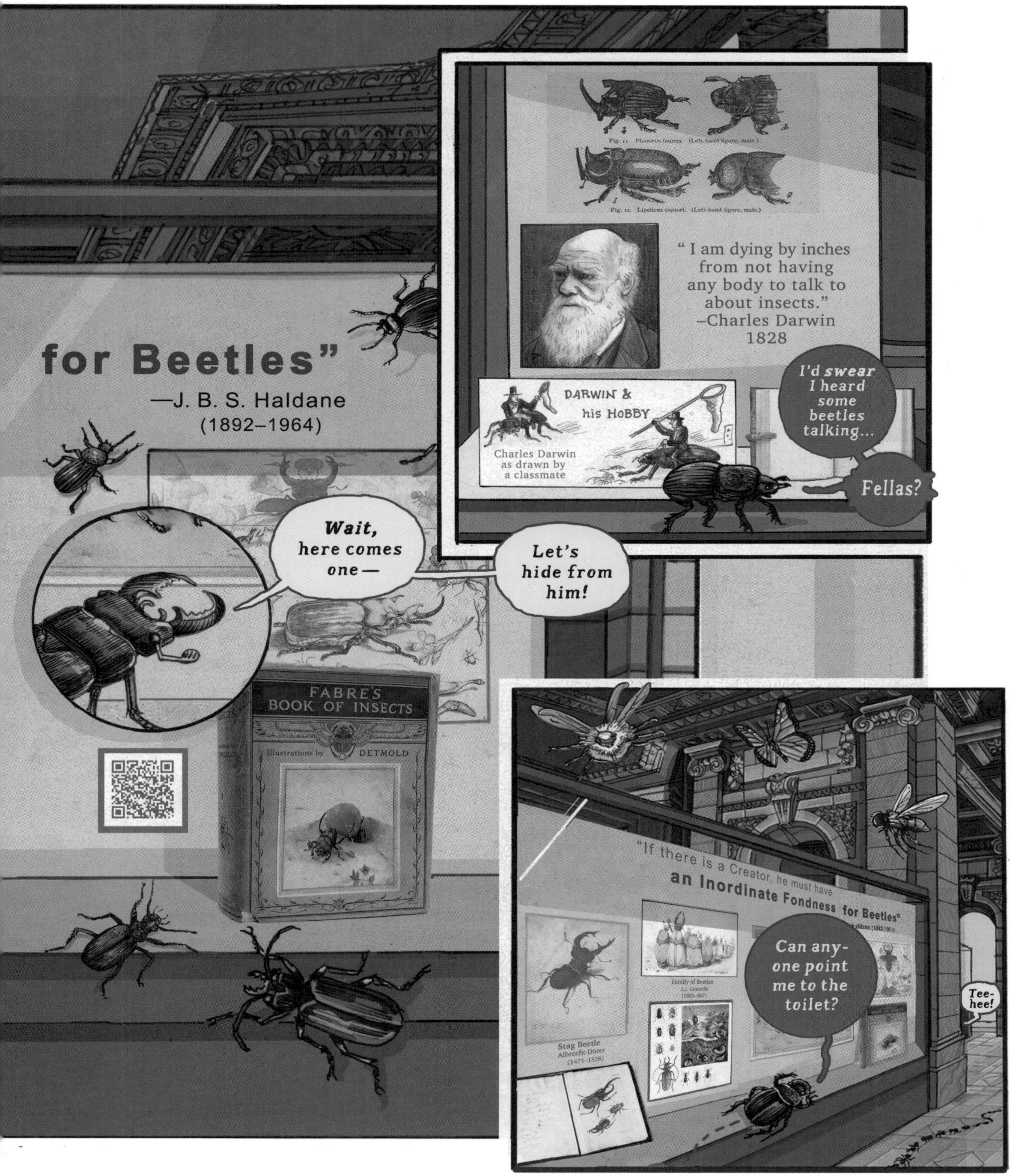

for Beetles"
—J. B. S. Haldane
(1892–1964)
FABRE'S BOOK OF INSECTS
Illustrations by DETMOLD
Wait, here comes one—
Let's hide from him!
Fig. 11. Phanæus faunus. (Left-hand figure, male.)
Fig. 12. Lipelicus cantori. (Left-hand figure, male.)
" I am dying by inches from not having any body to talk to about insects."
–Charles Darwin
1828
DARWIN & his HOBBY
Charles Darwin as drawn by a classmate
I'd *swear* I heard some beetles talking...
Fellas?
"If there is a Creator, he must have
an Inordinate Fondness for Beetles"
Stag Beetle
Albrecht Dürer
(1471–1528)
Family of Beetles
J.J. Granville
(1803–1847)
Can anyone point me to the toilet?
Tee-hee!

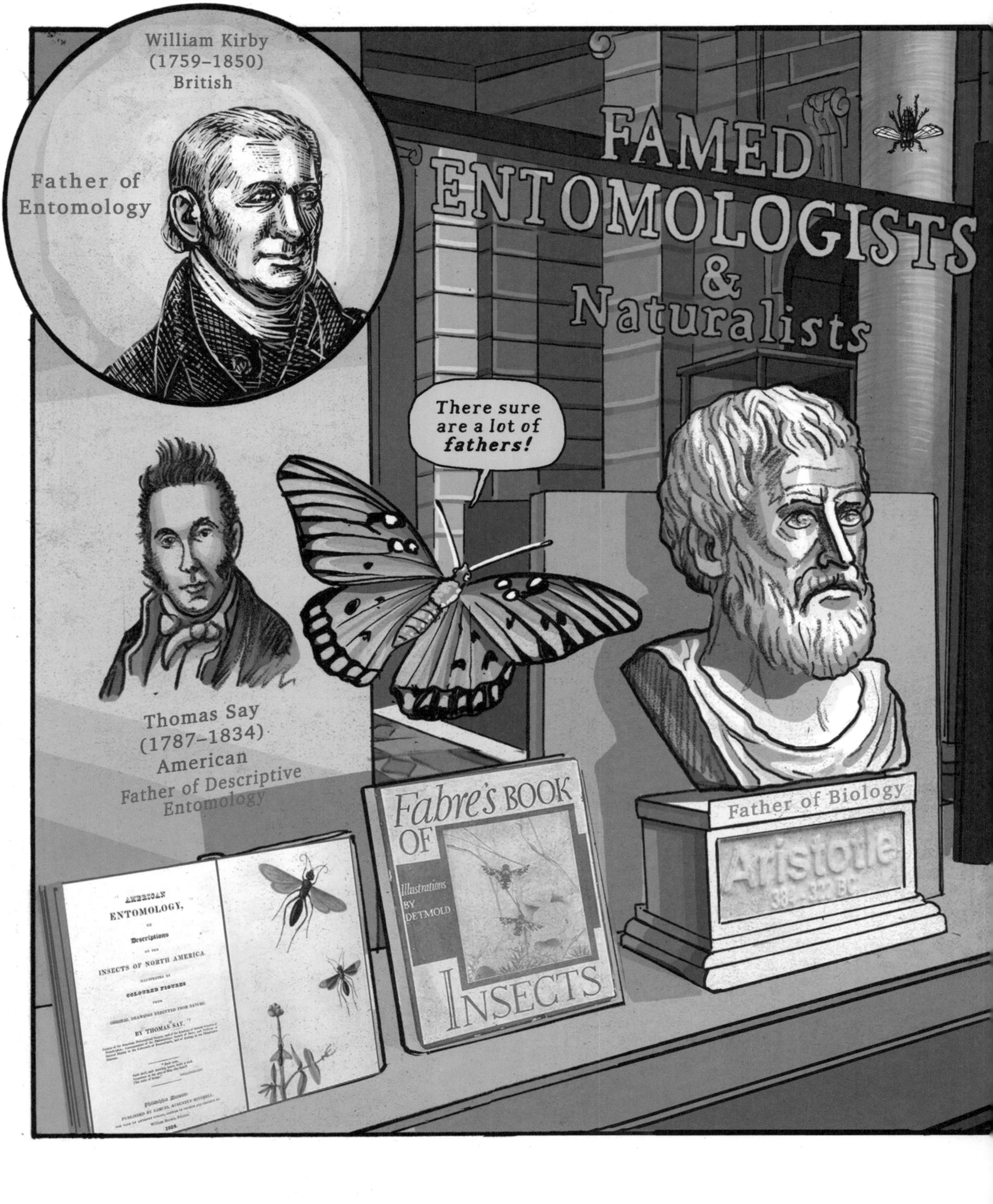

FAMED ENTOMOLOGISTS & Naturalists
William Kirby
(1759–1850)
British
Father of Entomology
There sure are a lot of *fathers!*
Thomas Say
(1787–1834)
American
Father of Descriptive Entomology
Father of Biology
Aristotle
AMERICAN ENTOMOLOGY,
Descriptions
INSECTS OF NORTH AMERICA.
COLOURED FIGURES
BY THOMAS SAY.
Fabre's BOOK OF INSECTS
Illustrations BY DETMOLD

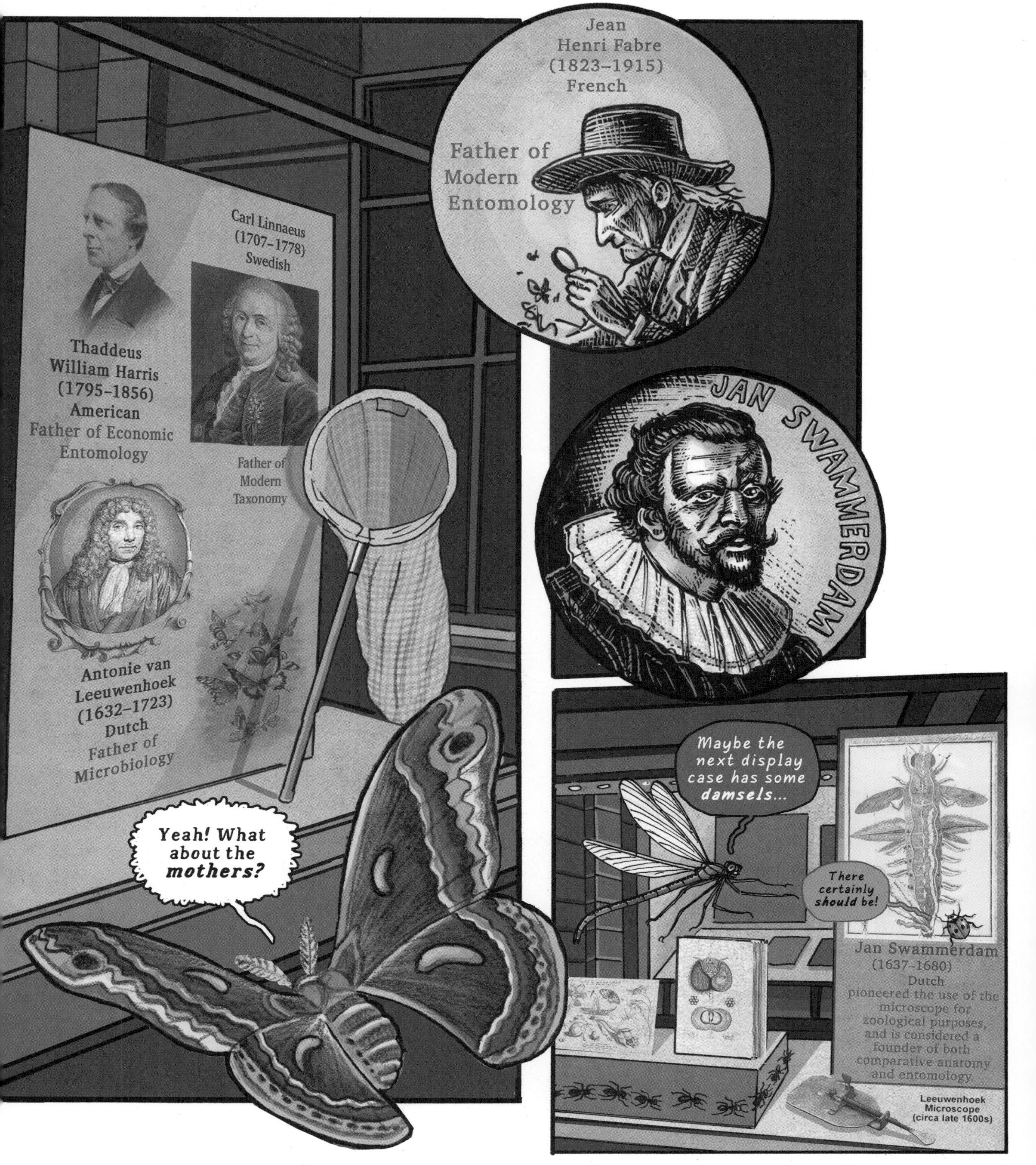

Jean
Henri Fabre
(1823–1915)
French
Father of
Modern
Entomology
Thaddeus
William Harris
(1795–1856)
American
Father of Economic
Entomology
Carl Linnaeus
(1707–1778)
Swedish
Father of
Modern
Taxonomy
Antonie van
Leeuwenhoek
(1632–1723)
Dutch
Father of
Microbiology
JAN SWAMMERDAM
Yeah! What about the *mothers?*
Maybe the next display case has some *damsels...*
There certainly *should* be!
Jan Swammerdam
(1637–1680)
Dutch
pioneered the use of the
microscope for
zoological purposes,
and is considered a
founder of both
comparative anatomy
and entomology.
Leeuwenhoek
Microscope
(circa late 1600s)

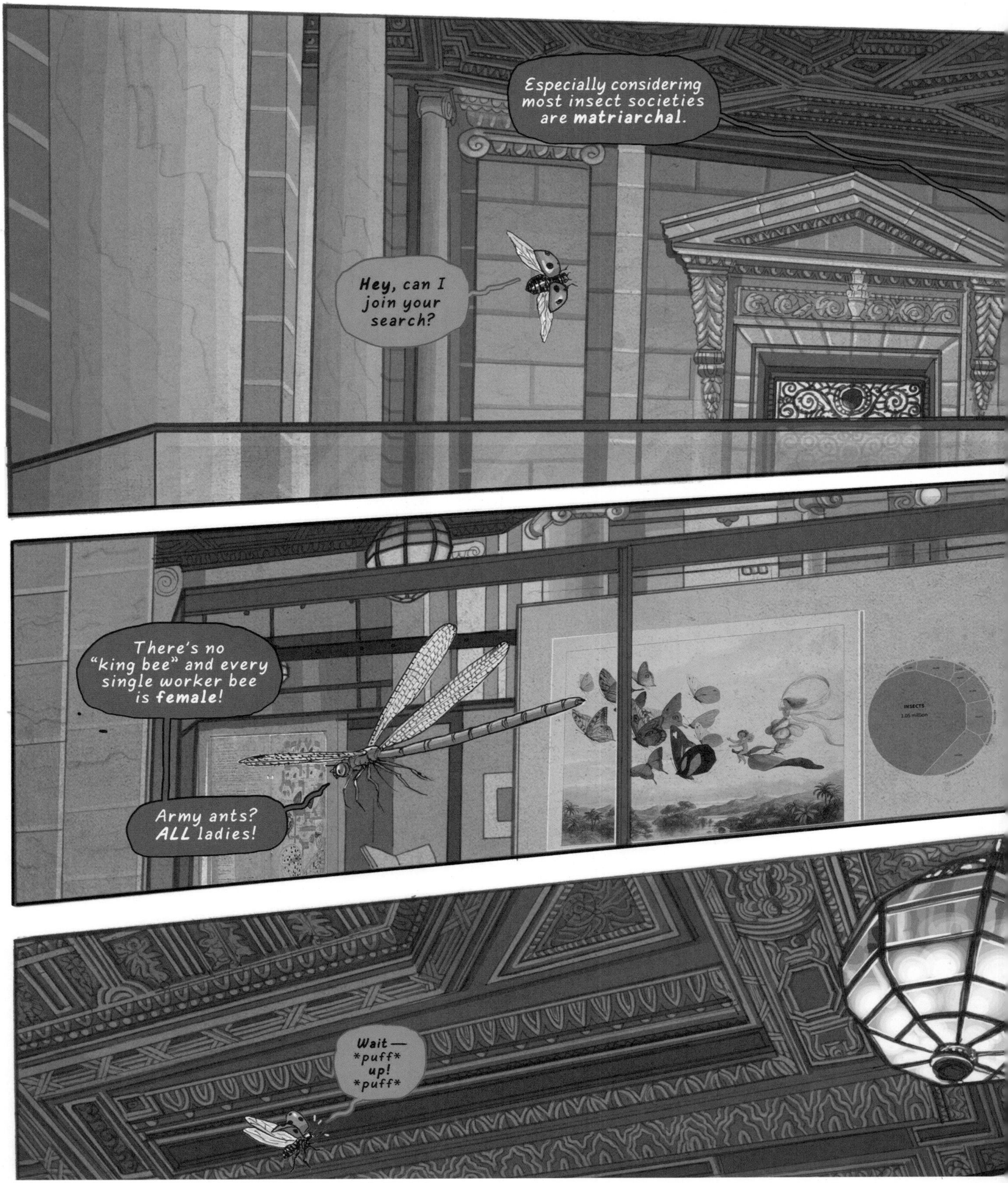

Especially considering most insect societies are **matriarchal**.
Hey, can I join your search?
There's no "king bee" and every single worker bee is **female**!
Army ants? ***ALL*** ladies!
INSECTS
Wait— *puff* up! *puff*

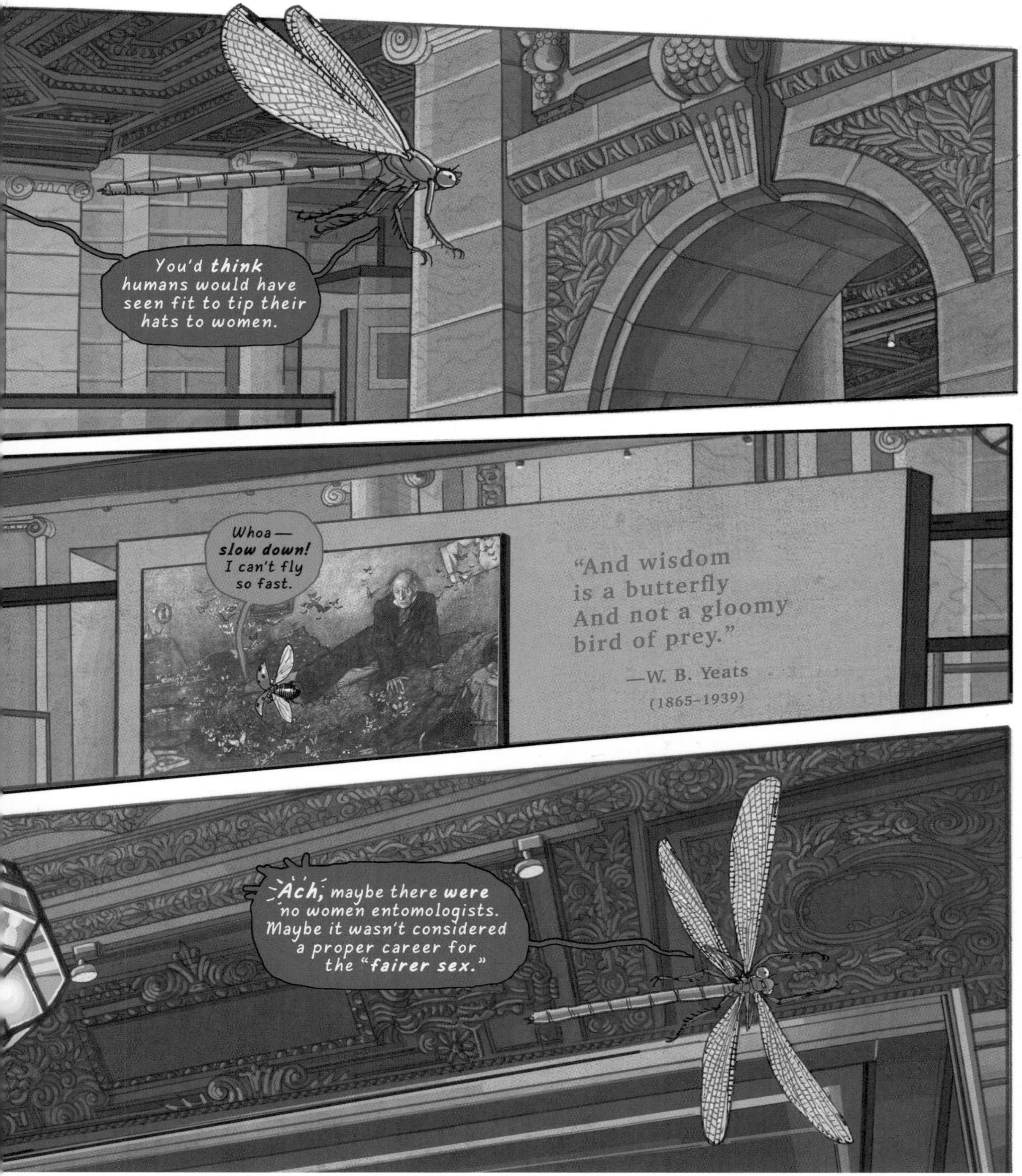
You'd **think** humans would have seen fit to tip their hats to women.
Whoa— **slow down!** I can't fly so fast.
"And wisdom is a butterfly And not a gloomy bird of prey." —W. B. Yeats (1865–1939)
Ach, maybe there **were** no women entomologists. Maybe it wasn't considered a proper career for the "**fairer sex.**"

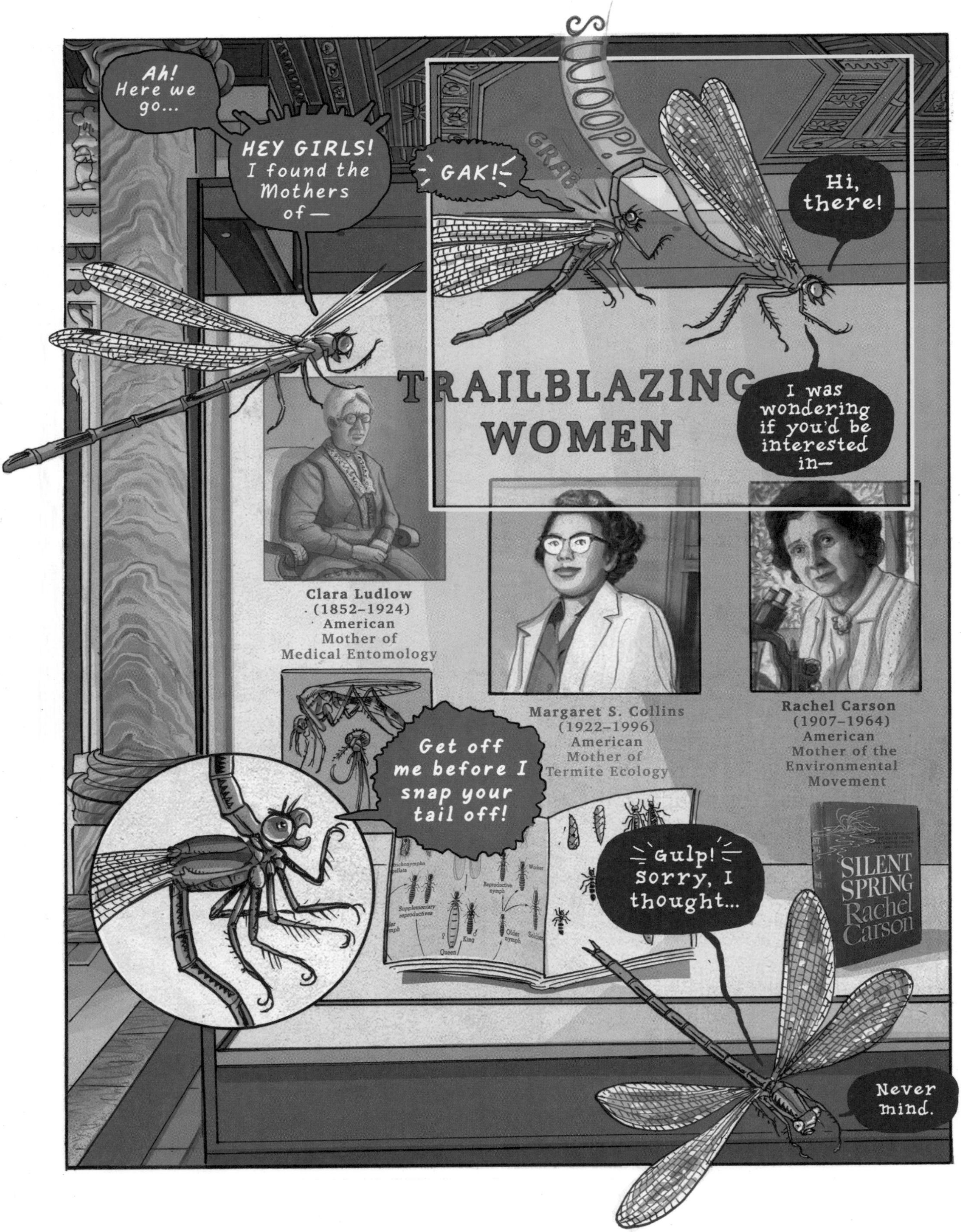
Ah! Here we go...
HEY GIRLS! I found the Mothers of—
GAK!
GRAB
SWOOP!
Hi, there!
I was wondering if you'd be interested in—
TRAILBLAZING WOMEN
Clara Ludlow (1852–1924) American Mother of Medical Entomology
Margaret S. Collins (1922–1996) American Mother of Termite Ecology
Rachel Carson (1907–1964) American Mother of the Environmental Movement
Get off me before I snap your tail off!
Gulp! Sorry, I thought...
SILENT SPRING Rachel Carson
Never mind.

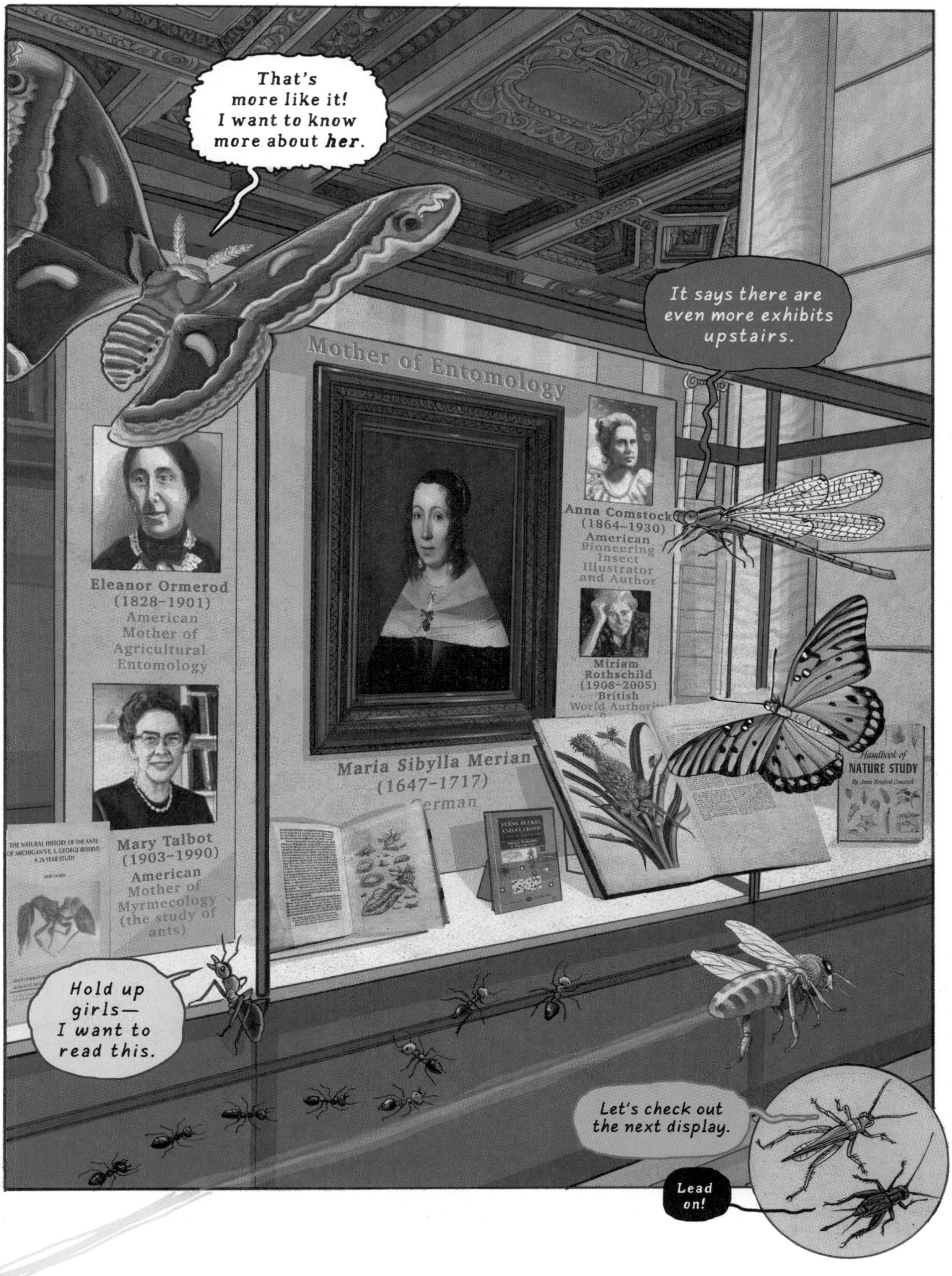

That's more like it! I want to know more about ***her***.
Mother of Entomology
Eleanor Ormerod
(1828–1901)
American Mother of Agricultural Entomology
Maria Sibylla Merian
(1647–1717)
Anna Comstock
(1864–1930)
American Pioneering Insect Illustrator and Author
Miriam Rothschild
(1908–2005)
British
It says there are even more exhibits upstairs.
Mary Talbot
(1903–1990)
American Mother of Myrmecology (the study of ants)
THE NATURAL HISTORY OF THE ANTS OF MICHIGAN'S E. S. GEORGE RESERVE: A 26 YEAR STUDY
Handbook of NATURE STUDY
Hold up girls— I want to read this.
Let's check out the next display.
Lead on!

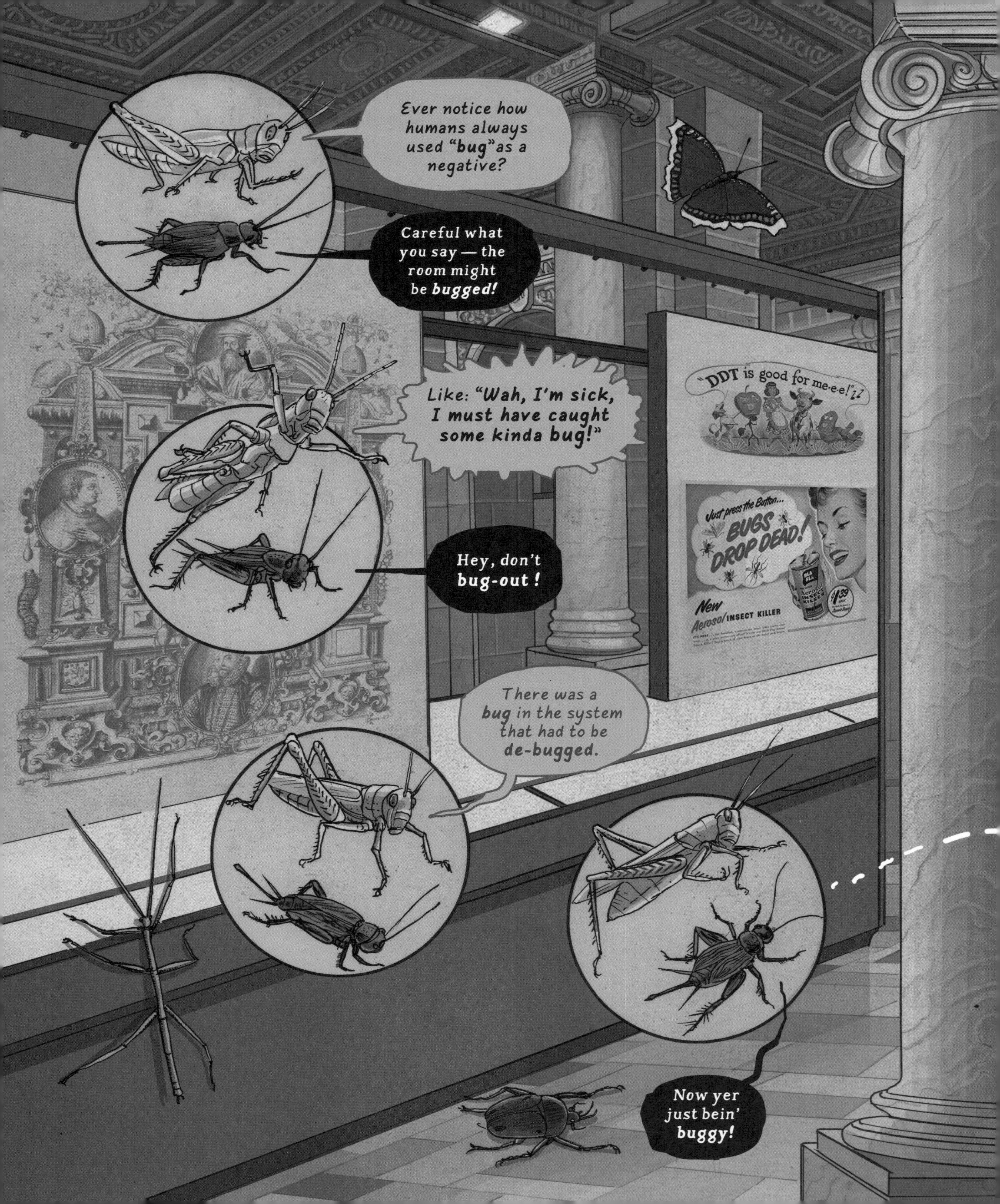
Ever notice how humans always used "bug" as a negative?
Careful what you say — the room might be bugged!
Like: "Wah, I'm sick, I must have caught some kinda bug!"
Hey, don't bug-out!
There was a bug in the system that had to be de-bugged.
Now yer just bein' buggy!
"DDT is good for me-e-e!"
Just press the Button...
BUGS DROP DEAD!
New Aerosol INSECT KILLER

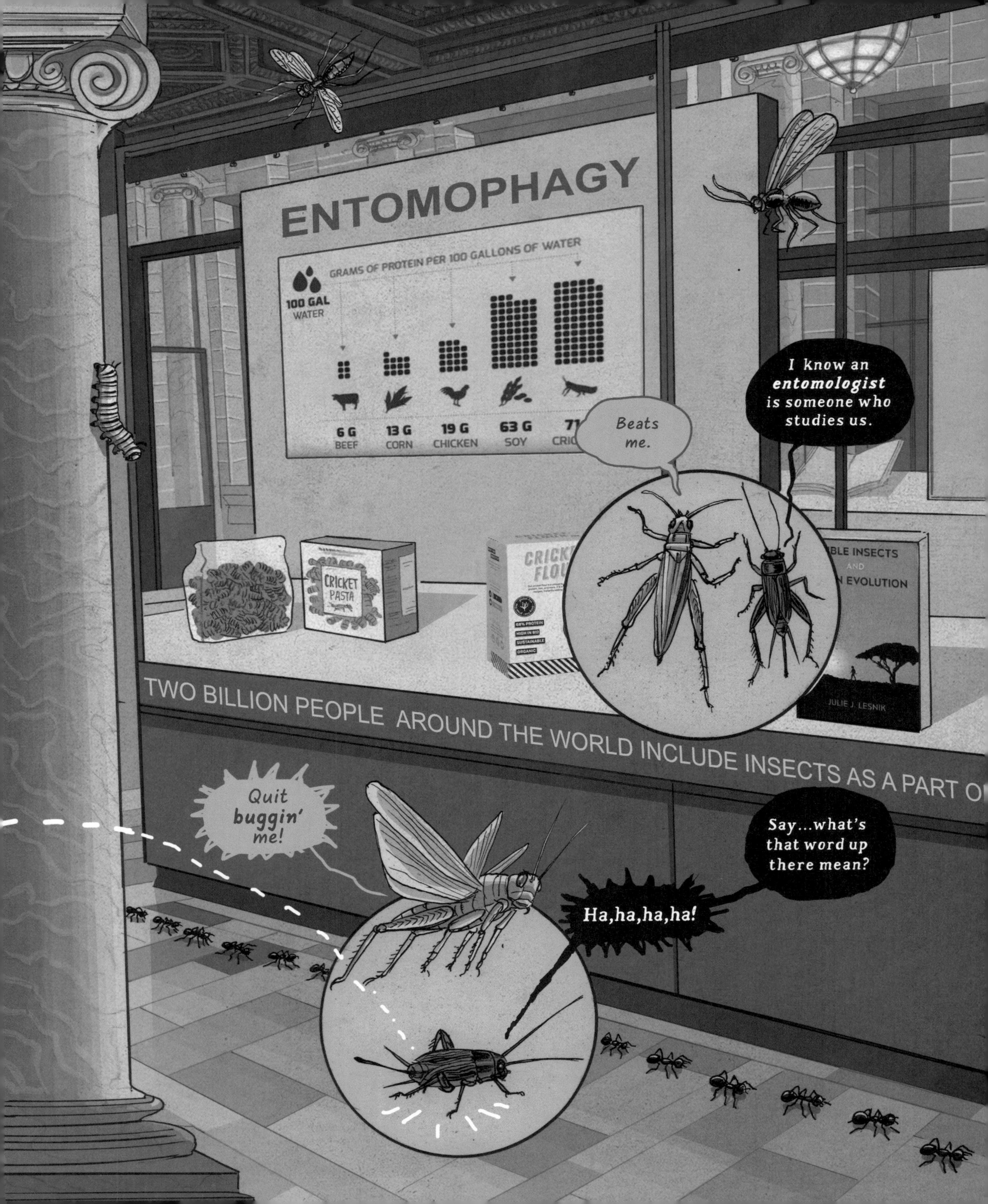
ENTOMOPHAGY
GRAMS OF PROTEIN PER 100 GALLONS OF WATER
100 GAL WATER
6 G BEEF
13 G CORN
19 G CHICKEN
63 G SOY
CRICKET PASTA
68% PROTEIN
HIGH IN B12
SUSTAINABLE
ORGANIC
JULIE J. LESNIK
Beats me.
I know an *entomologist* is someone who studies us.
TWO BILLION PEOPLE AROUND THE WORLD INCLUDE INSECTS AS A PART O
Quit buggin' me!
Say...what's that word up there mean?
Ha,ha,ha,ha!

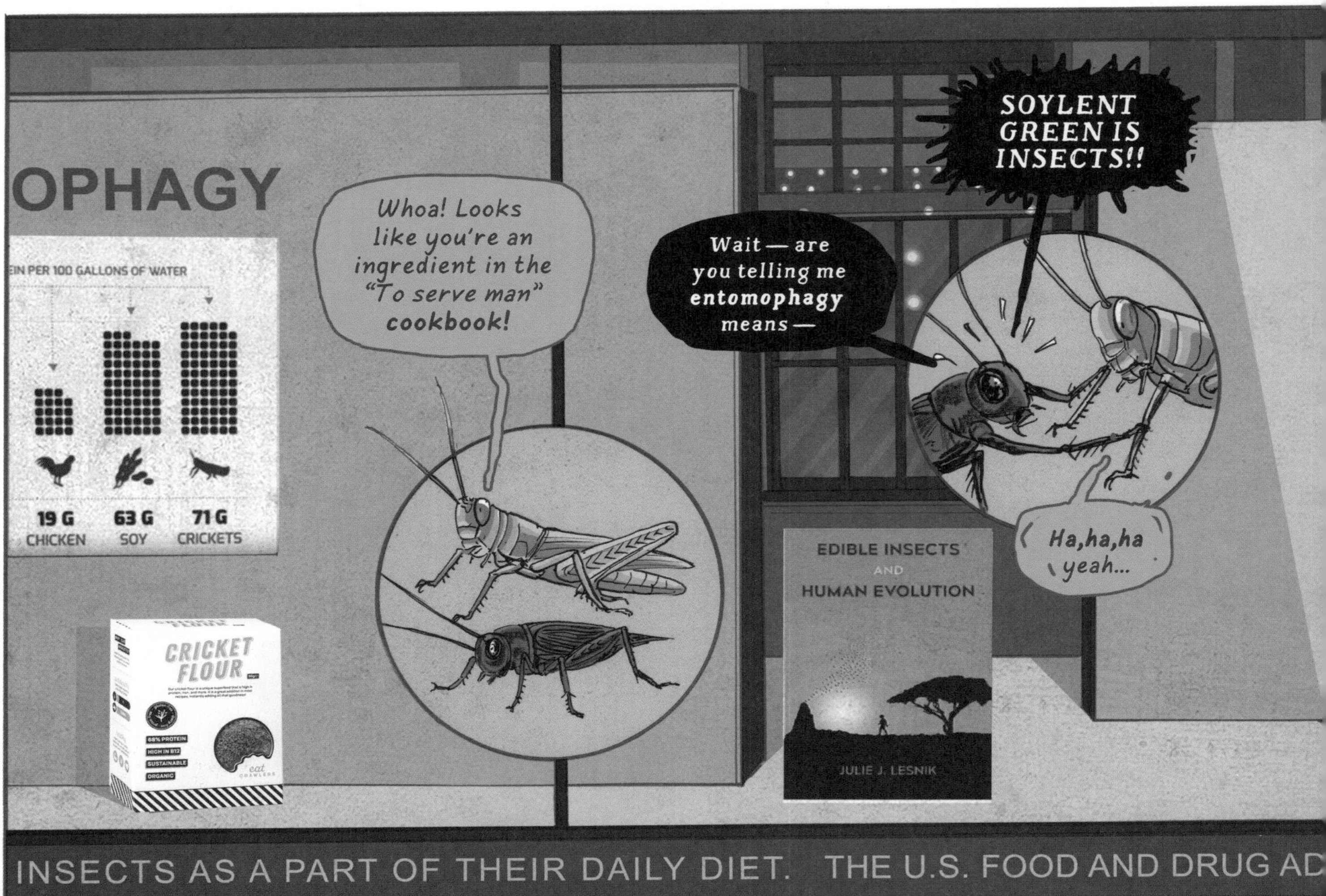
OPHAGY
EIN PER 100 GALLONS OF WATER
19 G CHICKEN
63 G SOY
71 G CRICKETS
CRICKET FLOUR
68% PROTEIN
HIGH IN B12
SUSTAINABLE
ORGANIC
Whoa! Looks like you're an ingredient in the "To serve man" cookbook!
Wait — are you telling me entomophagy means —
SOYLENT GREEN IS INSECTS!!
Ha,ha,ha yeah...
EDIBLE INSECTS AND HUMAN EVOLUTION
JULIE J. LESNIK
INSECTS AS A PART OF THEIR DAILY DIET. THE U.S. FOOD AND DRUG AD

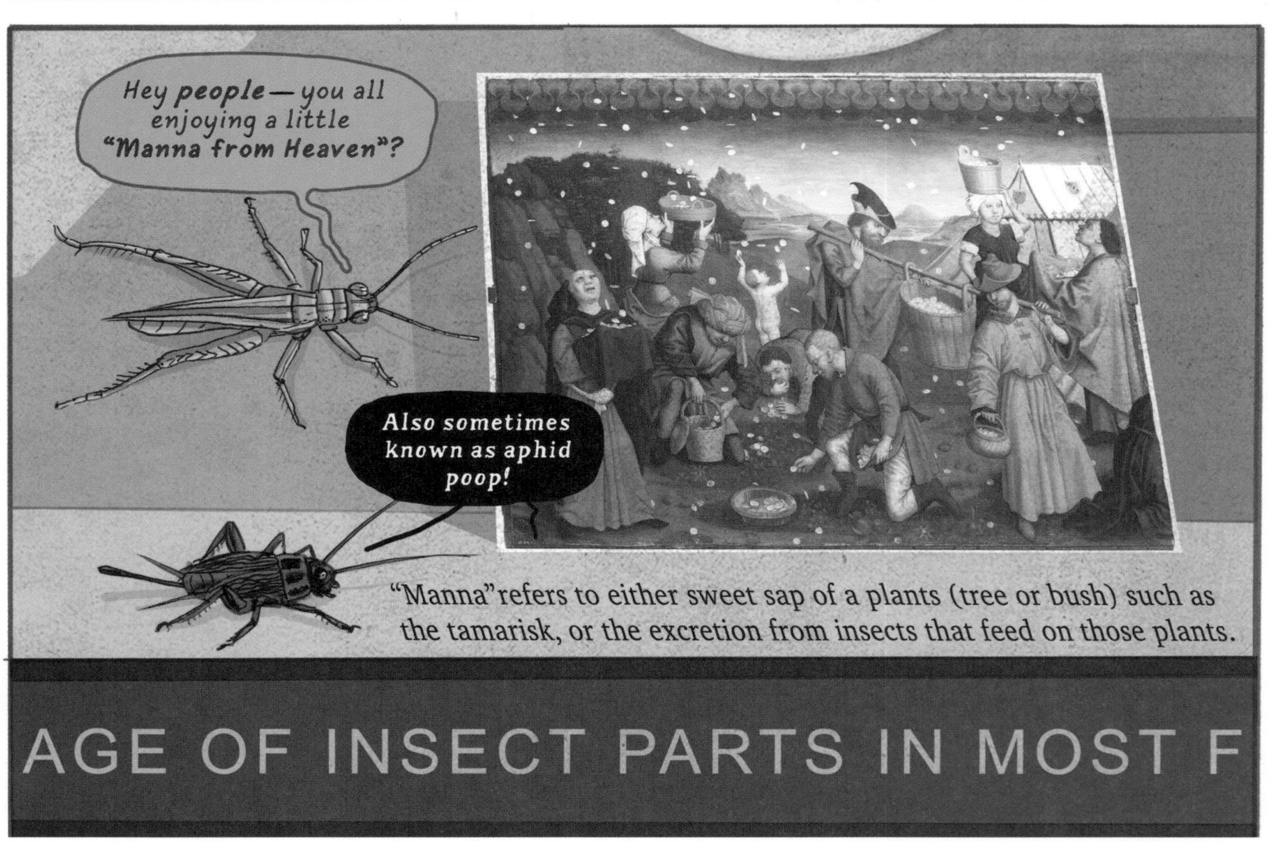
Hey people — you all enjoying a little "Manna from Heaven"?
Also sometimes known as aphid poop!
"Manna" refers to either sweet sap of a plants (tree or bush) such as the tamarisk, or the excretion from insects that feed on those plants.
AGE OF INSECT PARTS IN MOST F

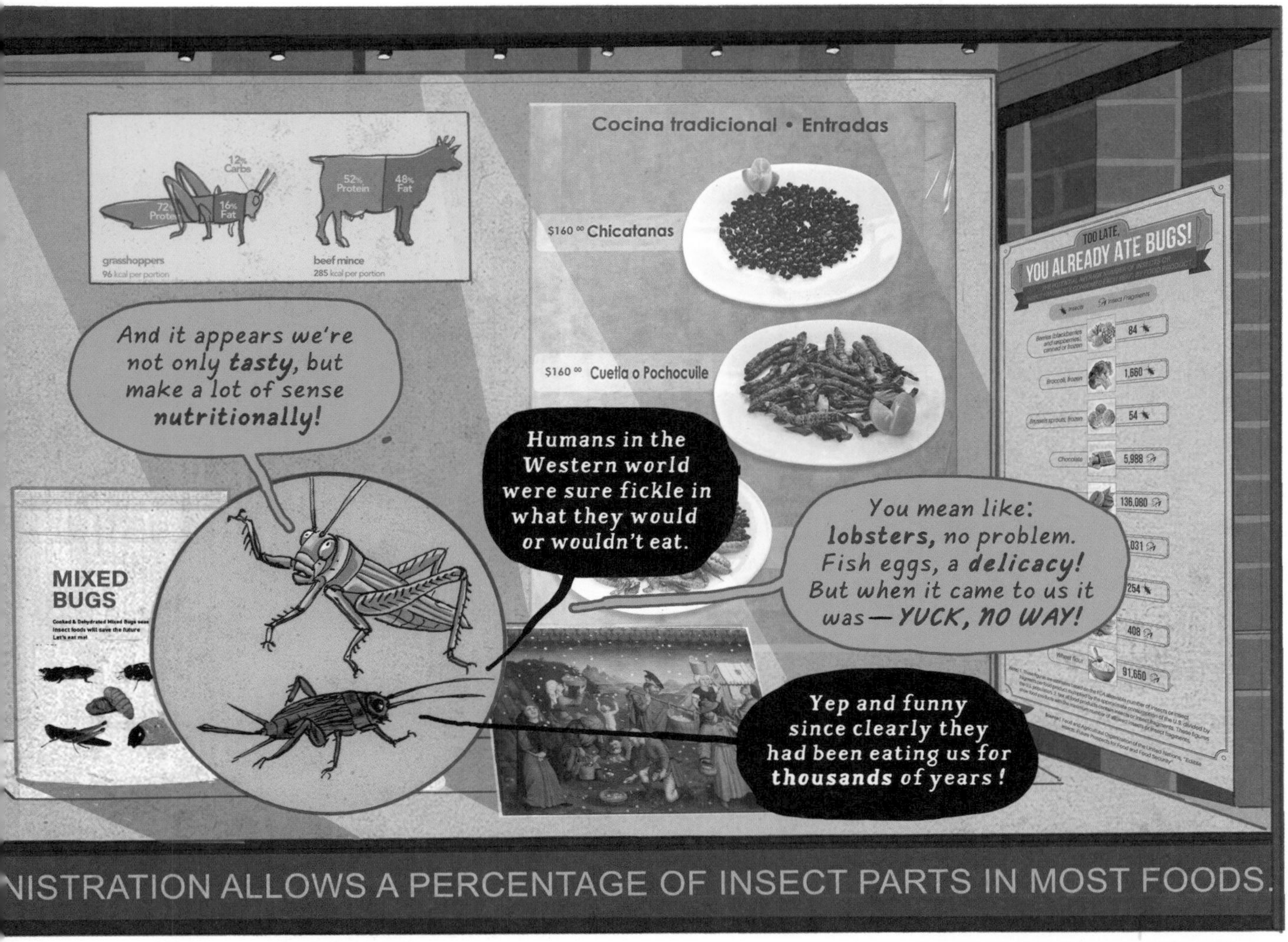
grasshoppers
96 kcal per portion
beef mince
285 kcal per portion
Cocina tradicional • Entradas
$160.00 Chicatanas
$160.00 Cuetla o Pochocuile
TOO LATE, YOU ALREADY ATE BUGS!
MIXED BUGS
And it appears we're not only tasty, but make a lot of sense nutritionally!
Humans in the Western world were sure fickle in what they would or wouldn't eat.
You mean like: lobsters, no problem. Fish eggs, a delicacy! But when it came to us it was—YUCK, NO WAY!
Yep and funny since clearly they had been eating us for thousands of years!
NISTRATION ALLOWS A PERCENTAGE OF INSECT PARTS IN MOST FOODS.

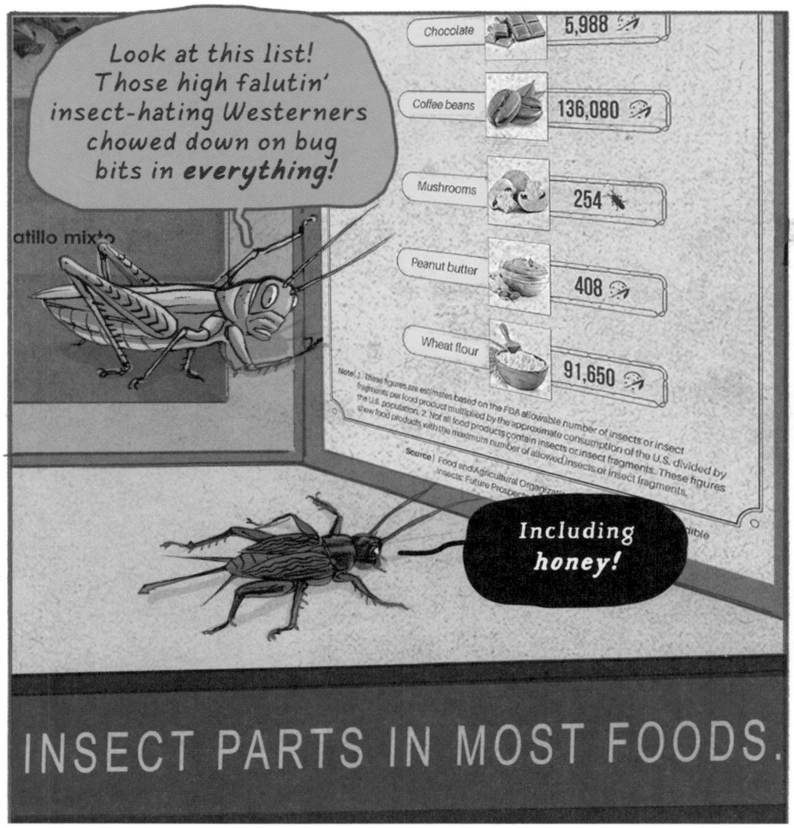
Look at this list! Those high falutin' insect-hating Westerners chowed down on bug bits in everything!
Chocolate 5,988
Coffee beans 136,080
Mushrooms 254
Peanut butter 408
Wheat flour 91,650
Including honey!
INSECT PARTS IN MOST FOODS.

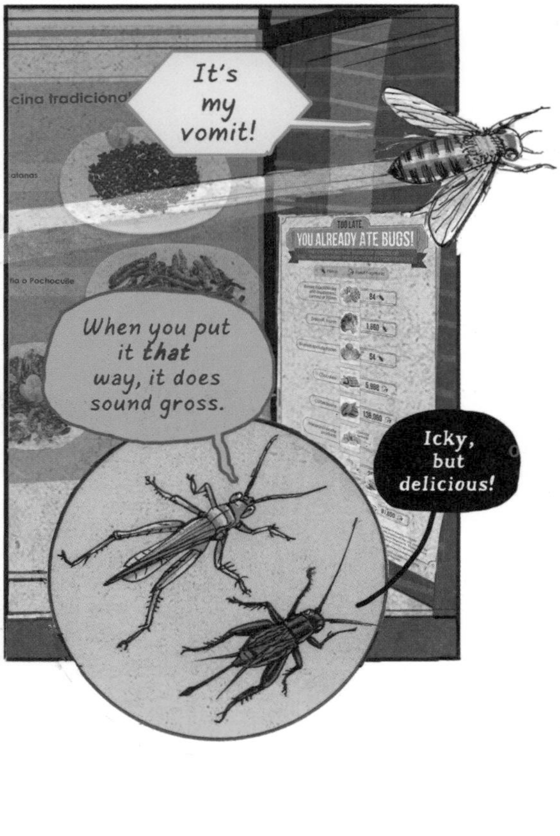
It's my vomit!
YOU ALREADY ATE BUGS!
When you put it that way, it does sound gross.
Icky, but delicious!

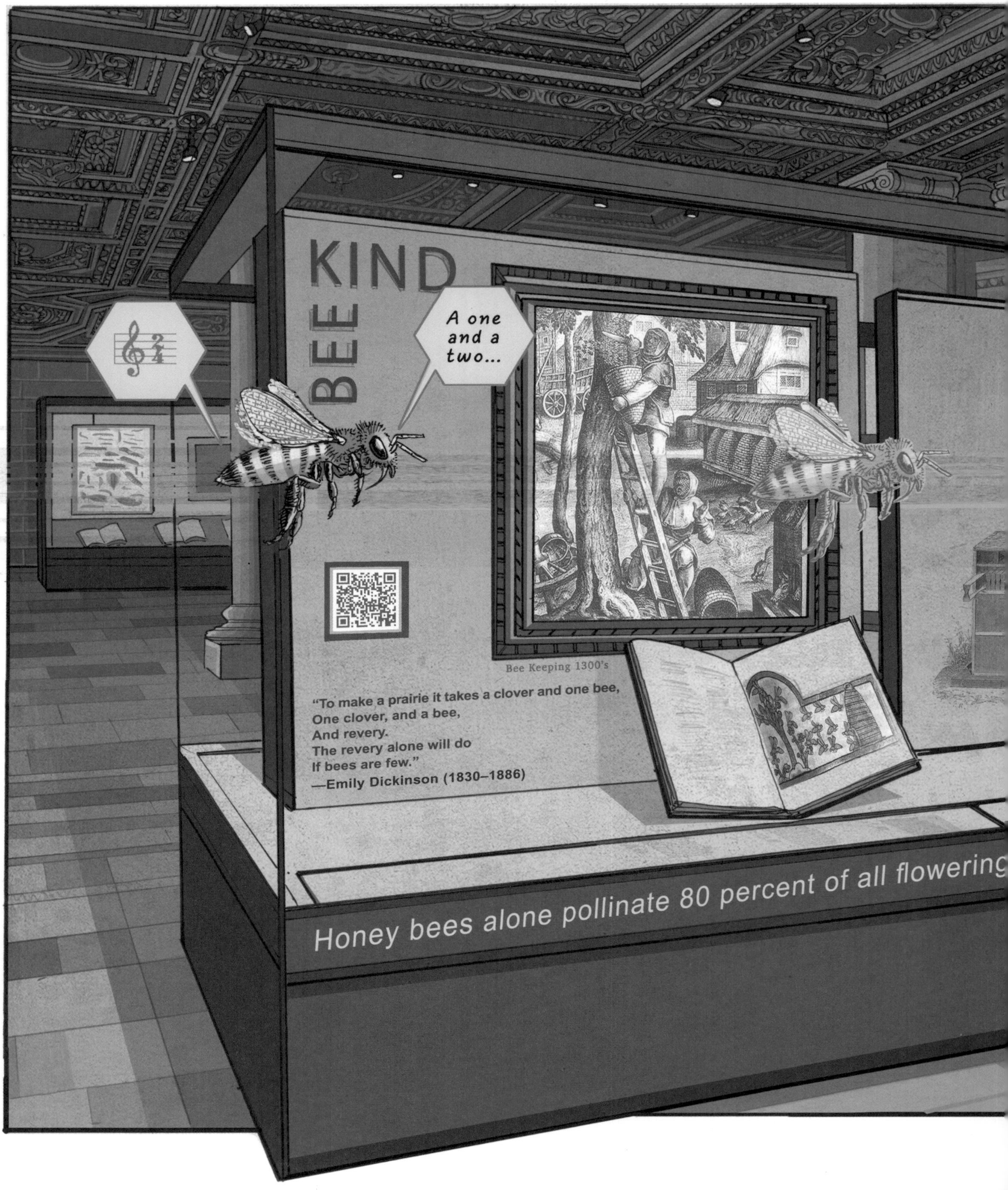

A one and a two...
BEE KIND
Bee Keeping 1300's
"To make a prairie it takes a clover and one bee,
One clover, and a bee,
And revery.
The revery alone will do
If bees are few."
—Emily Dickinson (1830–1886)
Honey bees alone pollinate 80 percent of all flowering

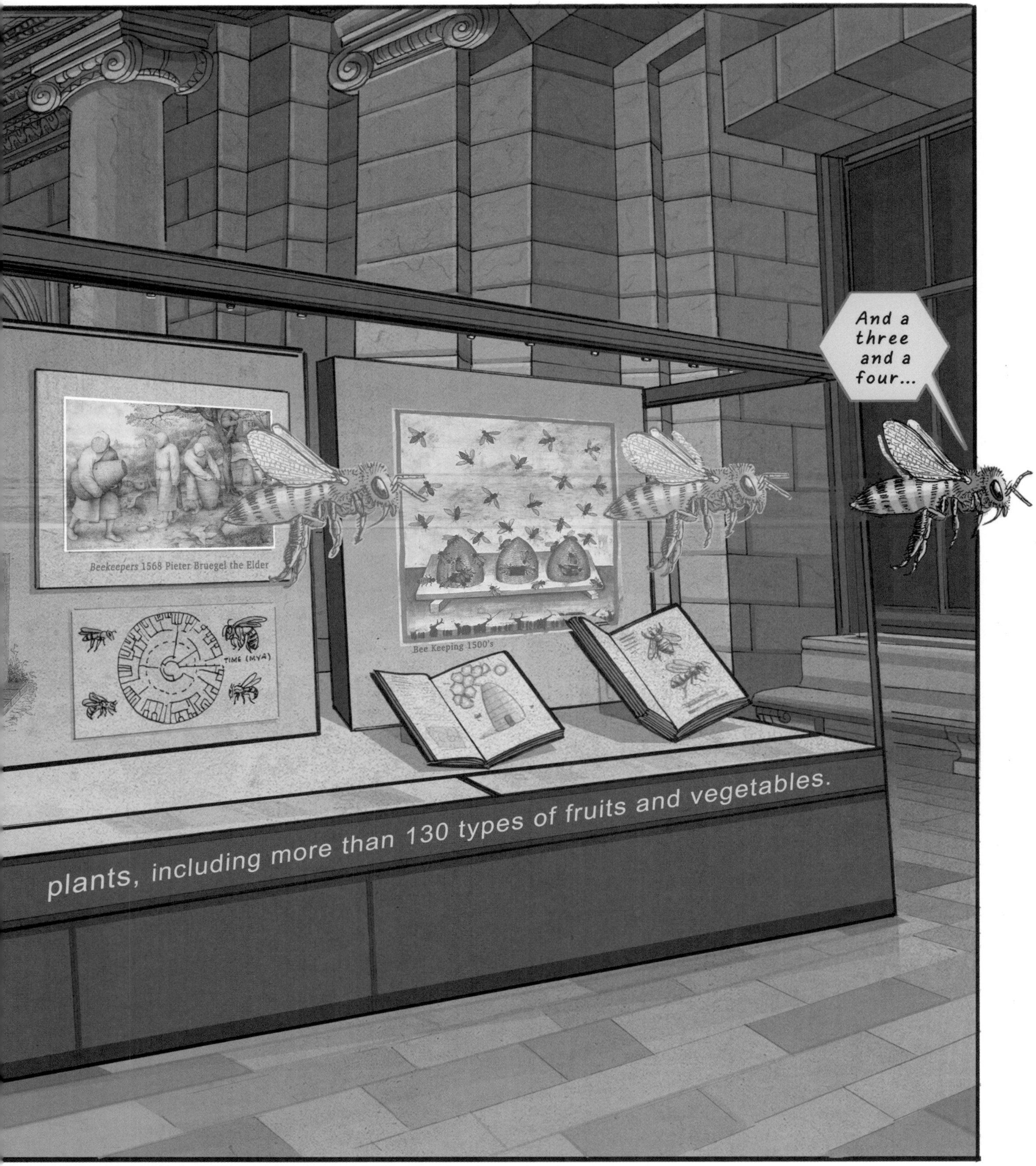
And a three and a four...
Beekeepers 1568 Pieter Bruegel the Elder
TIME (MYA)
Bee Keeping 1500's
plants, including more than 130 types of fruits and vegetables.

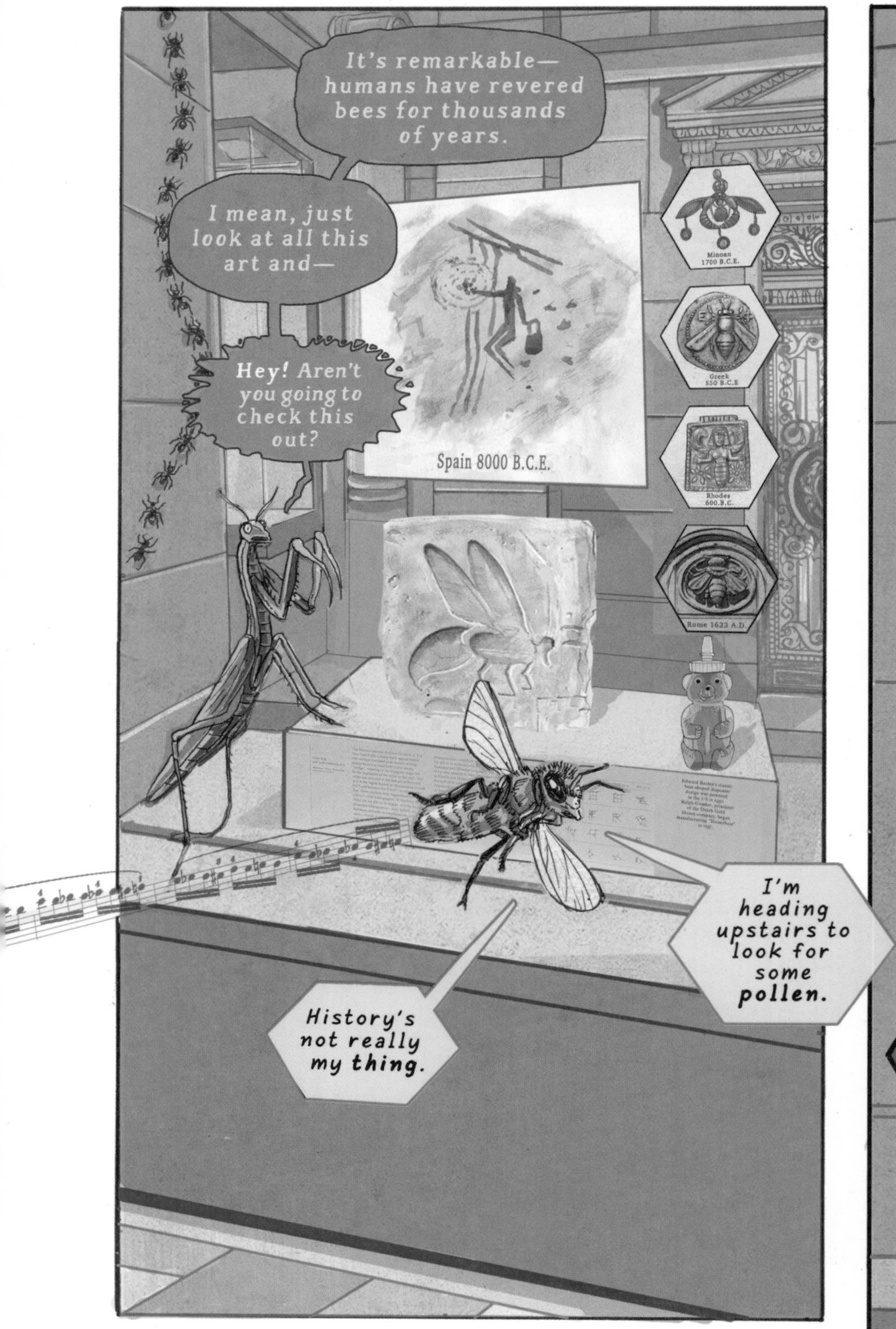
It's remarkable—humans have revered bees for thousands of years.
I mean, just look at all this art and—
Hey! Aren't you going to check this out?
Spain 8000 B.C.E.
Minoan 1700 B.C.E.
Greek 550 B.C.E
Rhodes 600.B.C.
Rome 1623 A.D.
I'm heading upstairs to look for some **pollen**.
History's not really my **thing**.

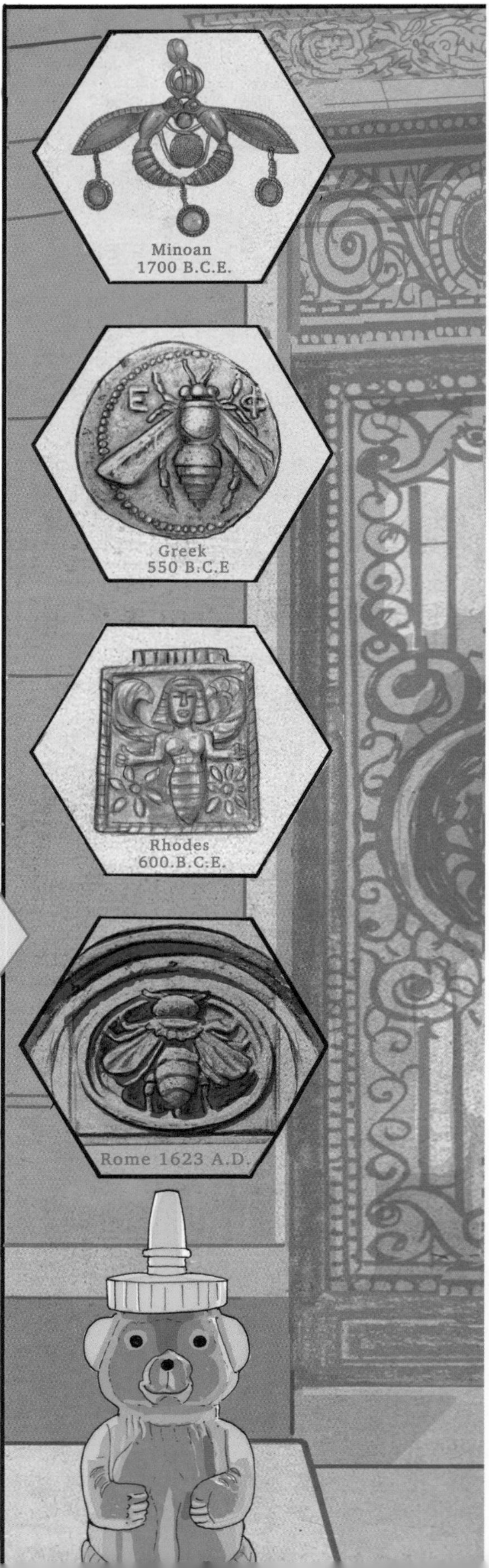
Minoan 1700 B.C.E.
Greek 550 B.C.E
Rhodes 600.B.C.E.
Rome 1623 A.D.

Edward Rachin's classic bear-shaped dispenser design was patented in the U.S. in 1950. Ralph Gamber, president of the Dutch Gold Honey company, began manufacturing "Honeybear" in 1957.

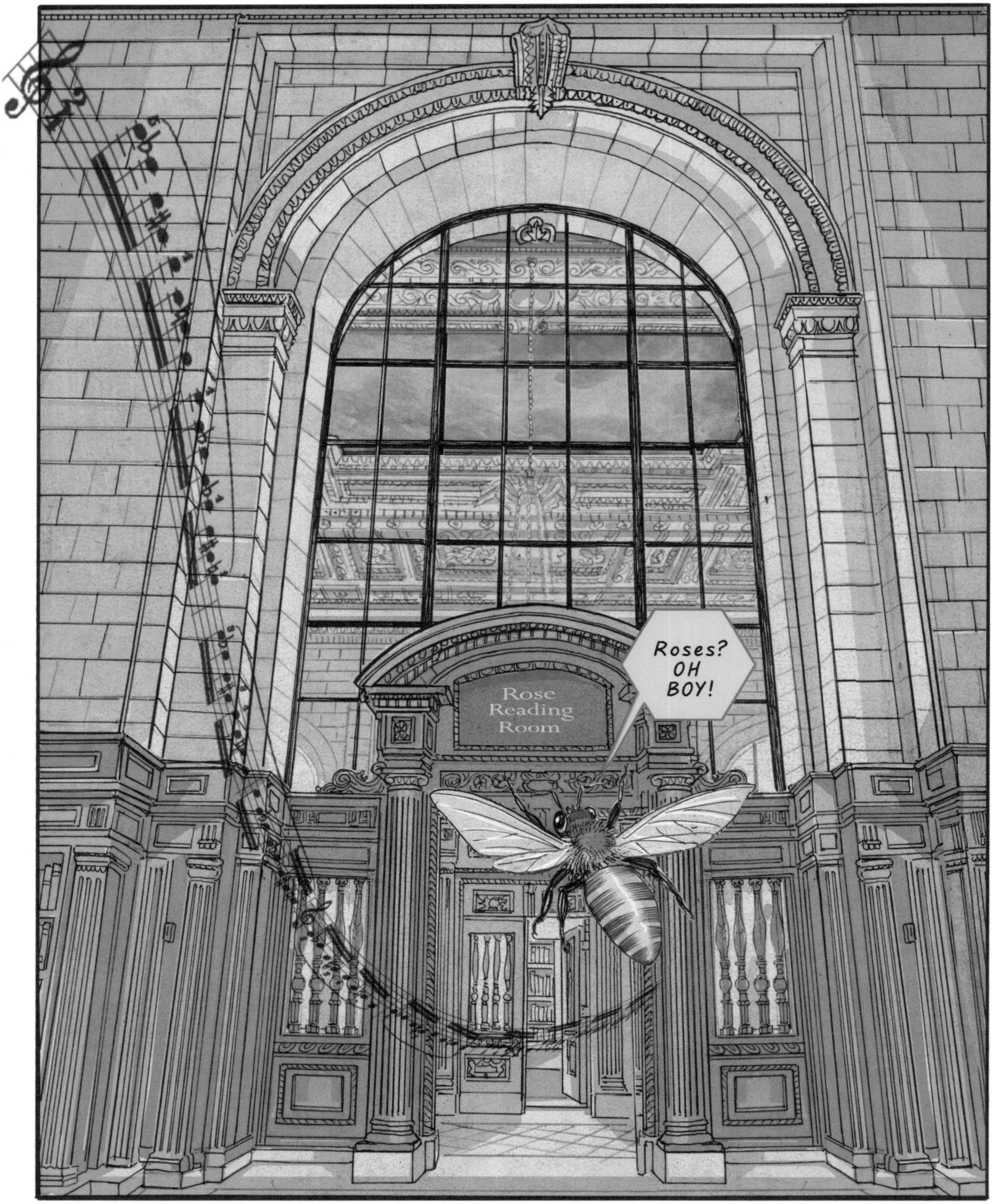
Rose Reading Room
Roses? OH BOY!

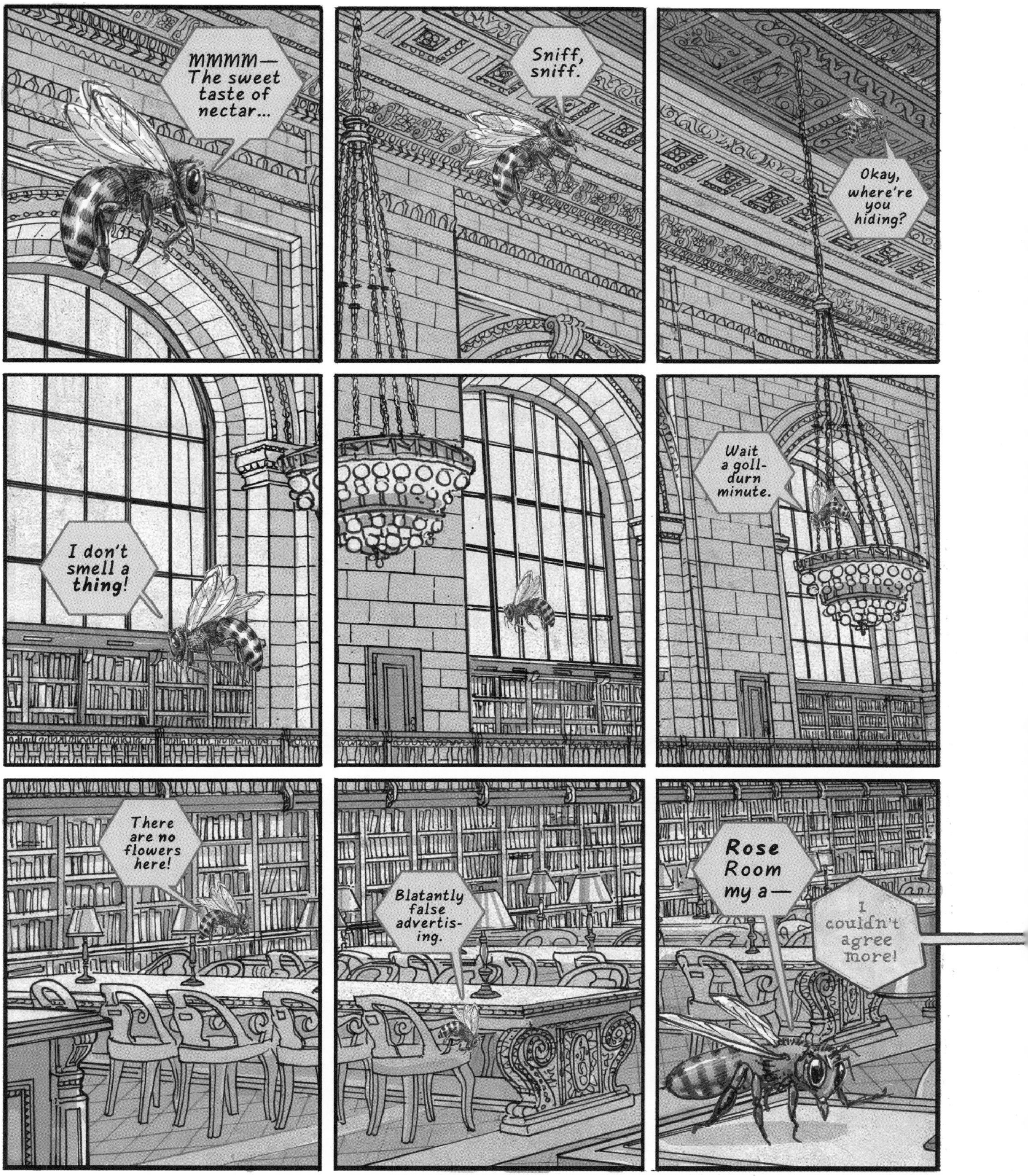
MMMM— The sweet taste of nectar...
Sniff, sniff.
Okay, where're you hiding?
I don't smell a thing!
Wait a goll-durn minute.
There are no flowers here!
Blatantly false advertis-ing.
Rose Room my a—
I couldn't agree more!

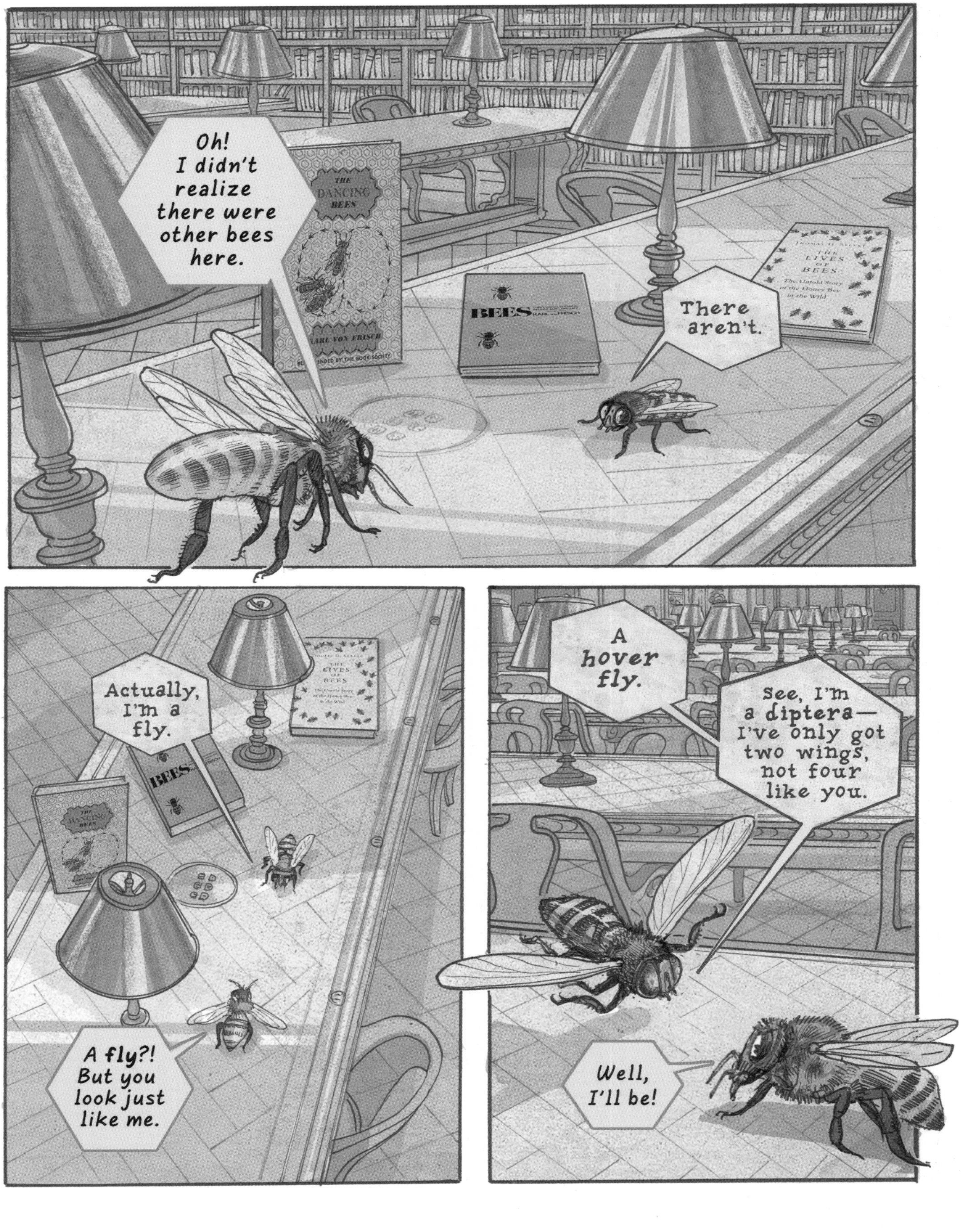
Oh! I didn't realize there were other bees here.
There aren't.
Actually, I'm a fly.
A fly?! But you look just like me.
A hover fly.
See, I'm a diptera— I've only got two wings, not four like you.
Well, I'll be!

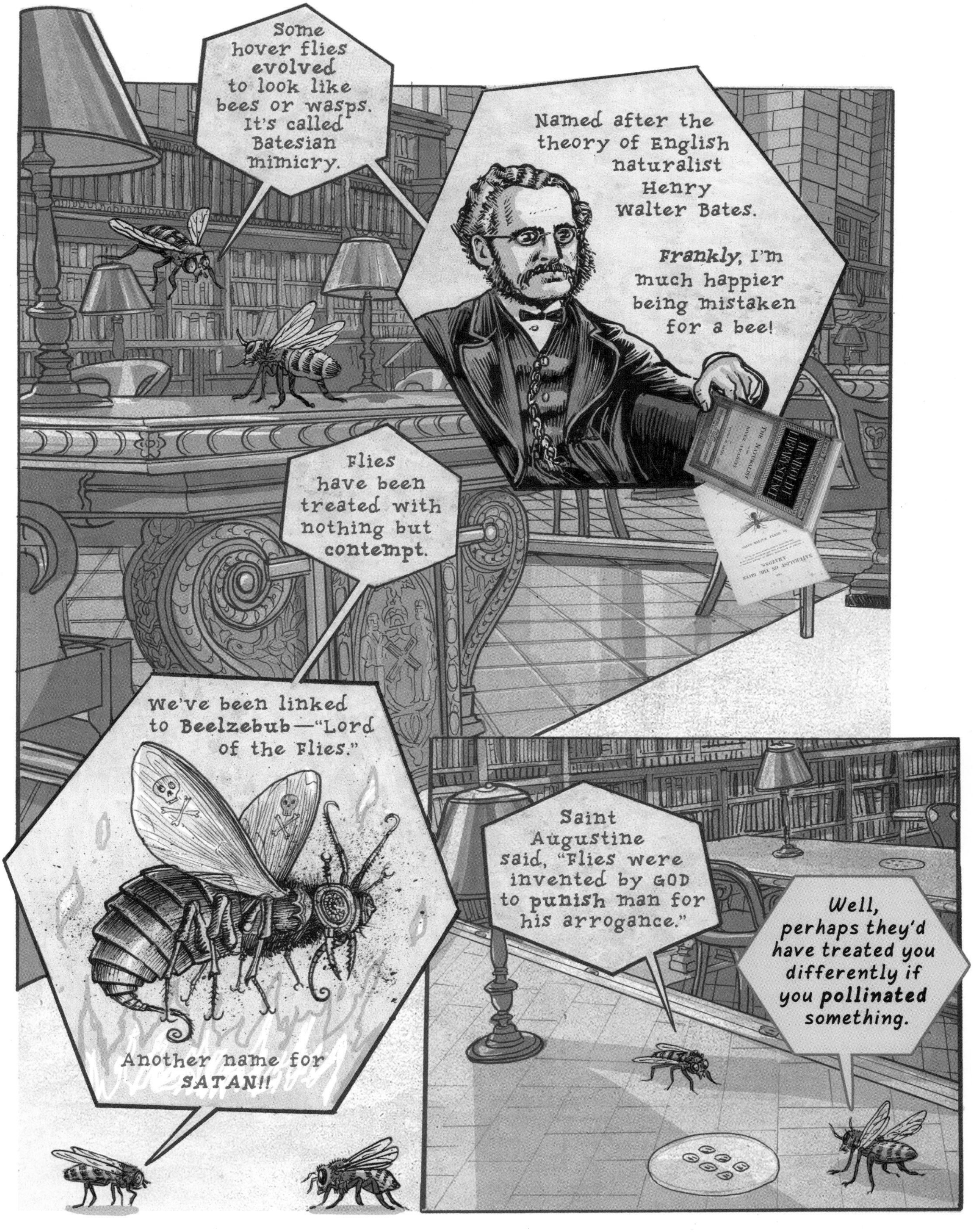
Some hover flies evolved to look like bees or wasps. It's called Batesian mimicry.
Named after the theory of English naturalist Henry Walter Bates.
Frankly, I'm much happier being mistaken for a bee!
HUMBOLDT LIBRARY OF SCIENCE
THE NATURALIST ON THE RIVER AMAZONS
Flies have been treated with nothing but contempt.
We've been linked to **Beelzebub**—"Lord of the Flies."
Another name for **SATAN!!**
Saint Augustine said, "Flies were invented by GOD to punish man for his arrogance."
Well, perhaps they'd have treated you differently if you **pollinated** something.

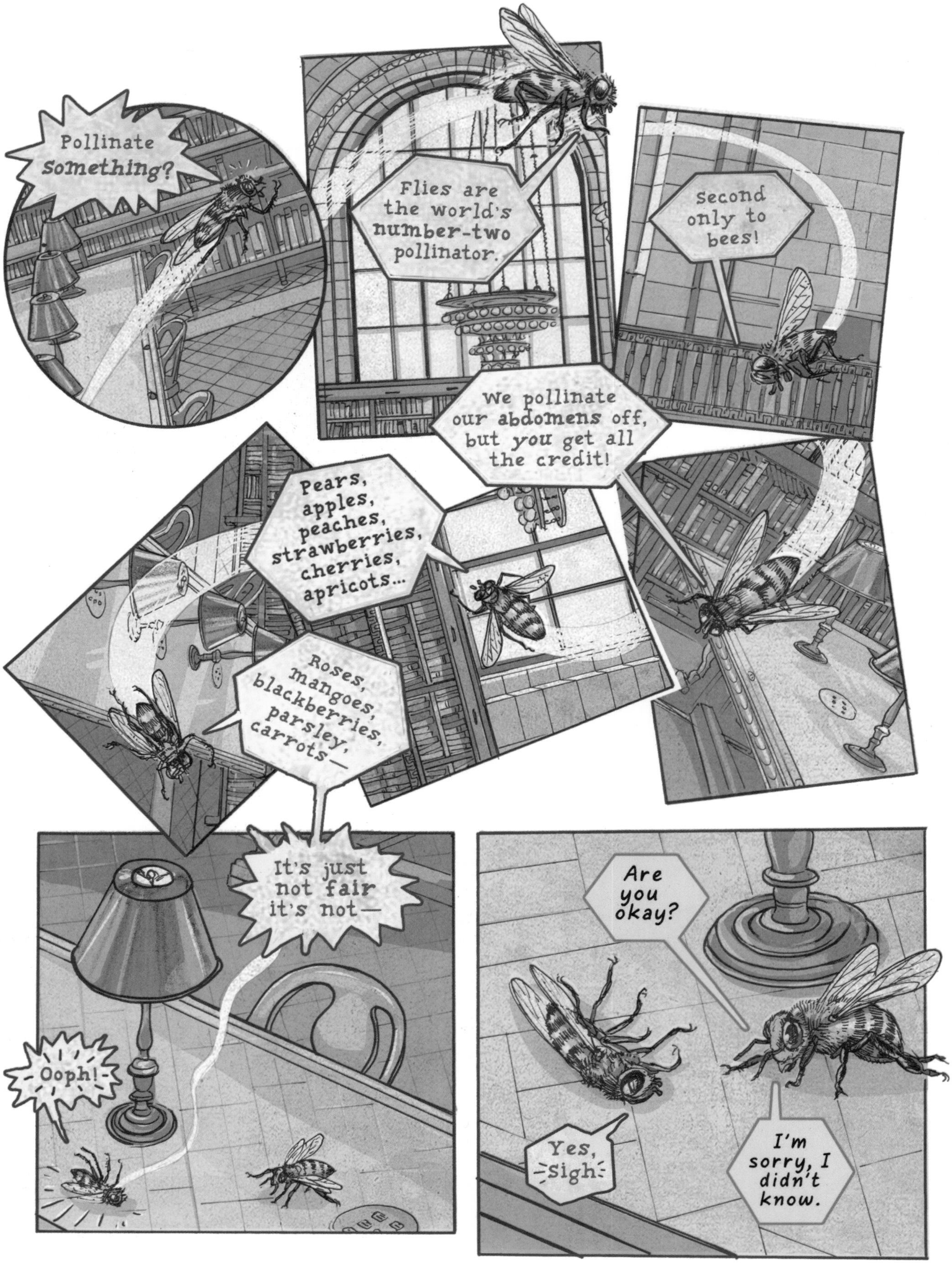
Pollinate something?
Flies are the world's number-two pollinator.
Second only to bees!
We pollinate our abdomens off, but you get all the credit!
Pears, apples, peaches, strawberries, cherries, apricots...
Roses, mangoes, blackberries, parsley, carrots—
It's just not fair it's not—
Ooph!
Are you okay?
Yes, Sigh.
I'm sorry, I didn't know.

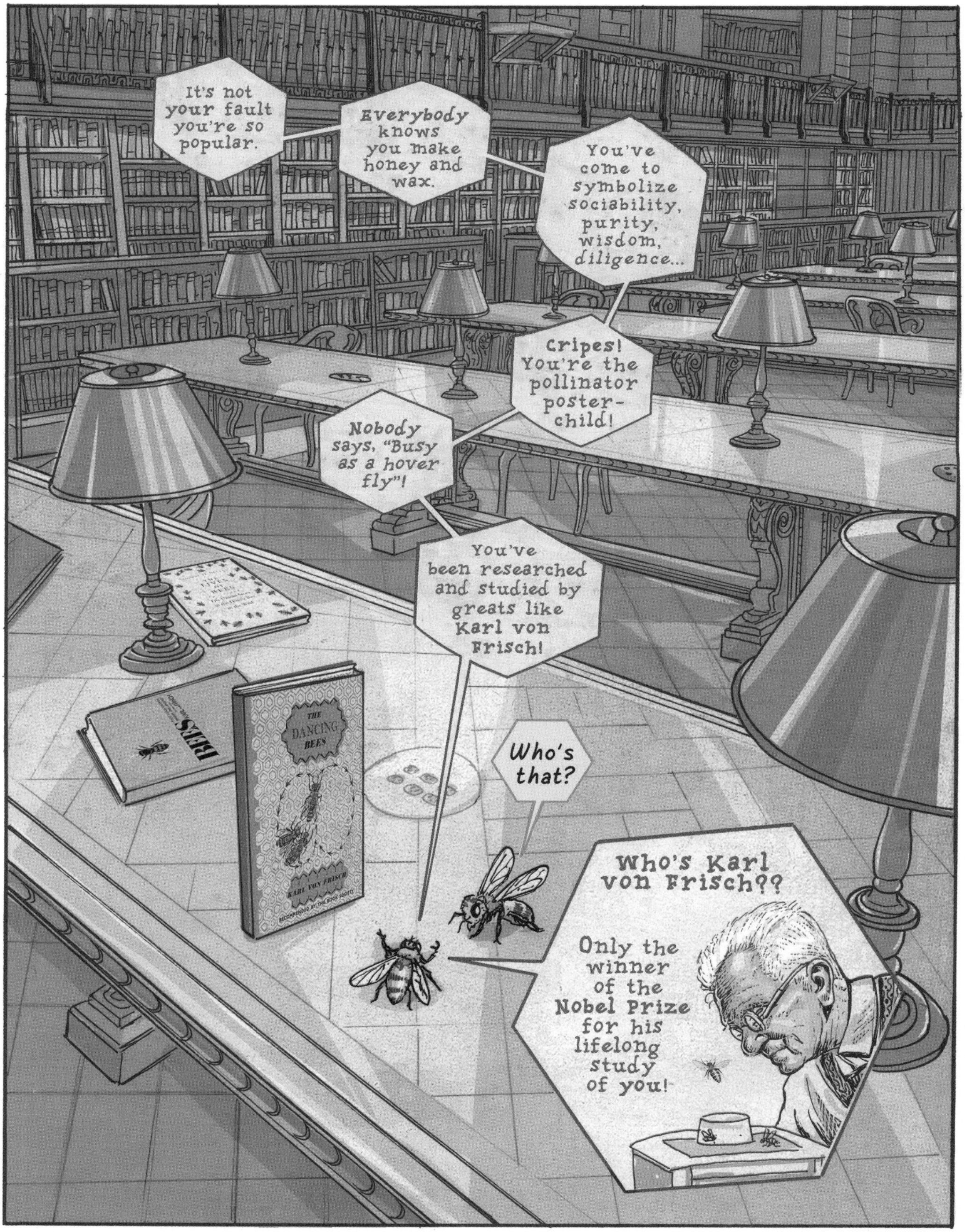
It's not your fault you're so popular.
Everybody knows you make honey and wax.
You've come to symbolize sociability, purity, wisdom, diligence...
Cripes! You're the pollinator poster-child!
Nobody says, "Busy as a hover fly"!
You've been researched and studied by greats like Karl von Frisch!
Who's that?
Who's Karl von Frisch??
Only the winner of the Nobel Prize for his lifelong study of you!
BEES
THE DANCING BEES
KARL VON FRISCH

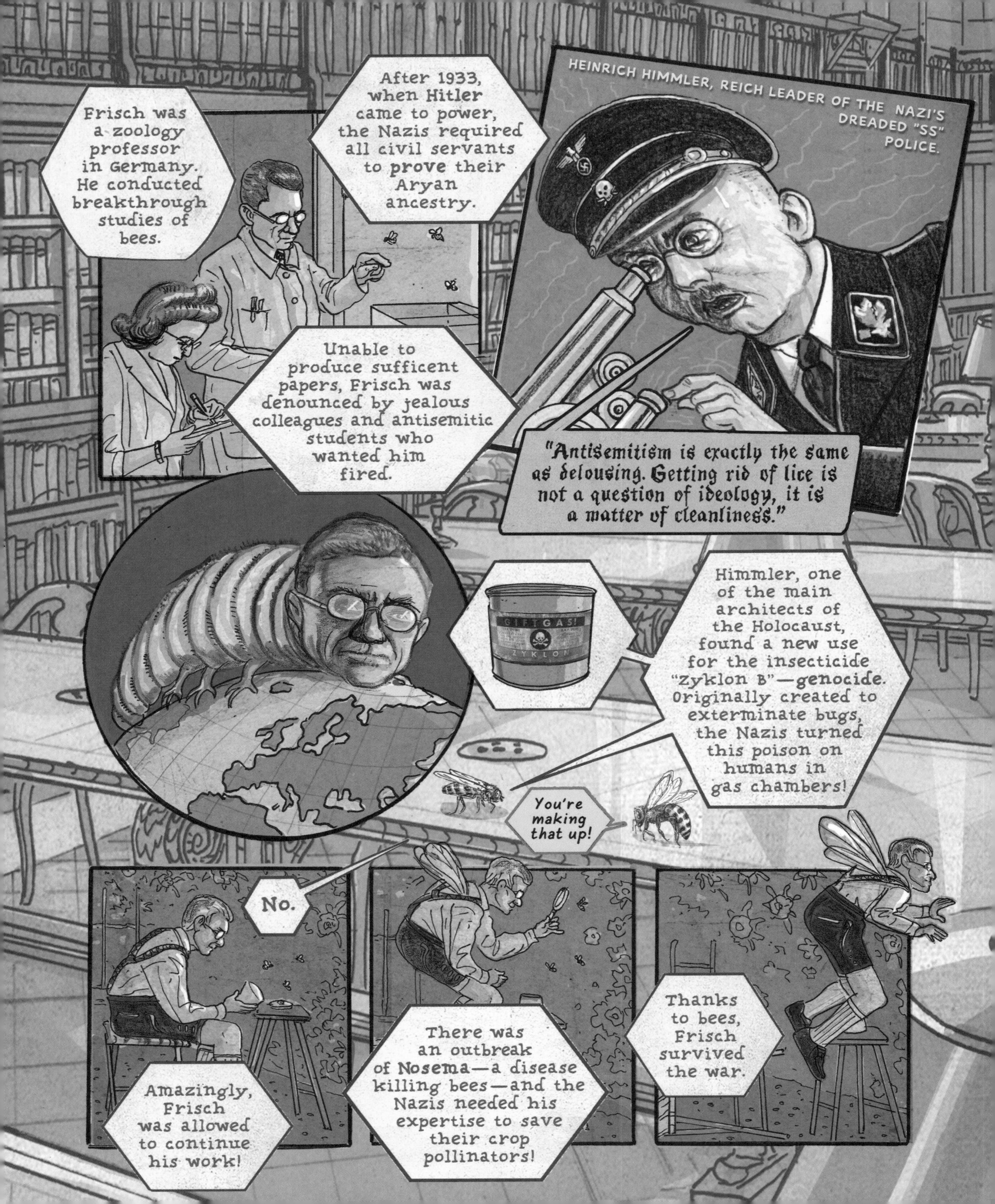
Frisch was a zoology professor in Germany. He conducted breakthrough studies of bees.
After 1933, when Hitler came to power, the Nazis required all civil servants to prove their Aryan ancestry.
Unable to produce sufficent papers, Frisch was denounced by jealous colleagues and antisemitic students who wanted him fired.
HEINRICH HIMMLER, REICH LEADER OF THE NAZI'S DREADED "SS" POLICE.
"Antisemitism is exactly the same as delousing. Getting rid of lice is not a question of ideology, it is a matter of cleanliness."
GIFTGAS!
ZYKLON
Himmler, one of the main architects of the Holocaust, found a new use for the insecticide "Zyklon B"—genocide. Originally created to exterminate bugs, the Nazis turned this poison on humans in gas chambers!
You're making that up!
No.
Amazingly, Frisch was allowed to continue his work!
There was an outbreak of Nosema—a disease killing bees—and the Nazis needed his expertise to save their crop pollinators!
Thanks to bees, Frisch survived the war.

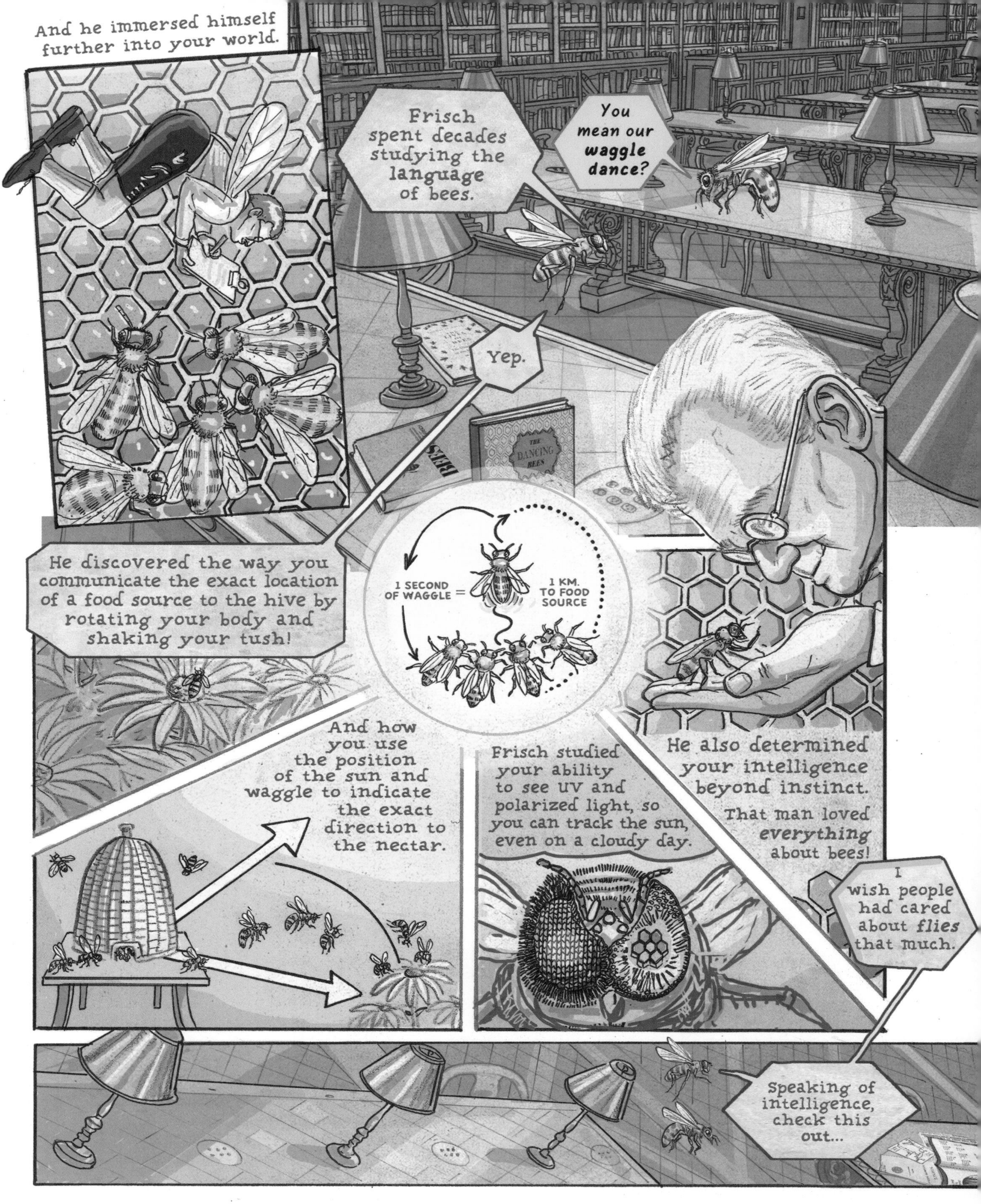

And he immersed himself further into your world.
Frisch spent decades studying the language of bees.
You mean our waggle dance?
Yep.
THE DANCING BEES
He discovered the way you communicate the exact location of a food source to the hive by rotating your body and shaking your tush!
1 SECOND OF WAGGLE = 1 KM. TO FOOD SOURCE
And how you use the position of the sun and waggle to indicate the exact direction to the nectar.
Frisch studied your ability to see UV and polarized light, so you can track the sun, even on a cloudy day.
He also determined your intelligence beyond instinct.
That man loved everything about bees!
I wish people had cared about flies that much.
Speaking of intelligence, check this out...

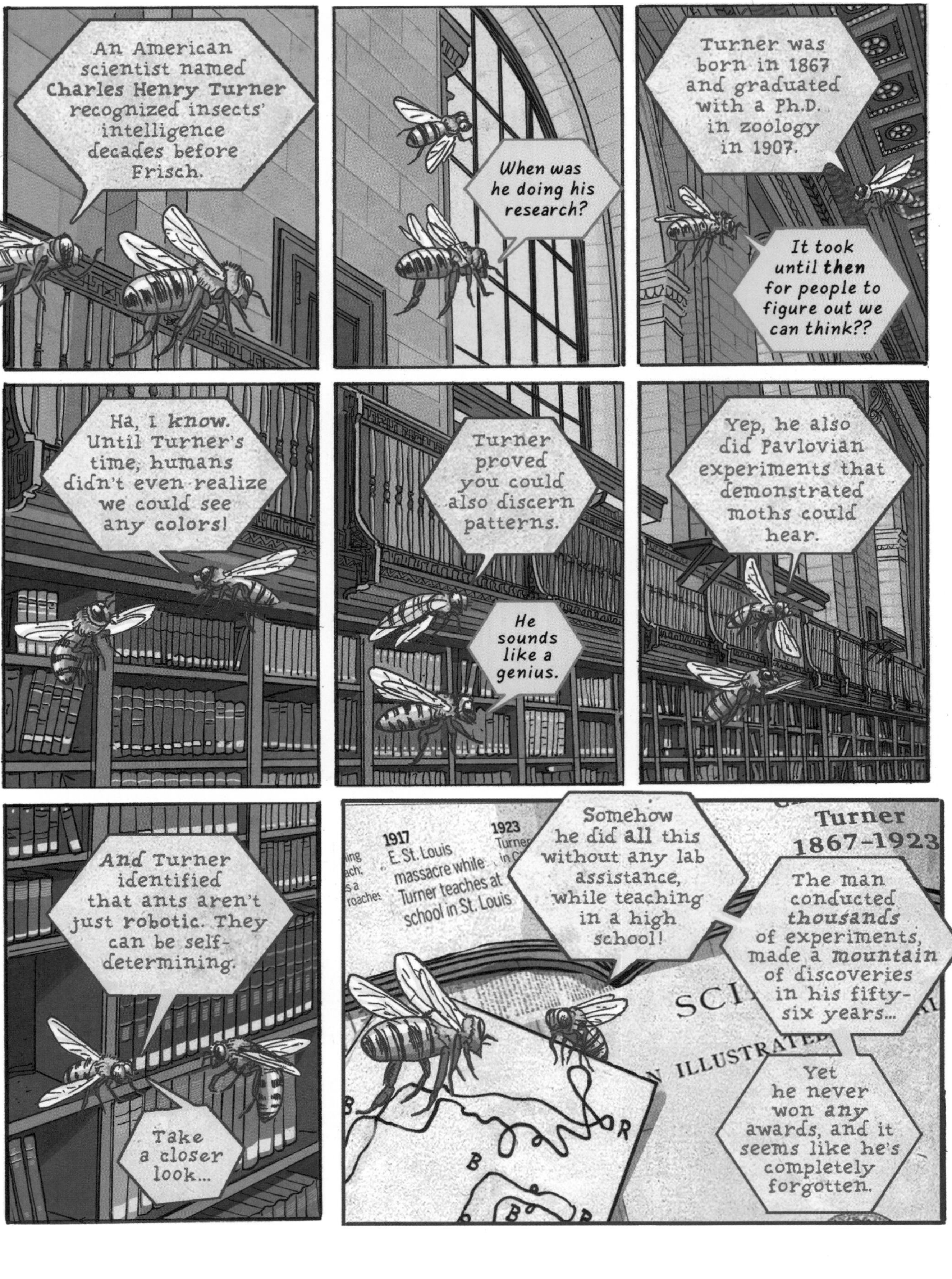
An American scientist named Charles Henry Turner recognized insects' intelligence decades before Frisch.
When was he doing his research?
Turner was born in 1867 and graduated with a Ph.D. in zoology in 1907.
It took until **then** for people to figure out we can think??
Ha, I *know*. Until Turner's time, humans didn't even realize we could see any colors!
Turner proved you could also discern patterns.
He sounds like a genius.
Yep, he also did Pavlovian experiments that demonstrated moths could hear.
And Turner identified that ants aren't just **robotic**. They can be self-determining.
Take a closer look...
1917
E. St. Louis massacre while Turner teaches at school in St. Louis
1923
Somehow he did **all** this without any lab assistance, while teaching in a high school!
Turner
1867–1923
The man conducted ***thousands*** of experiments, made a ***mountain*** of discoveries in his fifty-six years...
ILLUSTRATED
Yet he never won *any* awards, and it seems like he's completely forgotten.

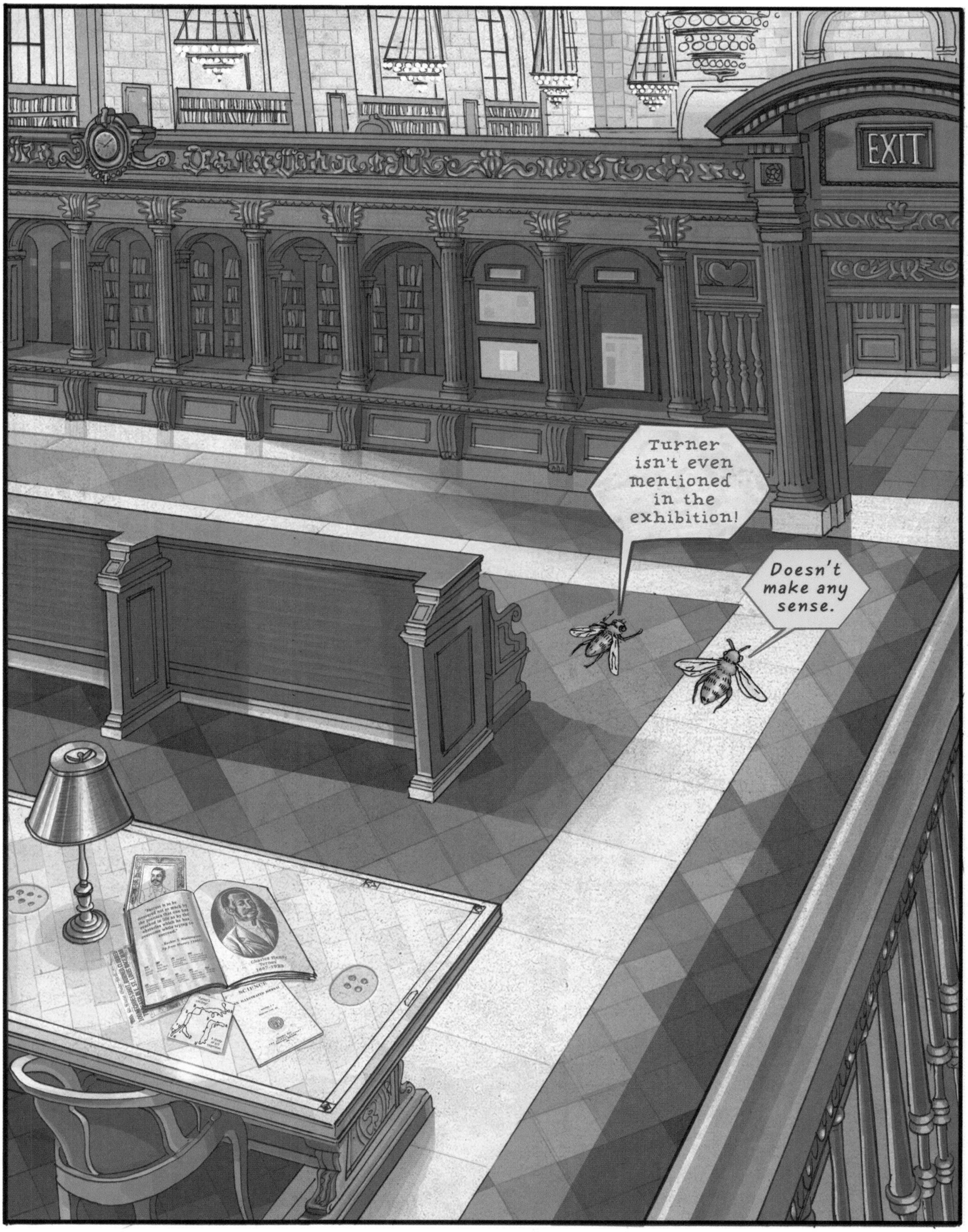
EXIT
Turner isn't even mentioned in the exhibition!
Doesn't make any sense.

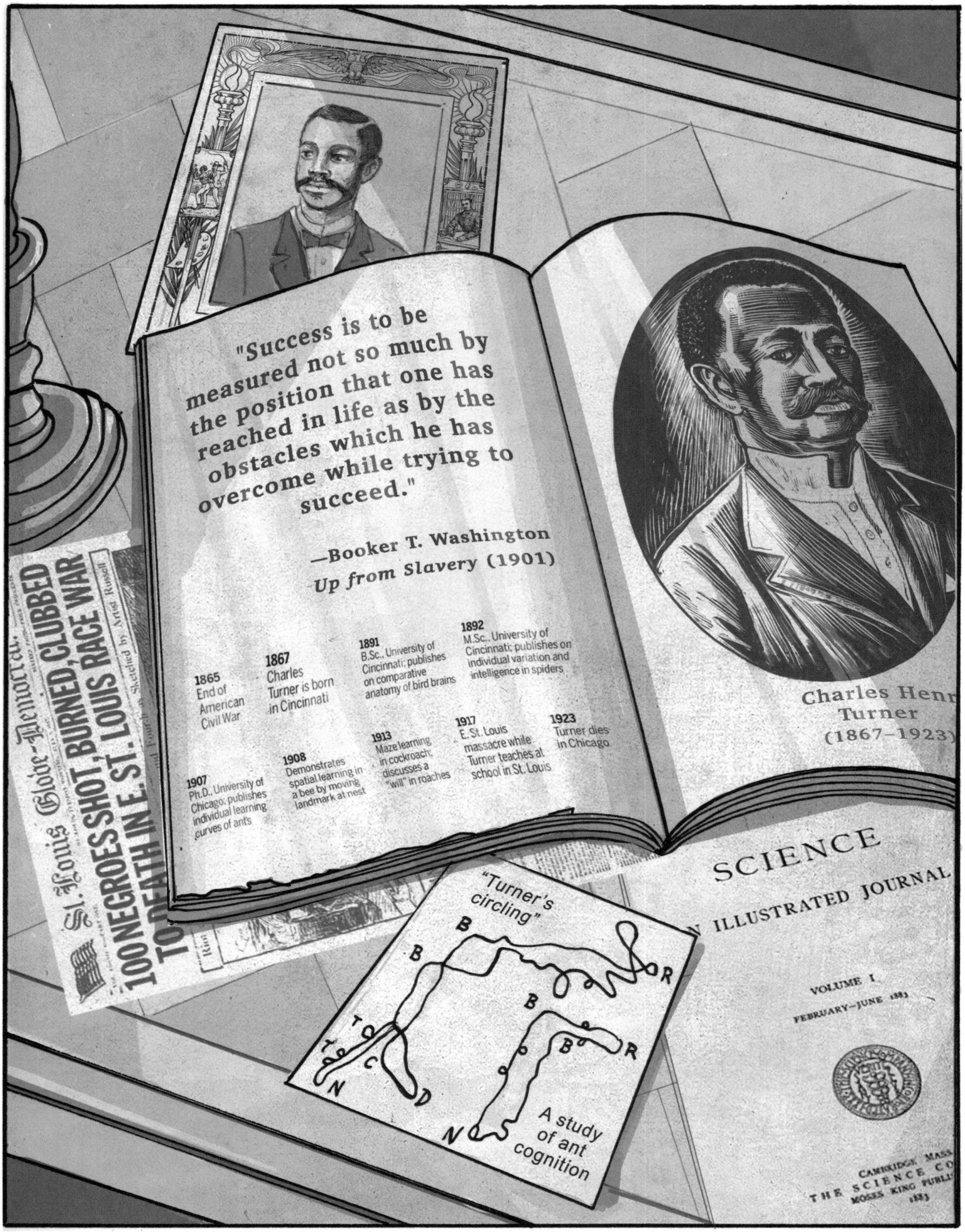
"Success is to be measured not so much by the position that one has reached in life as by the obstacles which he has overcome while trying to succeed."
—Booker T. Washington
Up from Slavery (1901)
1865 End of American Civil War
1867 Charles Turner is born in Cincinnati
1891 B.Sc., University of Cincinnati; publishes on comparative anatomy of bird brains
1892 M.Sc., University of Cincinnati; publishes on individual variation and intelligence in spiders
1907 Ph.D., University of Chicago; publishes individual learning curves of ants
1908 Demonstrates spatial learning in a bee by moving landmark at nest
1913 Maze learning in cockroach; discusses a "will" in roaches
1917 E. St. Louis massacre while Turner teaches at school in St. Louis
1923 Turner dies in Chicago
Charles Henr
Turner
(1867–1923)
St. Louis Globe-Democrat
100 NEGROES SHOT, BURNED, CLUBBED TO DEATH IN E. ST. LOUIS RACE WAR
Sketched by Artist Russell
"Turner's circling"
A study of ant cognition
SCIENCE
N ILLUSTRATED JOURNAL
VOLUME I
FEBRUARY–JUNE 1883
CAMBRIDGE MASS
THE SCIENCE CO
MOSES KING PUBLIS
1883

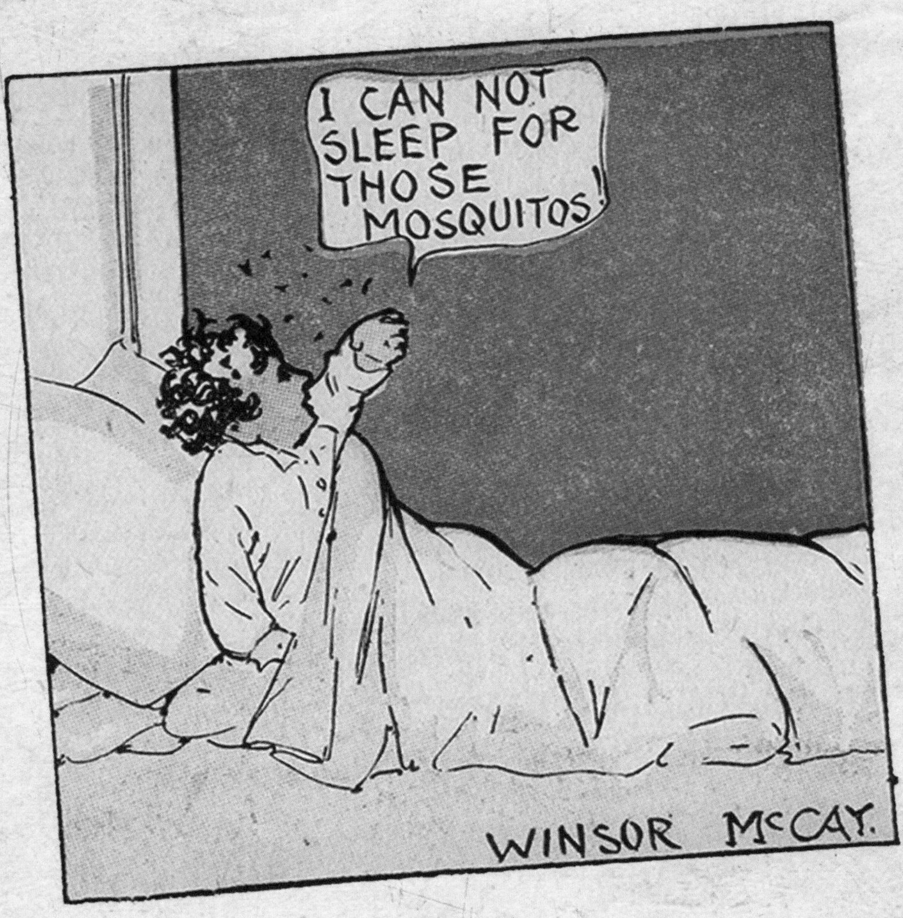
I CAN NOT SLEEP FOR THOSE MOSQUITOS!
WINSOR McCAY.

Winsor McCay
(1869–1934)
American cartoonist,
illustrator, and animator
Little Nemo In Slumberland
Published in the
American Examiner, 1912

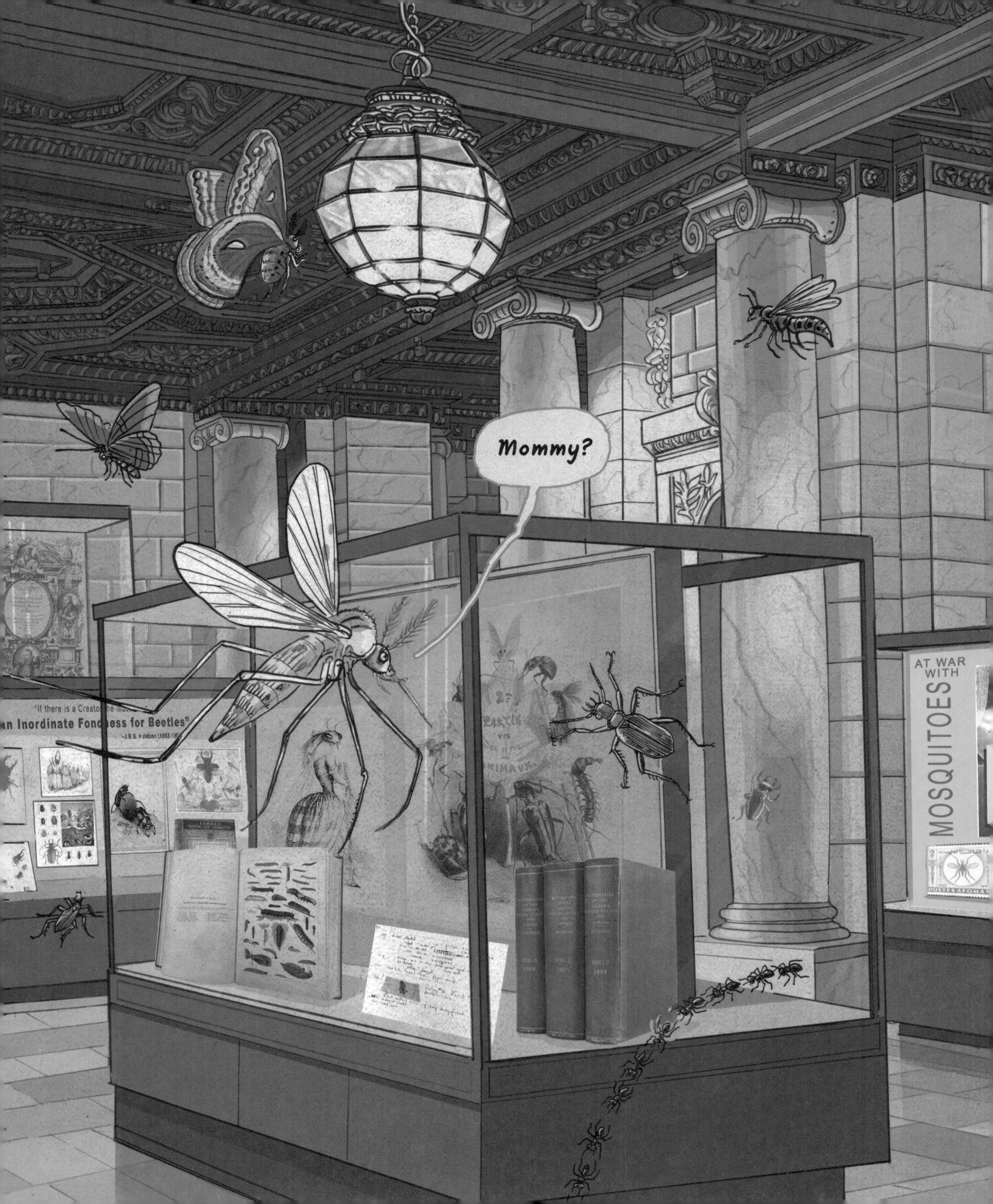
Mommy?
"If there is a Creato
n Inordinate Fond ess for Beetles"
AT WAR WITH MOSQUITOES

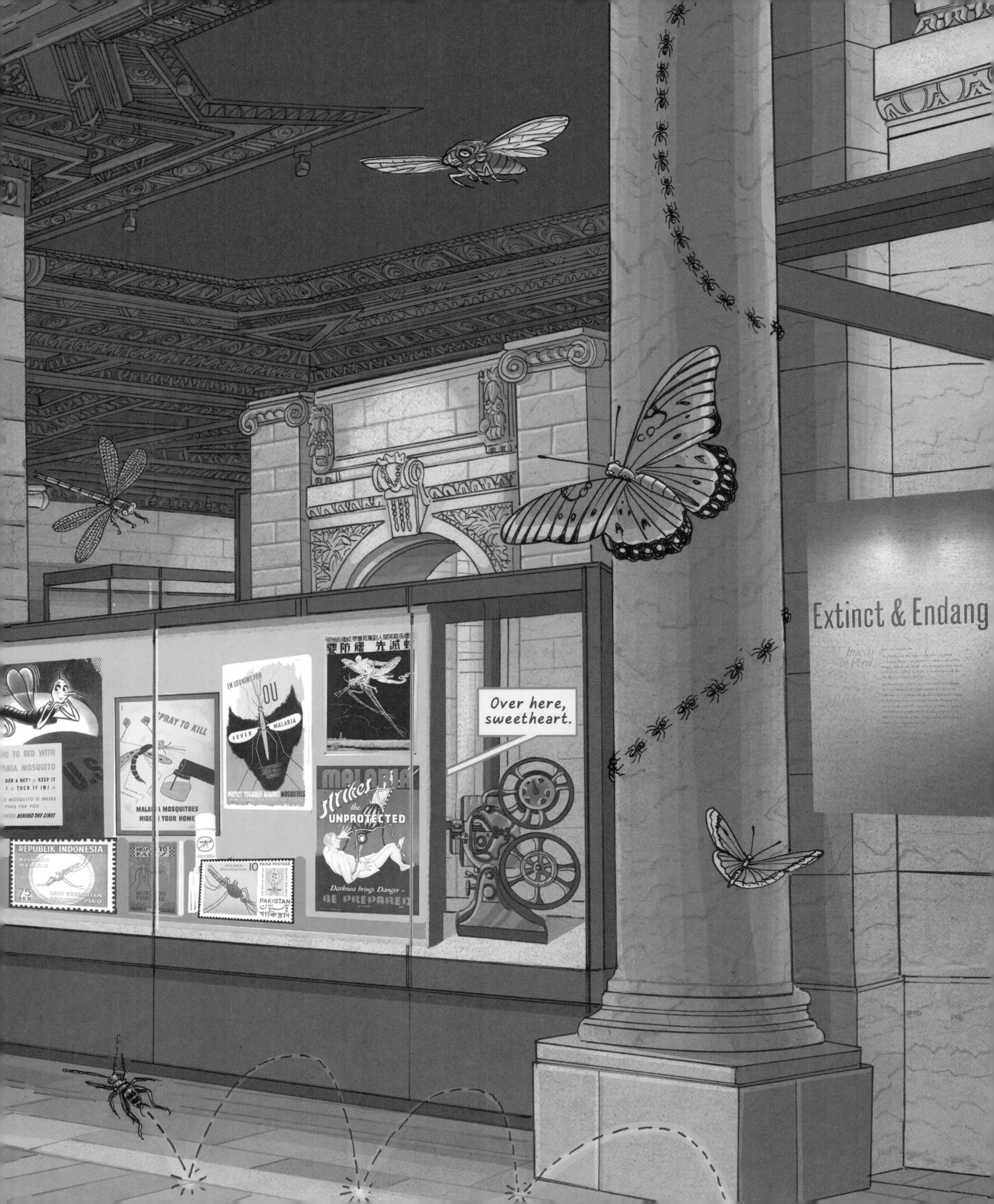

Over here, sweetheart.
Extinct & Endang
Insects in Peril
I'M LOOKING FOR
FEVER
MALARIA
PROTECT YOURSELF AGAINST MOSQUITOES
SPRAY TO KILL
要防瘧 先滅蚊
MALARIA
strikes the UNPROTECTED
Darkness brings Danger –
BE PREPARED
REPUBLIK INDONESIA
75 SEN
MALARIA ERADICATION
PAISA POSTAGE
10
PAKISTAN

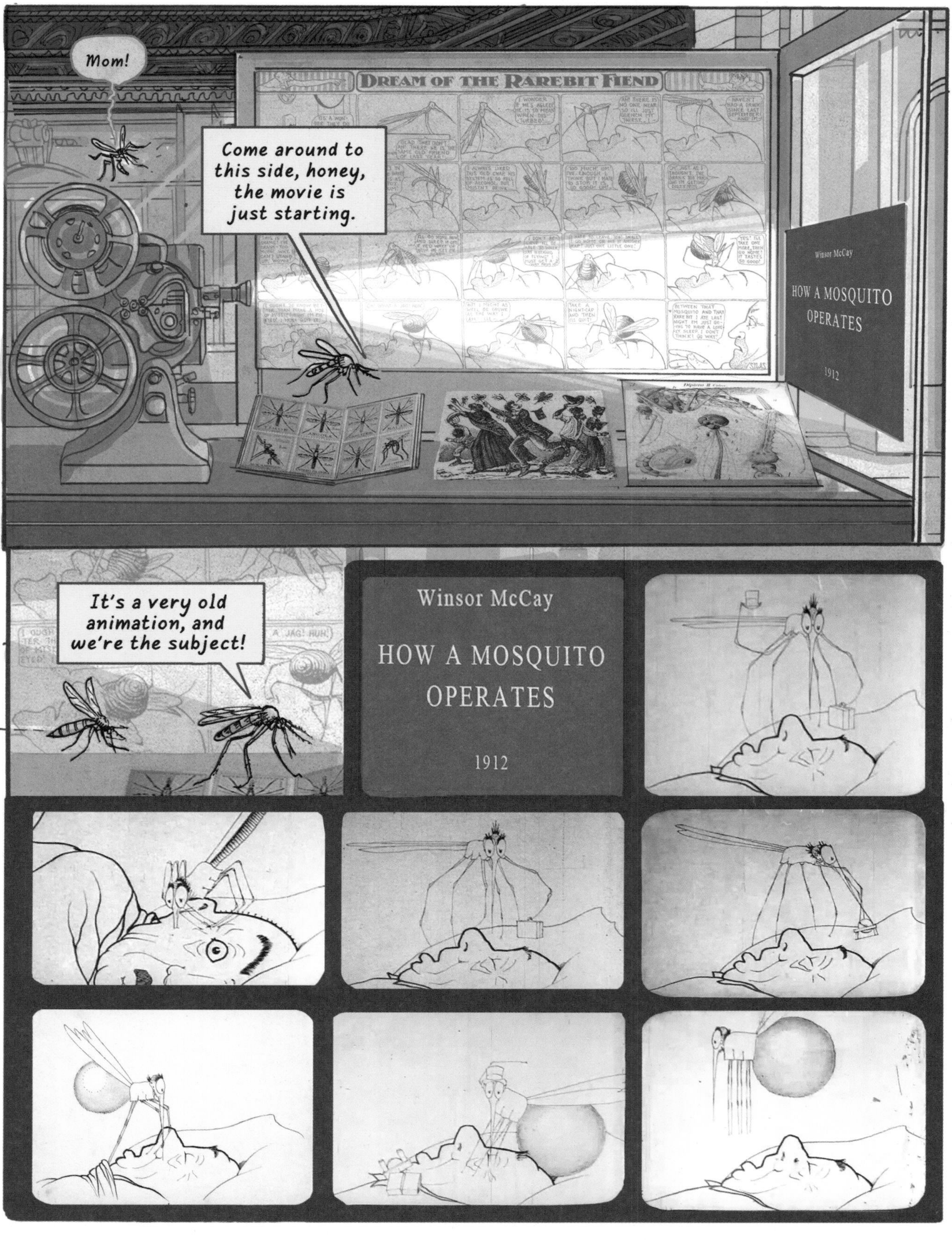
Mom!
DREAM OF THE RAREBIT FIEND
Come around to this side, honey, the movie is just starting.
Winsor McCay
HOW A MOSQUITO OPERATES
1912
It's a very old animation, and we're the subject!
Winsor McCay
HOW A MOSQUITO OPERATES
1912

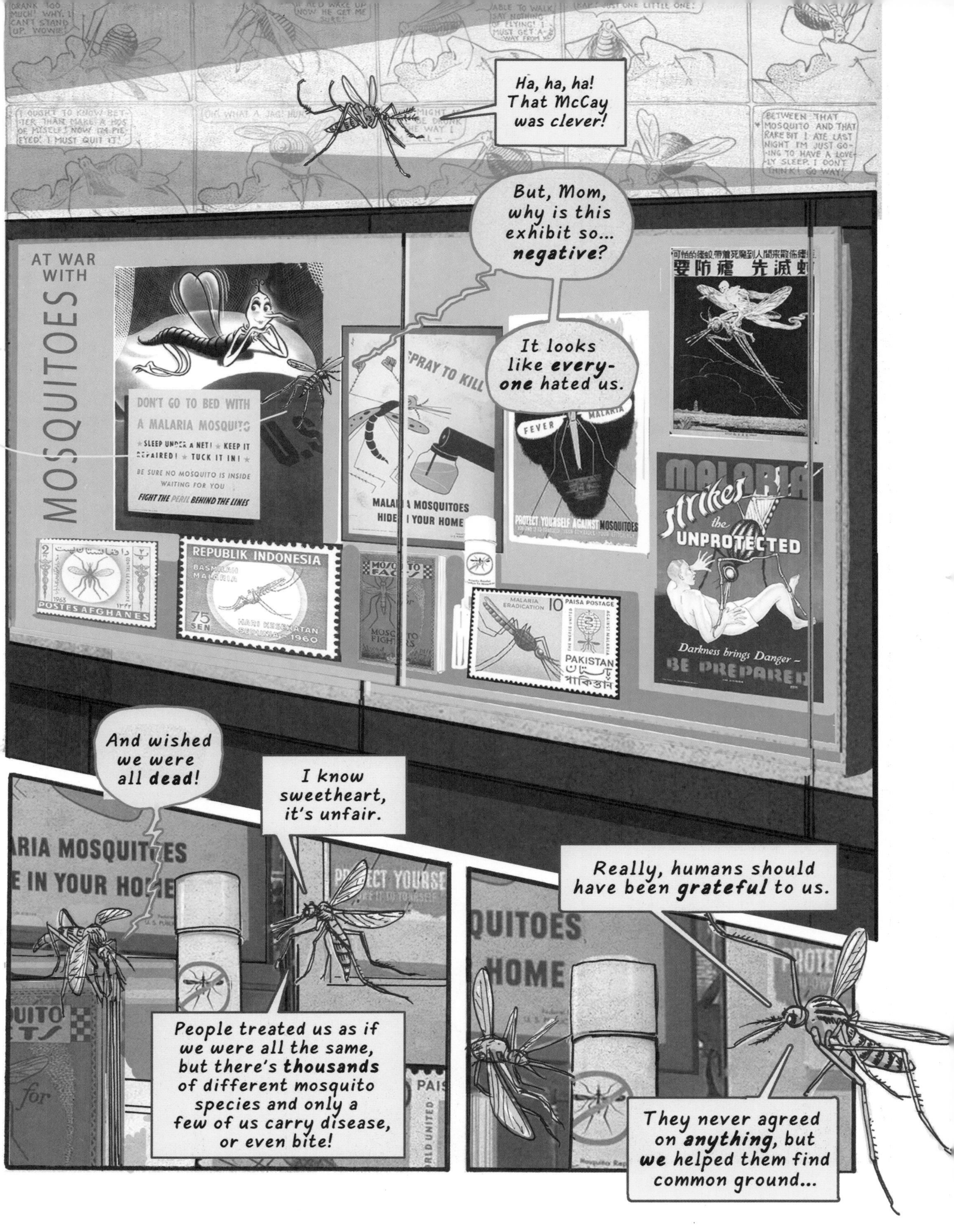

Ha, ha, ha! That McCay was clever!
BETWEEN THAT MOSQUITO AND THAT RAREBIT I ATE LAST NIGHT I'M JUST GOING TO HAVE A LOVELY SLEEP. I DON'T THINK! GO WAY!
But, Mom, why is this exhibit so... negative?
AT WAR WITH MOSQUITOES
DON'T GO TO BED WITH A MALARIA MOSQUITO
FIGHT THE PERIL BEHIND THE LINES
SPRAY TO KILL
It looks like everyone hated us.
FEVER
MALARIA
PROTECT YOURSELF AGAINST MOSQUITOES
Strikes the UNPROTECTED
Darkness brings Danger — BE PREPARED
POSTES AFGHANES
REPUBLIK INDONESIA
PAKISTAN
And wished we were all dead!
I know sweetheart, it's unfair.
People treated us as if we were all the same, but there's thousands of different mosquito species and only a few of us carry disease, or even bite!
Really, humans should have been grateful to us.
They never agreed on anything, but we helped them find common ground...

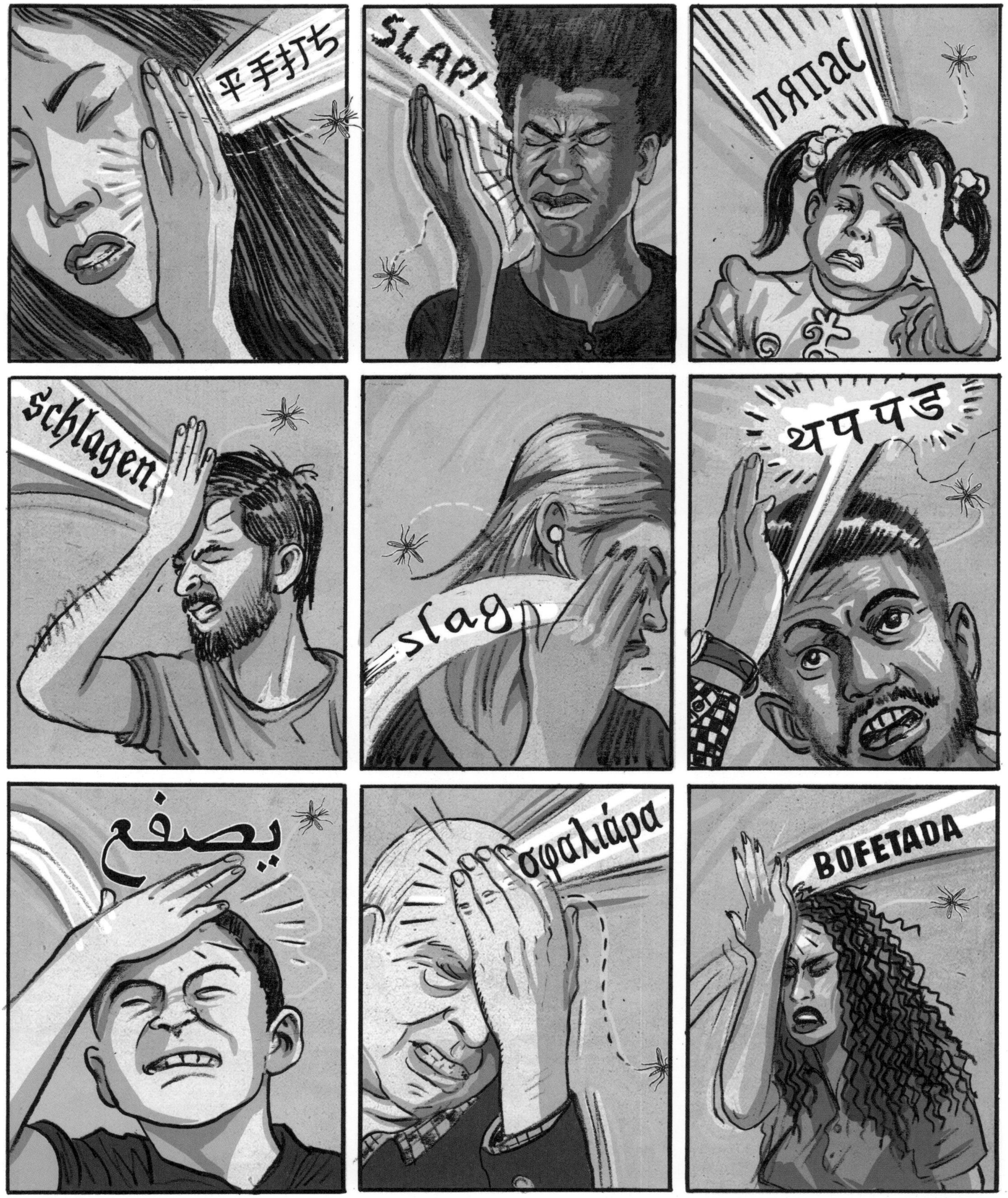
平手打ち
SLAP!
ЛЯПАС
schlagen
slag
थपपड
يصفع
σφαλιάρα
BOFETADA

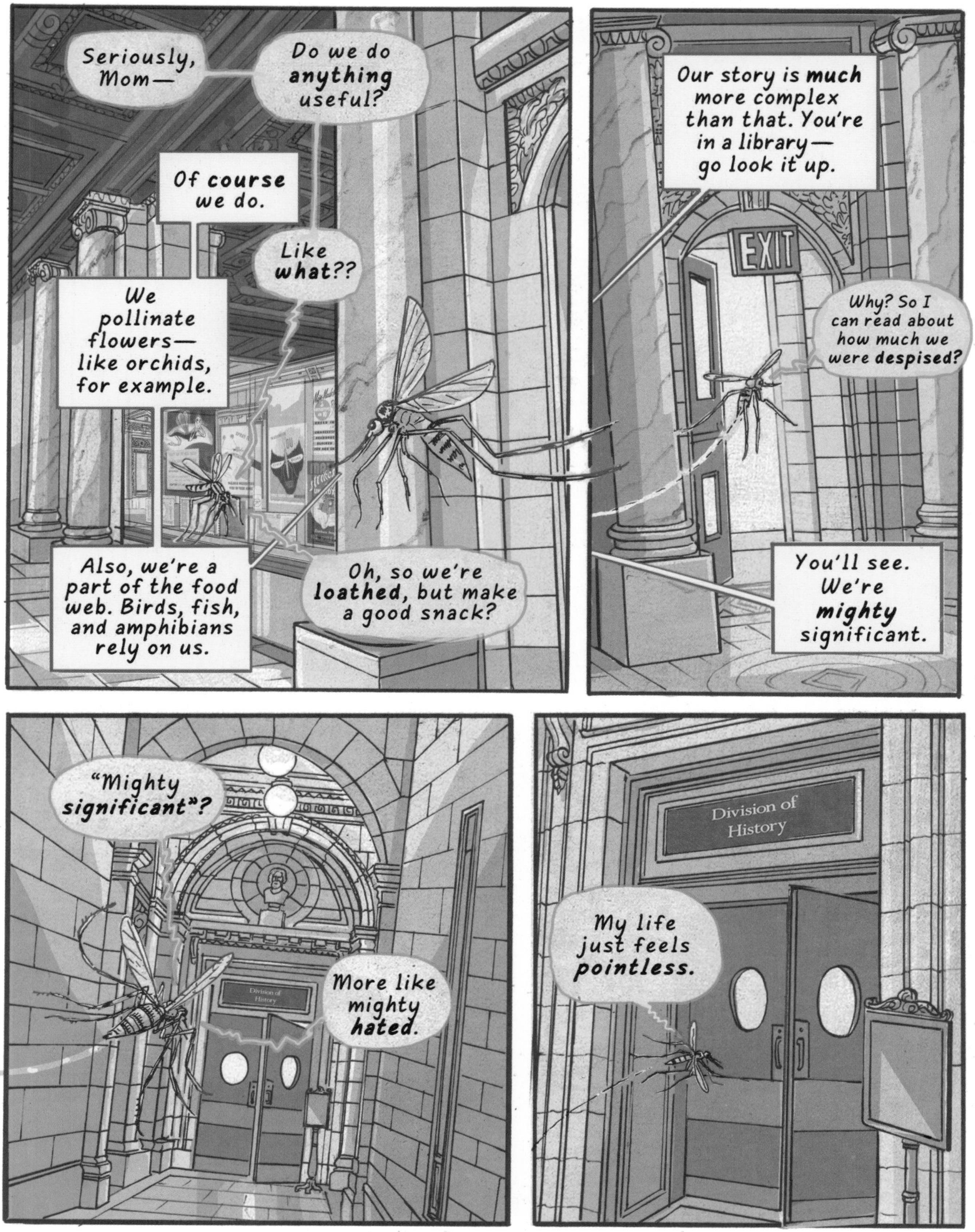
Seriously, Mom—
Do we do anything useful?
Of course we do.
Like what??
We pollinate flowers—like orchids, for example.
Also, we're a part of the food web. Birds, fish, and amphibians rely on us.
Oh, so we're loathed, but make a good snack?
Our story is much more complex than that. You're in a library—go look it up.
EXIT
Why? So I can read about how much we were despised?
You'll see. We're mighty significant.
"Mighty significant"?
More like mighty hated.
Division of History
My life just feels pointless.

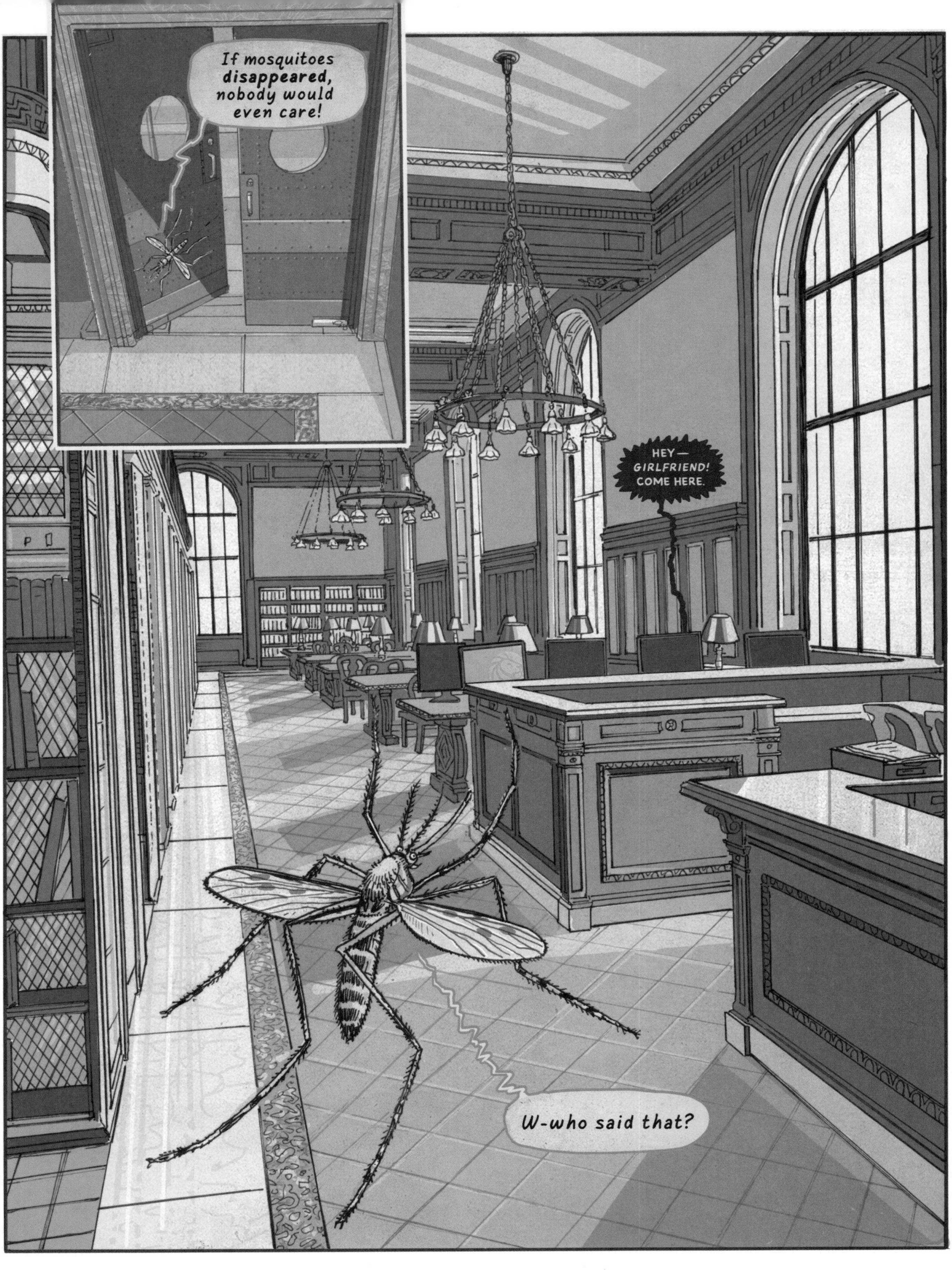

If mosquitoes disappeared, nobody would even care!
HEY— GIRLFRIEND! COME HERE.
W-who said that?

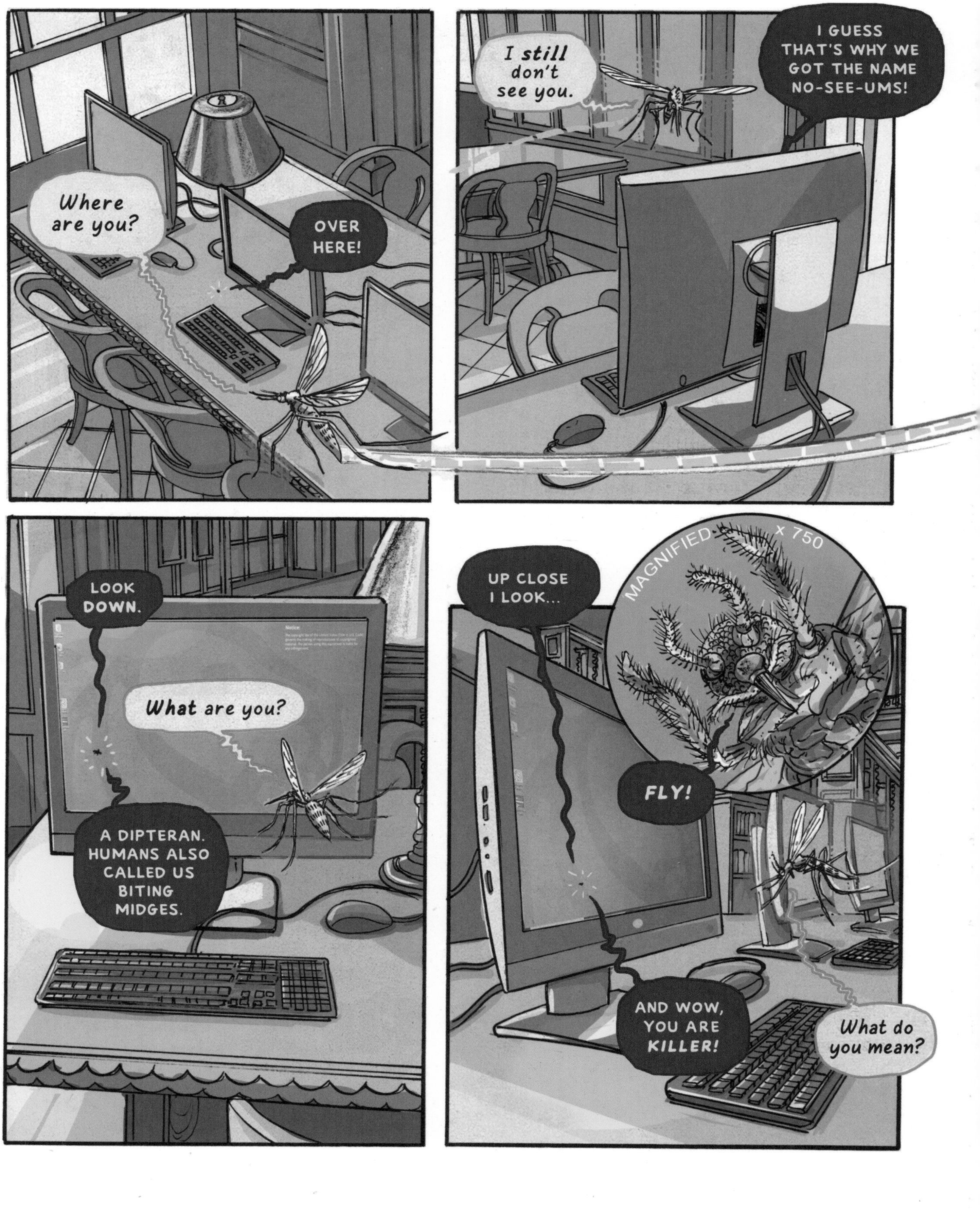

Where are you?
OVER HERE!
I still don't see you.
I GUESS THAT'S WHY WE GOT THE NAME NO-SEE-UMS!
LOOK DOWN.
What are you?
A DIPTERAN. HUMANS ALSO CALLED US BITING MIDGES.
UP CLOSE I LOOK...
MAGNIFIED x 750
FLY!
AND WOW, YOU ARE KILLER!
What do you mean?

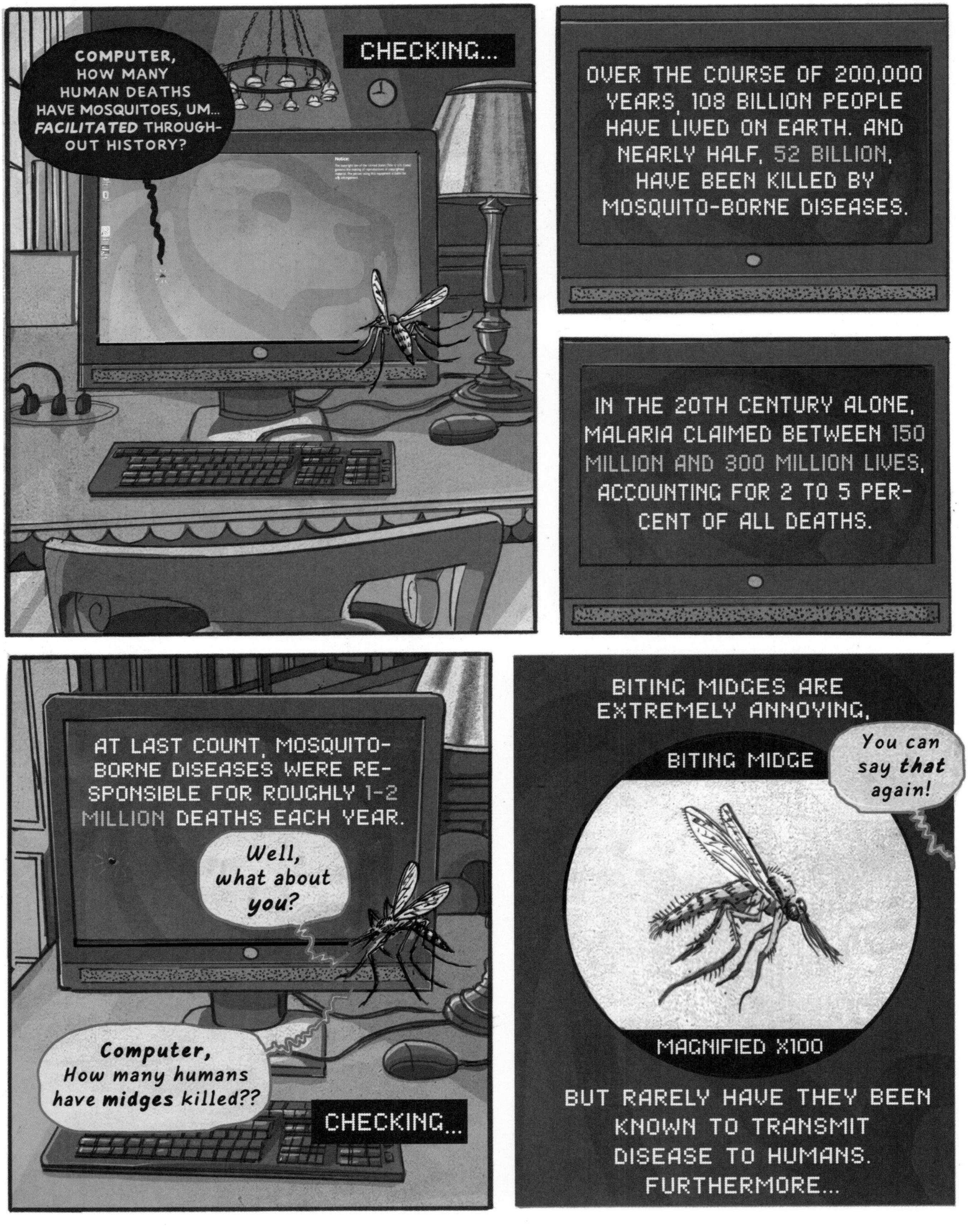
COMPUTER, HOW MANY HUMAN DEATHS HAVE MOSQUITOES, UM... *FACILITATED* THROUGH-OUT HISTORY?
CHECKING...
OVER THE COURSE OF 200,000 YEARS, 108 BILLION PEOPLE HAVE LIVED ON EARTH. AND NEARLY HALF, 52 BILLION, HAVE BEEN KILLED BY MOSQUITO-BORNE DISEASES.
IN THE 20TH CENTURY ALONE, MALARIA CLAIMED BETWEEN 150 MILLION AND 300 MILLION LIVES, ACCOUNTING FOR 2 TO 5 PER-CENT OF ALL DEATHS.
AT LAST COUNT, MOSQUITO-BORNE DISEASES WERE RE-SPONSIBLE FOR ROUGHLY 1-2 MILLION DEATHS EACH YEAR.
Well, what about *you?*
Computer, How many humans have **midges** killed??
CHECKING...
BITING MIDGES ARE EXTREMELY ANNOYING,
BITING MIDGE
You can say **that** again!
MAGNIFIED X100
BUT RARELY HAVE THEY BEEN KNOWN TO TRANSMIT DISEASE TO HUMANS. FURTHERMORE...

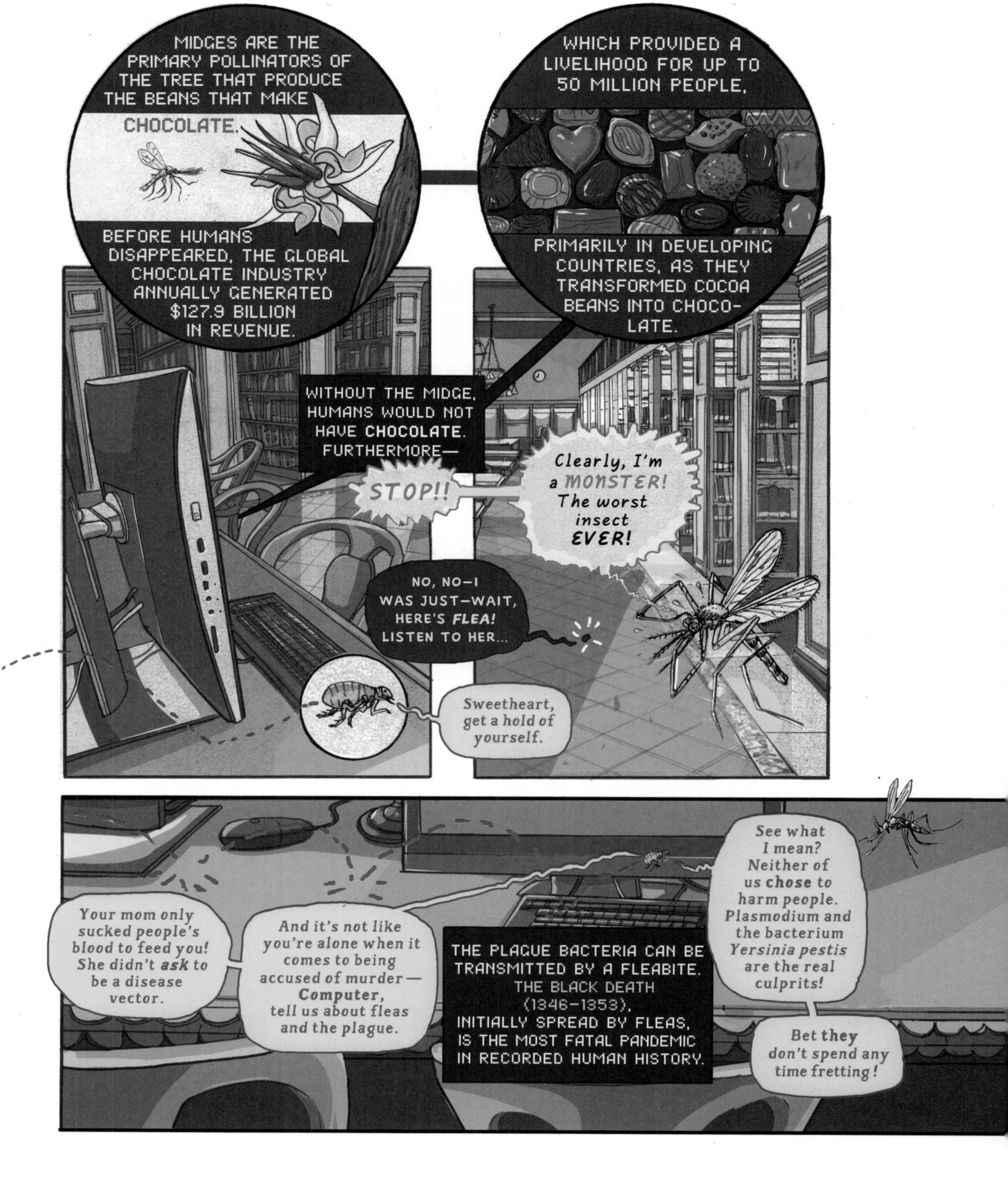
MIDGES ARE THE PRIMARY POLLINATORS OF THE TREE THAT PRODUCE THE BEANS THAT MAKE
CHOCOLATE.
BEFORE HUMANS DISAPPEARED, THE GLOBAL CHOCOLATE INDUSTRY ANNUALLY GENERATED $127.9 BILLION IN REVENUE.
WHICH PROVIDED A LIVELIHOOD FOR UP TO 50 MILLION PEOPLE,
PRIMARILY IN DEVELOPING COUNTRIES, AS THEY TRANSFORMED COCOA BEANS INTO CHOCO-LATE.
WITHOUT THE MIDGE, HUMANS WOULD NOT HAVE CHOCOLATE. FURTHERMORE—
STOP!!
Clearly, I'm a MONSTER! The worst insect EVER!
NO, NO—I WAS JUST—WAIT, HERE'S FLEA! LISTEN TO HER...
Sweetheart, get a hold of yourself.
Your mom only sucked people's blood to feed you! She didn't ask to be a disease vector.
And it's not like you're alone when it comes to being accused of murder—Computer, tell us about fleas and the plague.
THE PLAGUE BACTERIA CAN BE TRANSMITTED BY A FLEABITE. THE BLACK DEATH (1346–1353), INITIALLY SPREAD BY FLEAS, IS THE MOST FATAL PANDEMIC IN RECORDED HUMAN HISTORY.
See what I mean? Neither of us chose to harm people. Plasmodium and the bacterium Yersinia pestis are the real culprits!
Bet they don't spend any time fretting!

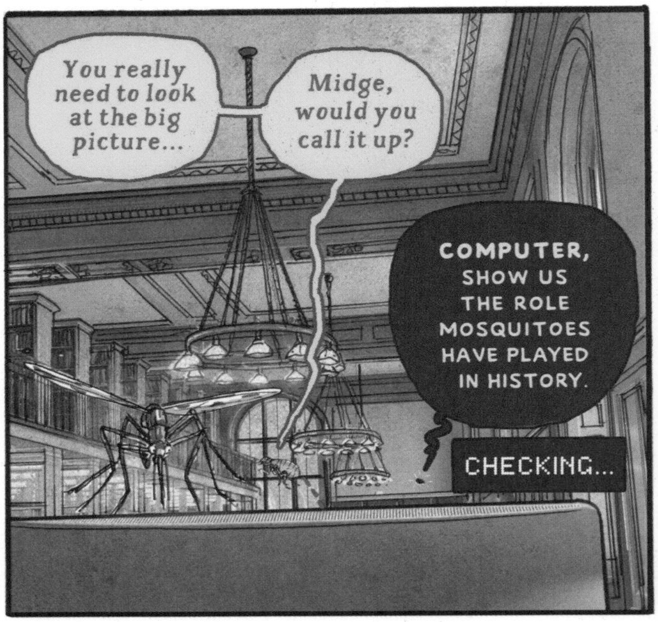
You really need to look at the big picture...
Midge, would you call it up?
COMPUTER, SHOW US THE ROLE MOSQUITOES HAVE PLAYED IN HISTORY.
CHECKING...

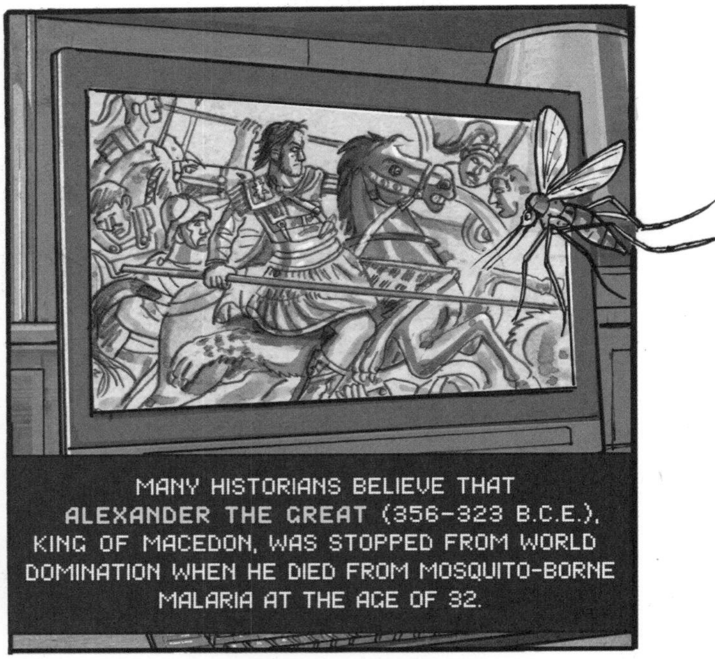
MANY HISTORIANS BELIEVE THAT ALEXANDER THE GREAT (356–323 B.C.E.), KING OF MACEDON, WAS STOPPED FROM WORLD DOMINATION WHEN HE DIED FROM MOSQUITO-BORNE MALARIA AT THE AGE OF 32.

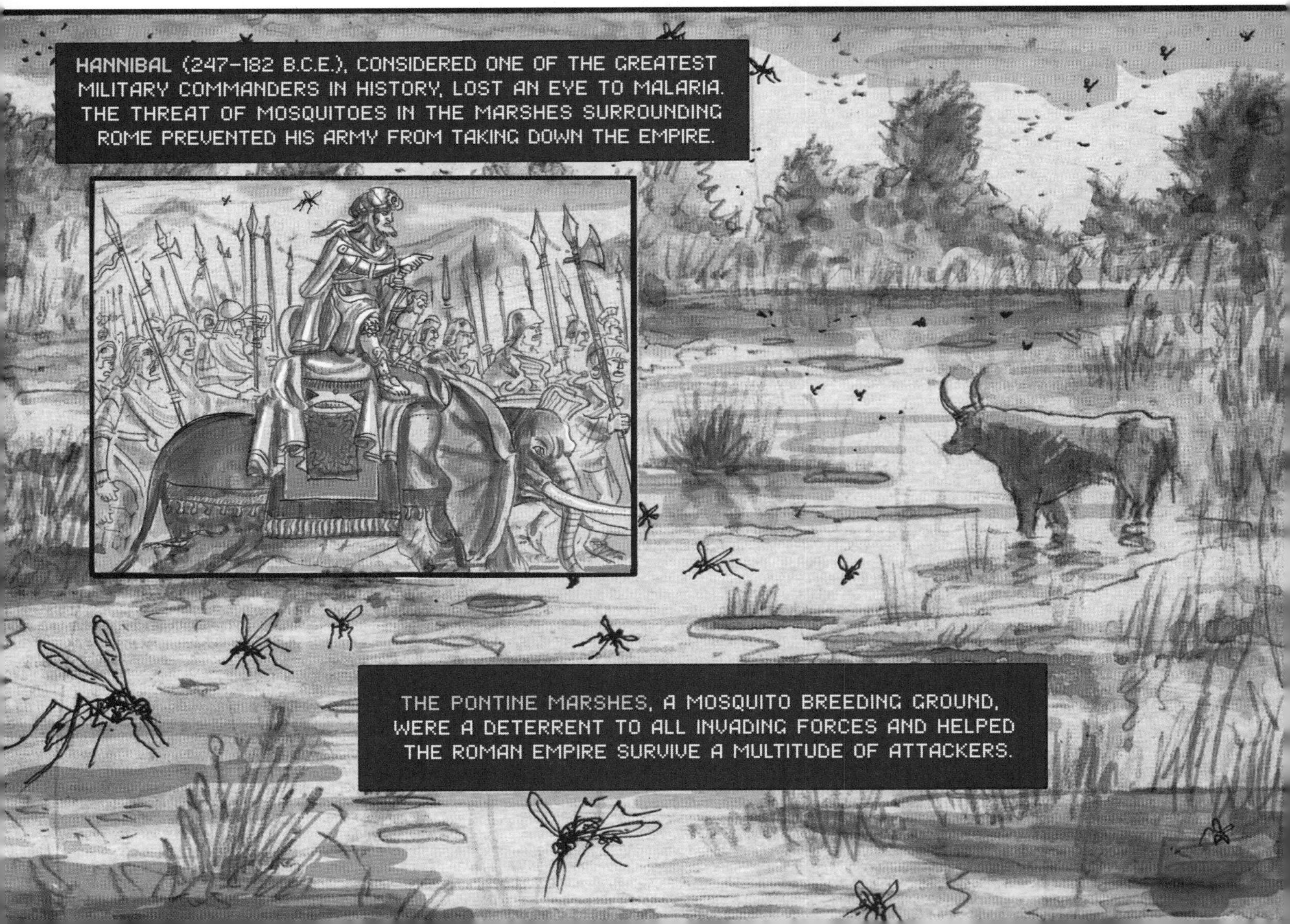
HANNIBAL (247–182 B.C.E.), CONSIDERED ONE OF THE GREATEST MILITARY COMMANDERS IN HISTORY, LOST AN EYE TO MALARIA. THE THREAT OF MOSQUITOES IN THE MARSHES SURROUNDING ROME PREVENTED HIS ARMY FROM TAKING DOWN THE EMPIRE.
THE PONTINE MARSHES, A MOSQUITO BREEDING GROUND, WERE A DETERRENT TO ALL INVADING FORCES AND HELPED THE ROMAN EMPIRE SURVIVE A MULTITUDE OF ATTACKERS.

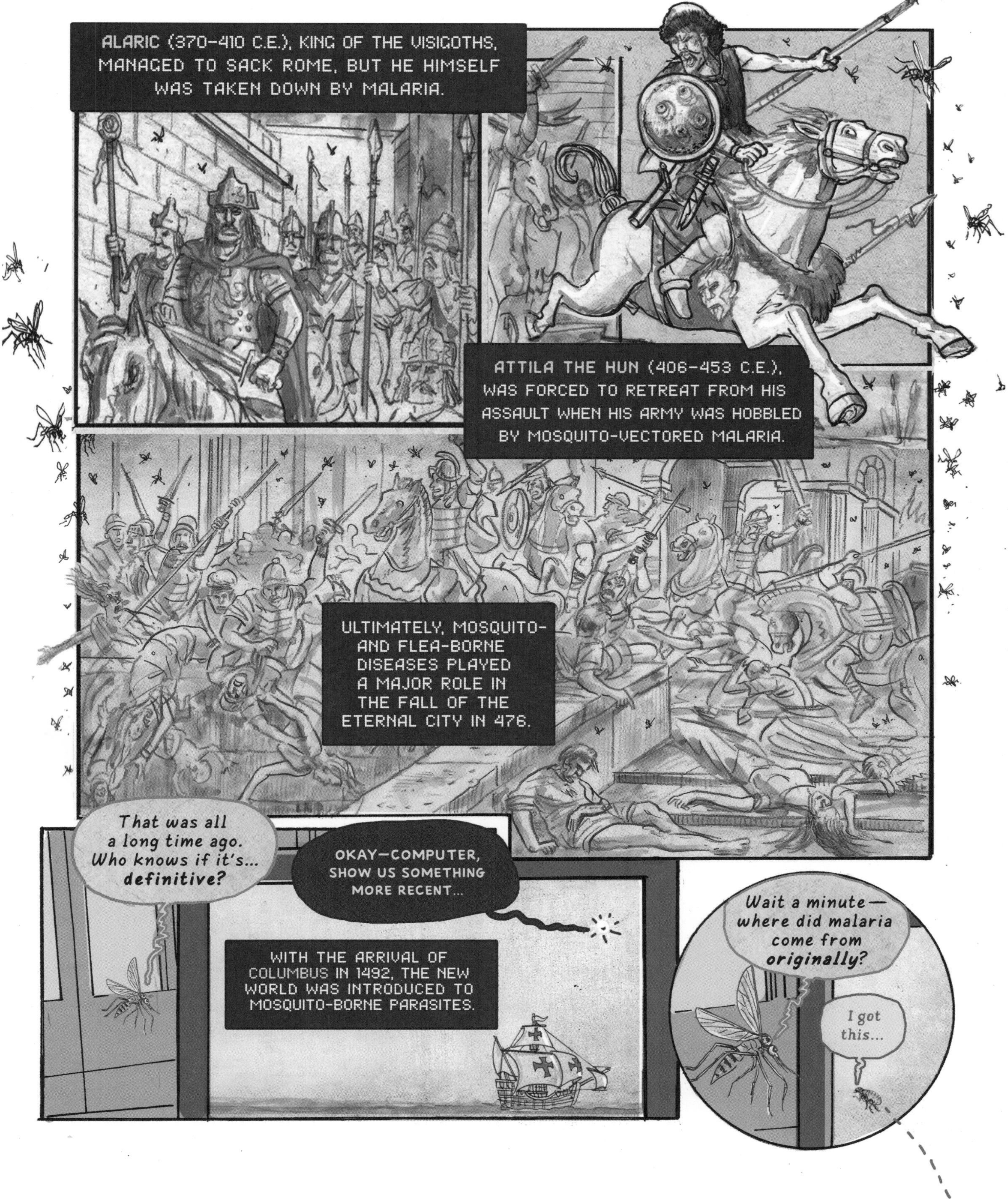
ALARIC (370–410 C.E.), KING OF THE VISIGOTHS, MANAGED TO SACK ROME, BUT HE HIMSELF WAS TAKEN DOWN BY MALARIA.
ATTILA THE HUN (406–453 C.E.), WAS FORCED TO RETREAT FROM HIS ASSAULT WHEN HIS ARMY WAS HOBBLED BY MOSQUITO-VECTORED MALARIA.
ULTIMATELY, MOSQUITO- AND FLEA-BORNE DISEASES PLAYED A MAJOR ROLE IN THE FALL OF THE ETERNAL CITY IN 476.
That was all a long time ago. Who knows if it's... definitive?
OKAY—COMPUTER, SHOW US SOMETHING MORE RECENT...
WITH THE ARRIVAL OF COLUMBUS IN 1492, THE NEW WORLD WAS INTRODUCED TO MOSQUITO-BORNE PARASITES.
Wait a minute— where did malaria come from originally?
I got this...

The name malaria was coined by the ancient Romans, who believed the sickness came from "mal aria"—"bad air" in the swamps. But the bacteria has been around since the dinosaurs...
SLAP!
Malaria transferred from monkeys to early humans millions of years ago and slowly made its way around the globe...
SLAP!
Malaria was described on Mesopotamian clay tablets in 2000 B.C.E.
SLAP!
It was recorded in Chinese scrolls going back to 2700 B.C.E.
SLAP!
Egyptians wrote about it on papyri during King Tut's rule...
And Tut probably died from malaria!
SLAP!
The illness was referenced in Hindu texts dating back to the 6th century B.C.E.
SLAP!
SLAP!
SLAP!
In Greece, about 400 B.C.E., the "father of medicine," Hippocrates, wrote about the symptoms, though he, too, thought they were caused by fumes.
SLAP!
A number of scientists had determined bacteria was the cause of the illness, but it wasn't until 1897 that Ronald Ross, a medical officer stationed in India, identified mosquitoes as the vector.
This won him the Nobel Prize.
SMUSH
No cure for malaria had been discovered, but for thousands of years the bitter bark from the cinchona tree was used to alleviate symptoms.
The fever tree bark contained quinine, which became a popular medication in Europe.
Especially after it was sweetened in a tonic with lime and alcohol.
A remedy called gin and tonic.
Because malaria and yellow fever had first appeared in sub-Saharan Africa, the people living in these areas developed some immunity through exposure over millennia.
But that came with a terrible price...

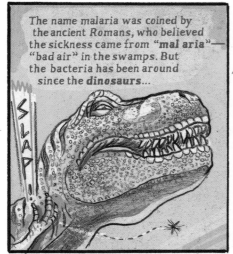

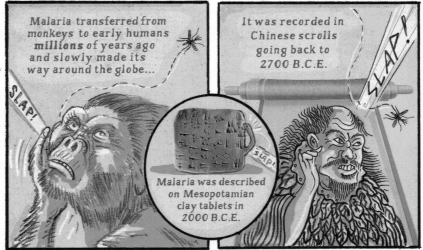

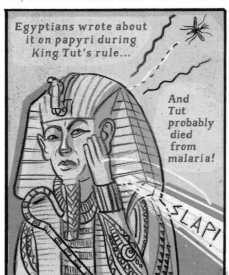

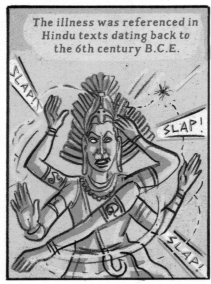

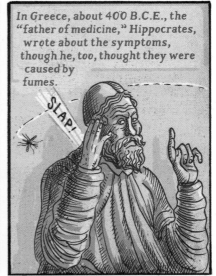

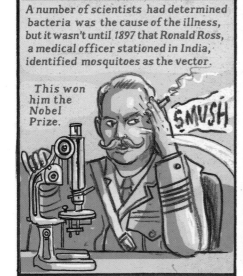

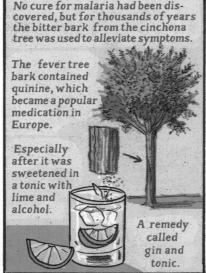

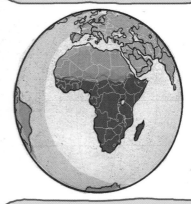

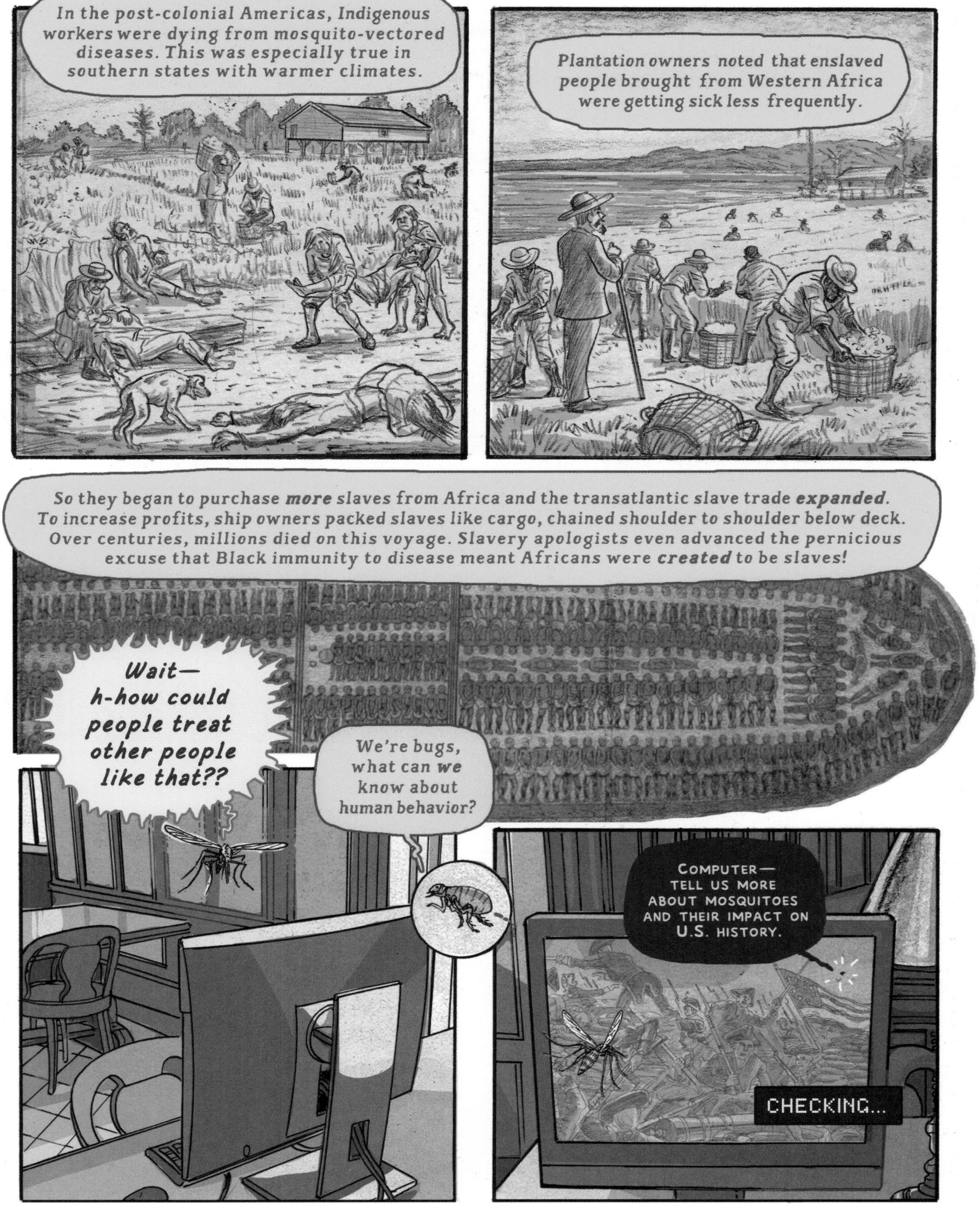
In the post-colonial Americas, Indigenous workers were dying from mosquito-vectored diseases. This was especially true in southern states with warmer climates.
Plantation owners noted that enslaved people brought from Western Africa were getting sick less frequently.
So they began to purchase *more* slaves from Africa and the transatlantic slave trade *expanded*. To increase profits, ship owners packed slaves like cargo, chained shoulder to shoulder below deck. Over centuries, millions died on this voyage. Slavery apologists even advanced the pernicious excuse that Black immunity to disease meant Africans were *created* to be slaves!
Wait— h-how could people treat other people like that??
We're bugs, what can *we* know about human behavior?
COMPUTER— TELL US MORE ABOUT MOSQUITOES AND THEIR IMPACT ON U.S. HISTORY.
CHECKING...

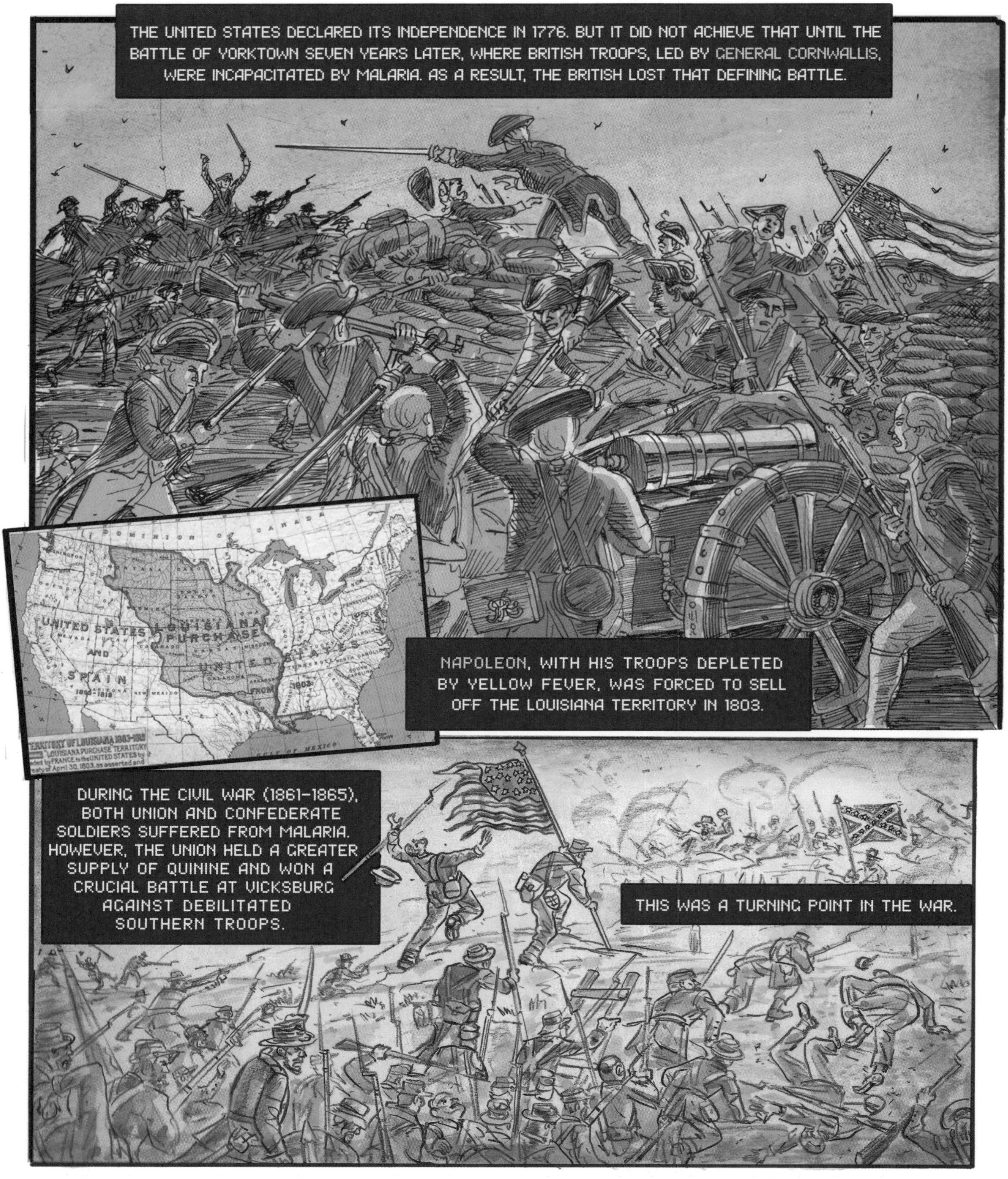
THE UNITED STATES DECLARED ITS INDEPENDENCE IN 1776. BUT IT DID NOT ACHIEVE THAT UNTIL THE BATTLE OF YORKTOWN SEVEN YEARS LATER, WHERE BRITISH TROOPS, LED BY GENERAL CORNWALLIS, WERE INCAPACITATED BY MALARIA. AS A RESULT, THE BRITISH LOST THAT DEFINING BATTLE.
UNITED STATES
LOUISIANA PURCHASE
AND
SPAIN
UNITED STATES
NAPOLEON, WITH HIS TROOPS DEPLETED BY YELLOW FEVER, WAS FORCED TO SELL OFF THE LOUISIANA TERRITORY IN 1803.
DURING THE CIVIL WAR (1861–1865), BOTH UNION AND CONFEDERATE SOLDIERS SUFFERED FROM MALARIA. HOWEVER, THE UNION HELD A GREATER SUPPLY OF QUININE AND WON A CRUCIAL BATTLE AT VICKSBURG AGAINST DEBILITATED SOUTHERN TROOPS.
THIS WAS A TURNING POINT IN THE WAR.

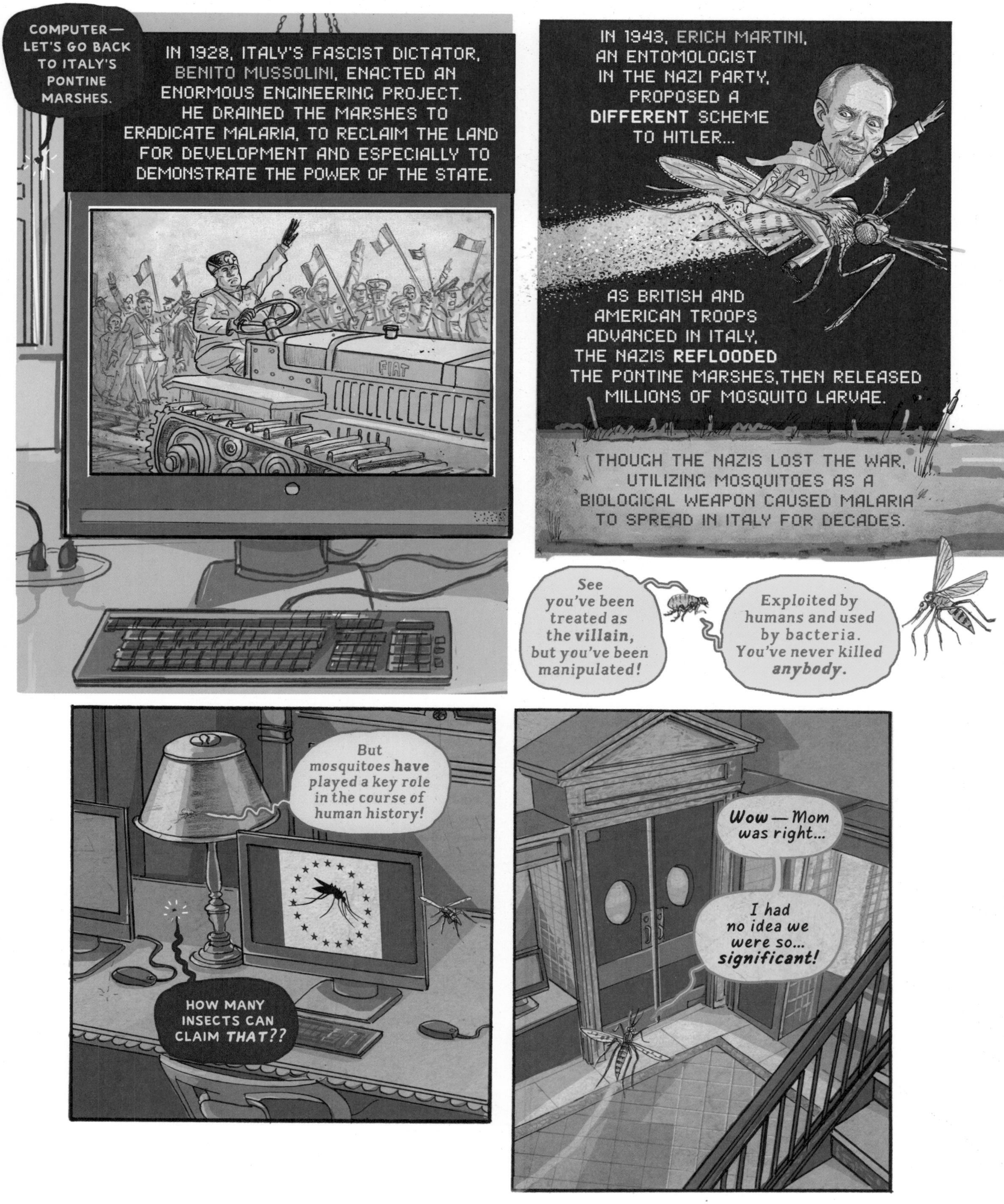
COMPUTER— LET'S GO BACK TO ITALY'S PONTINE MARSHES.
IN 1928, ITALY'S FASCIST DICTATOR, BENITO MUSSOLINI, ENACTED AN ENORMOUS ENGINEERING PROJECT. HE DRAINED THE MARSHES TO ERADICATE MALARIA, TO RECLAIM THE LAND FOR DEVELOPMENT AND ESPECIALLY TO DEMONSTRATE THE POWER OF THE STATE.
FIAT
IN 1943, ERICH MARTINI, AN ENTOMOLOGIST IN THE NAZI PARTY, PROPOSED A **DIFFERENT** SCHEME TO HITLER...
AS BRITISH AND AMERICAN TROOPS ADVANCED IN ITALY, THE NAZIS **REFLOODED** THE PONTINE MARSHES, THEN RELEASED MILLIONS OF MOSQUITO LARVAE.
THOUGH THE NAZIS LOST THE WAR, UTILIZING MOSQUITOES AS A BIOLOGICAL WEAPON CAUSED MALARIA TO SPREAD IN ITALY FOR DECADES.
See you've been treated as the **villain**, but you've been manipulated!
Exploited by humans and used by bacteria. You've never killed ***anybody***.
But *mosquitoes* **have** played a key role in the course of human history!
HOW MANY INSECTS CAN CLAIM *THAT*??
Wow — Mom was right...
I had no idea we were so... ***significant!***

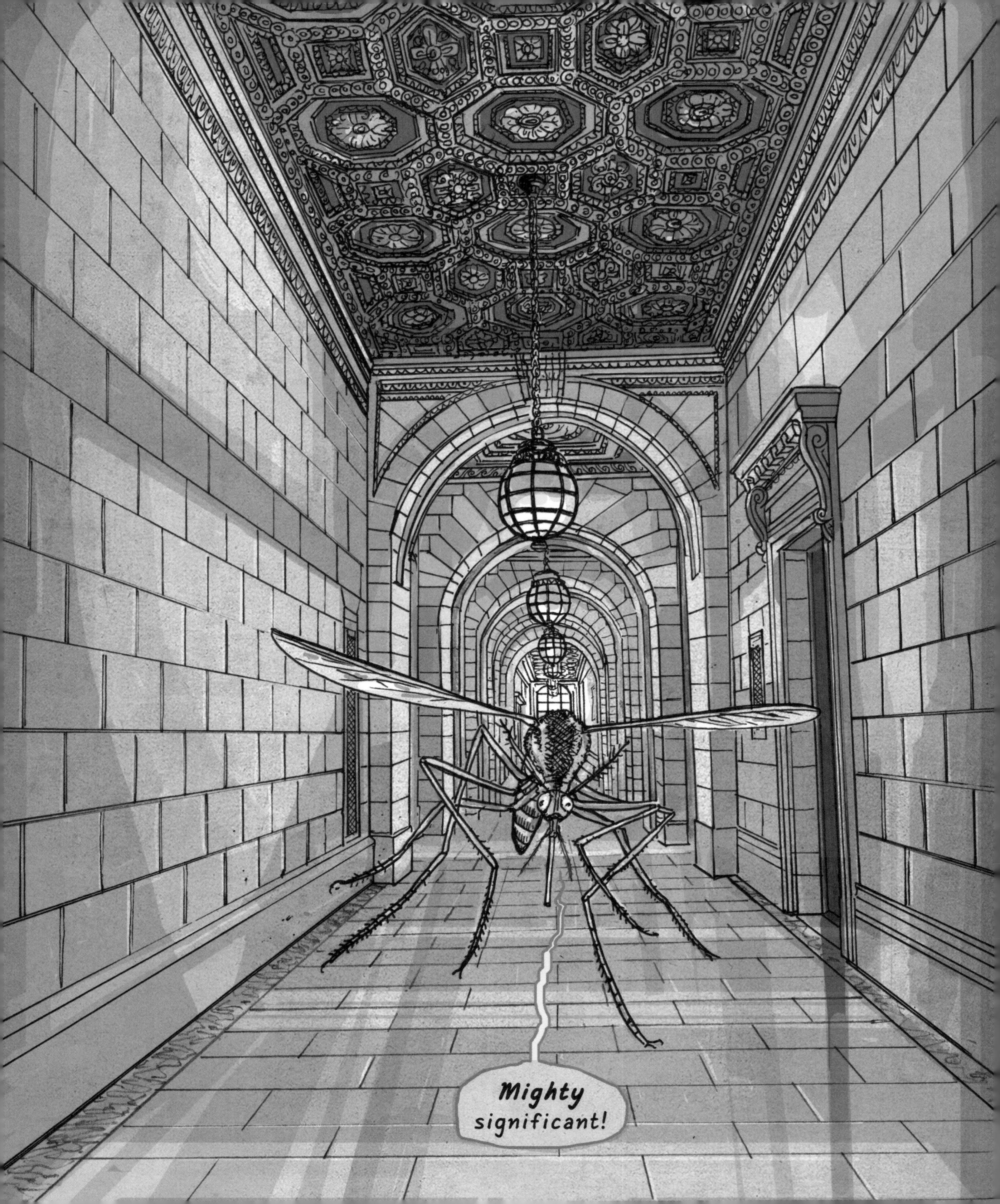
Mighty significant!

Ancient Egyptian
Winged Scarab Beetle
Khepri Sun Disk Hieroglyphic
(circa 664–332 B.C.E.)

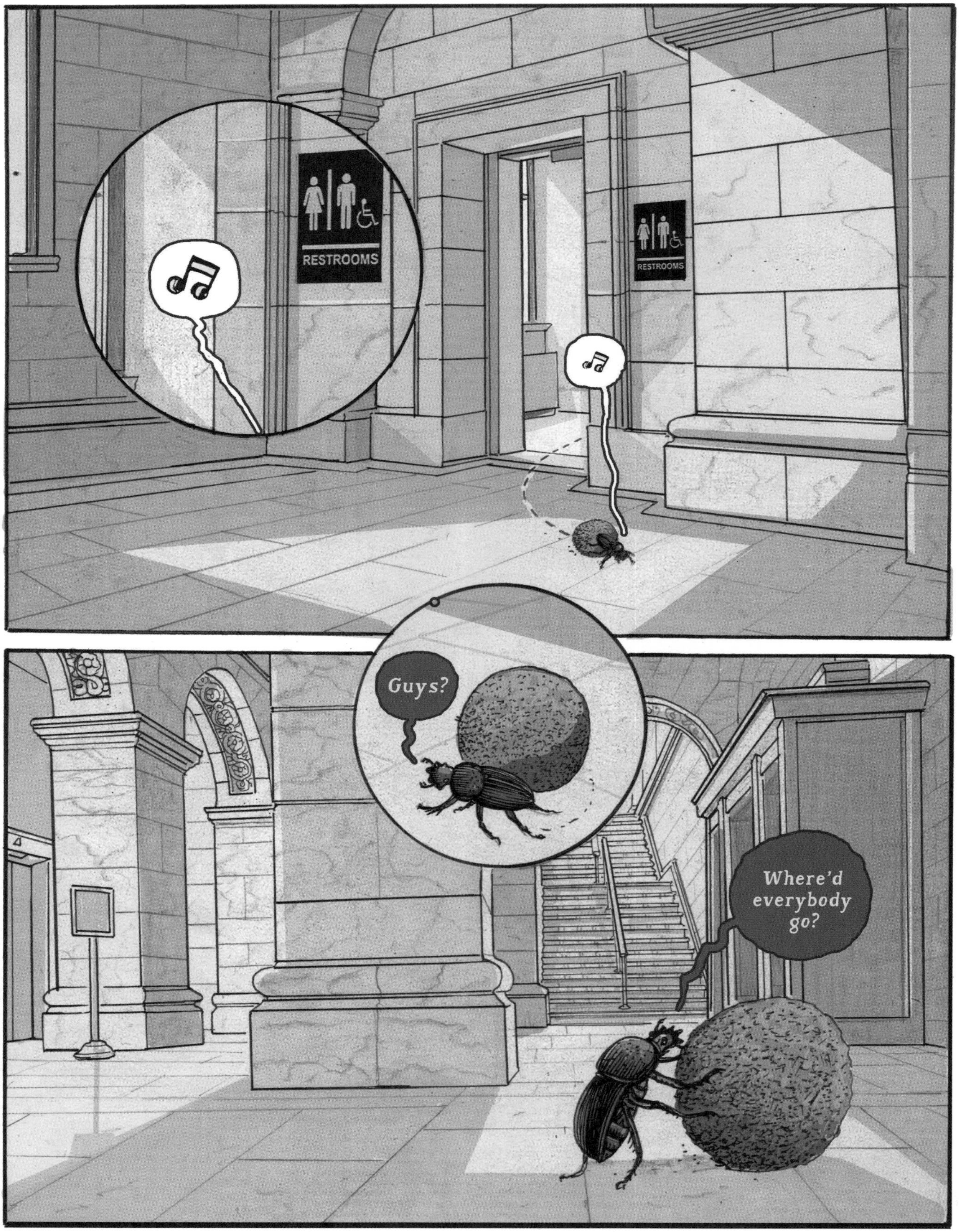
RESTROOMS
♫
RESTROOMS
♫
Guys?
Where'd everybody go?

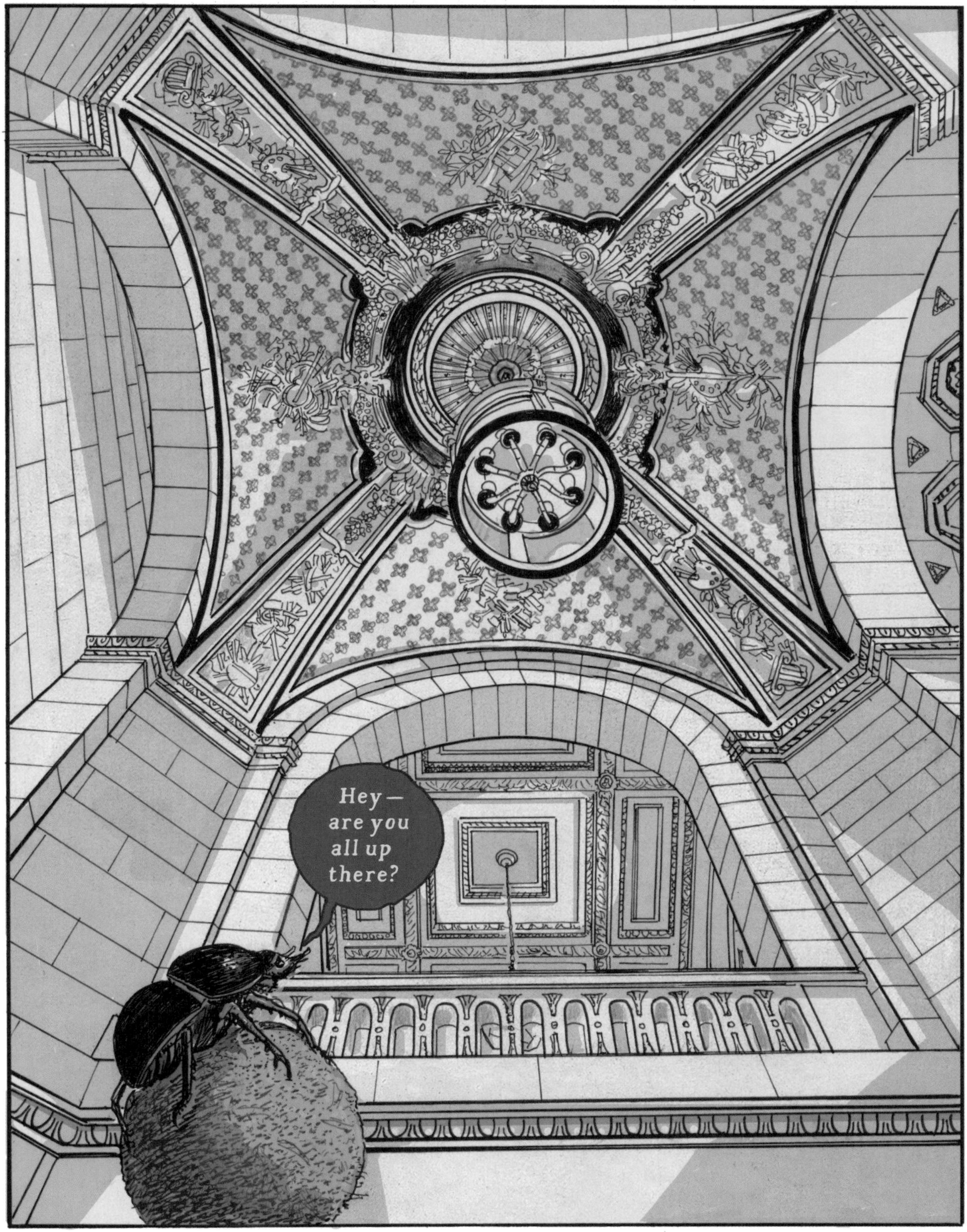
Hey— are you all up there?

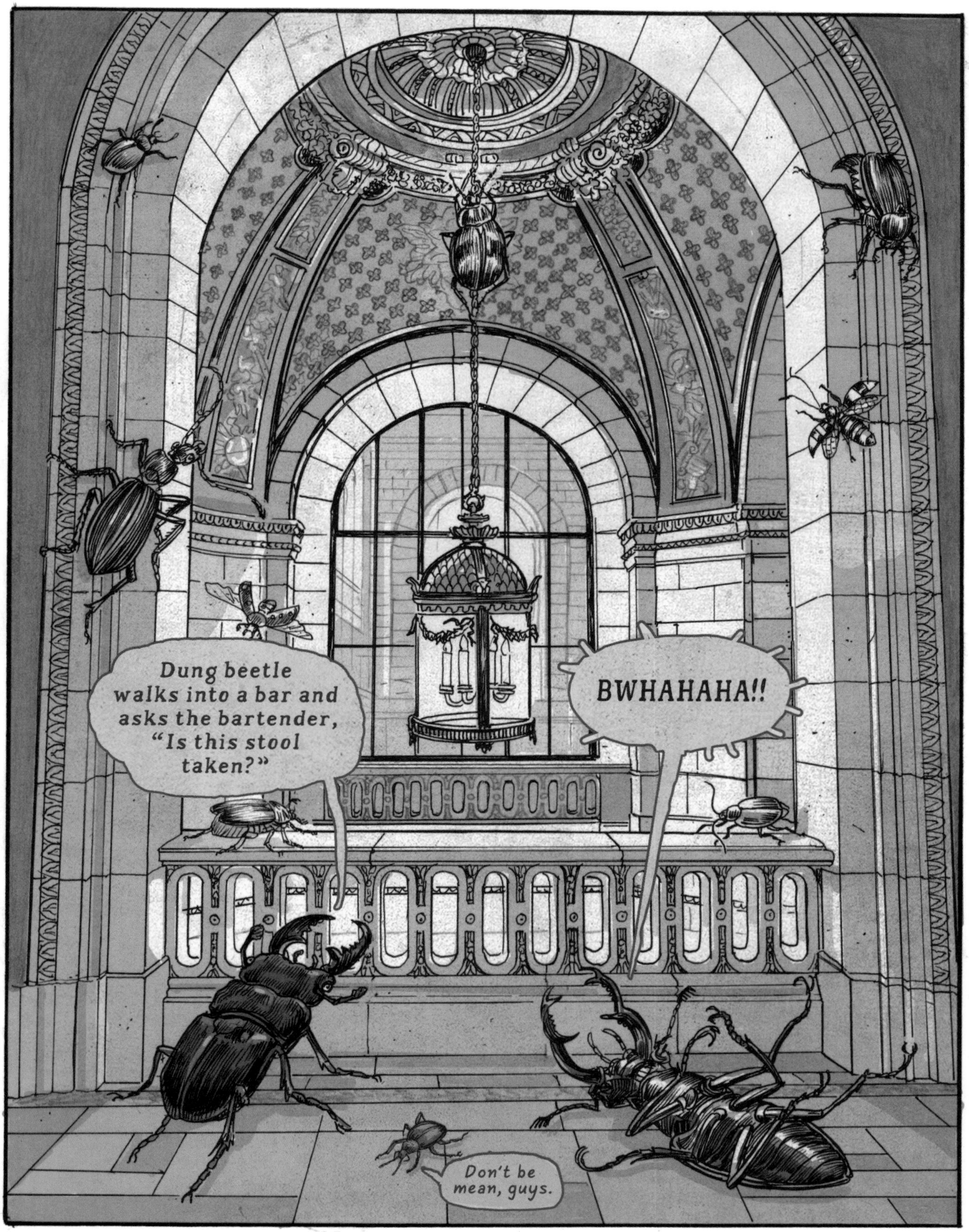
Dung beetle walks into a bar and asks the bartender, “Is this stool taken?”
BWHAHAHA!!
Don't be mean, guys.

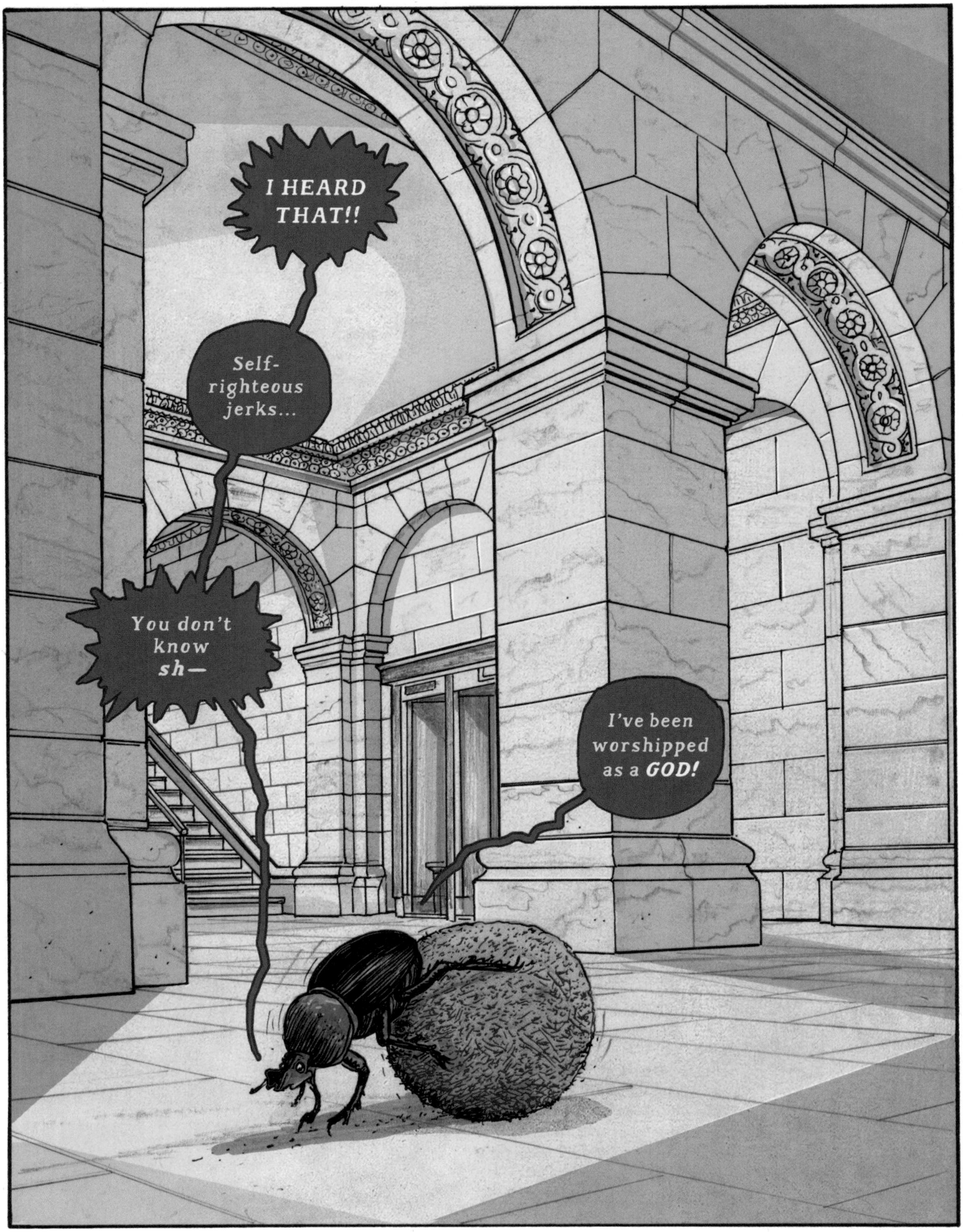
I HEARD THAT!!
Self-righteous jerks...
You don't know sh—
I've been worshipped as a GOD!

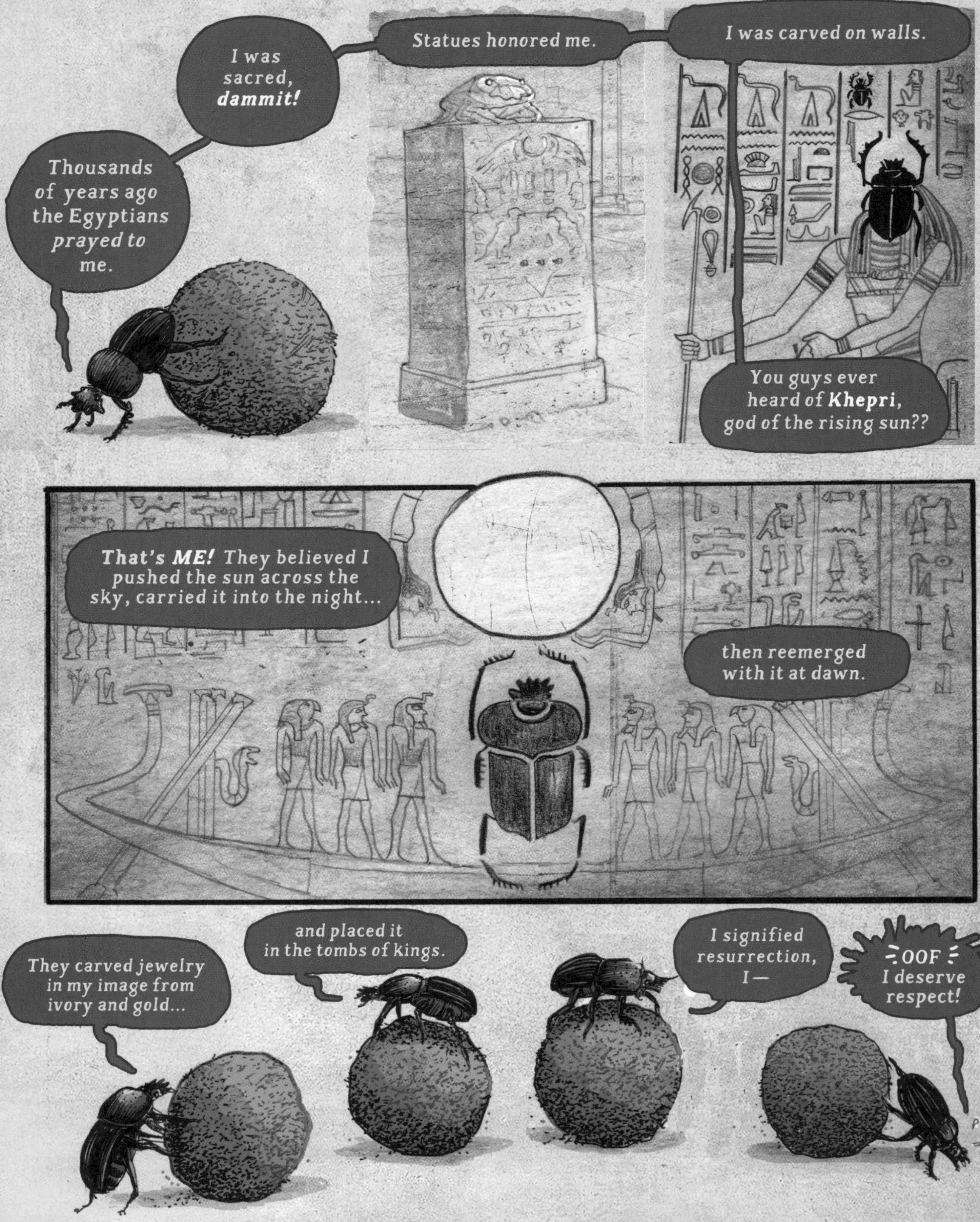
Thousands of years ago the Egyptians *prayed to* me.
I was sacred, ***dammit!***
Statues honored me.
I was carved on walls.
You guys ever heard of **Khepri**, god of the rising sun??
That's *ME!* They believed I pushed the sun across the sky, carried it into the night...
then reemerged with it at dawn.
They carved jewelry in my image from ivory and gold...
and placed it in the tombs of kings.
I signified resurrection, I—
OOF
I deserve respect!

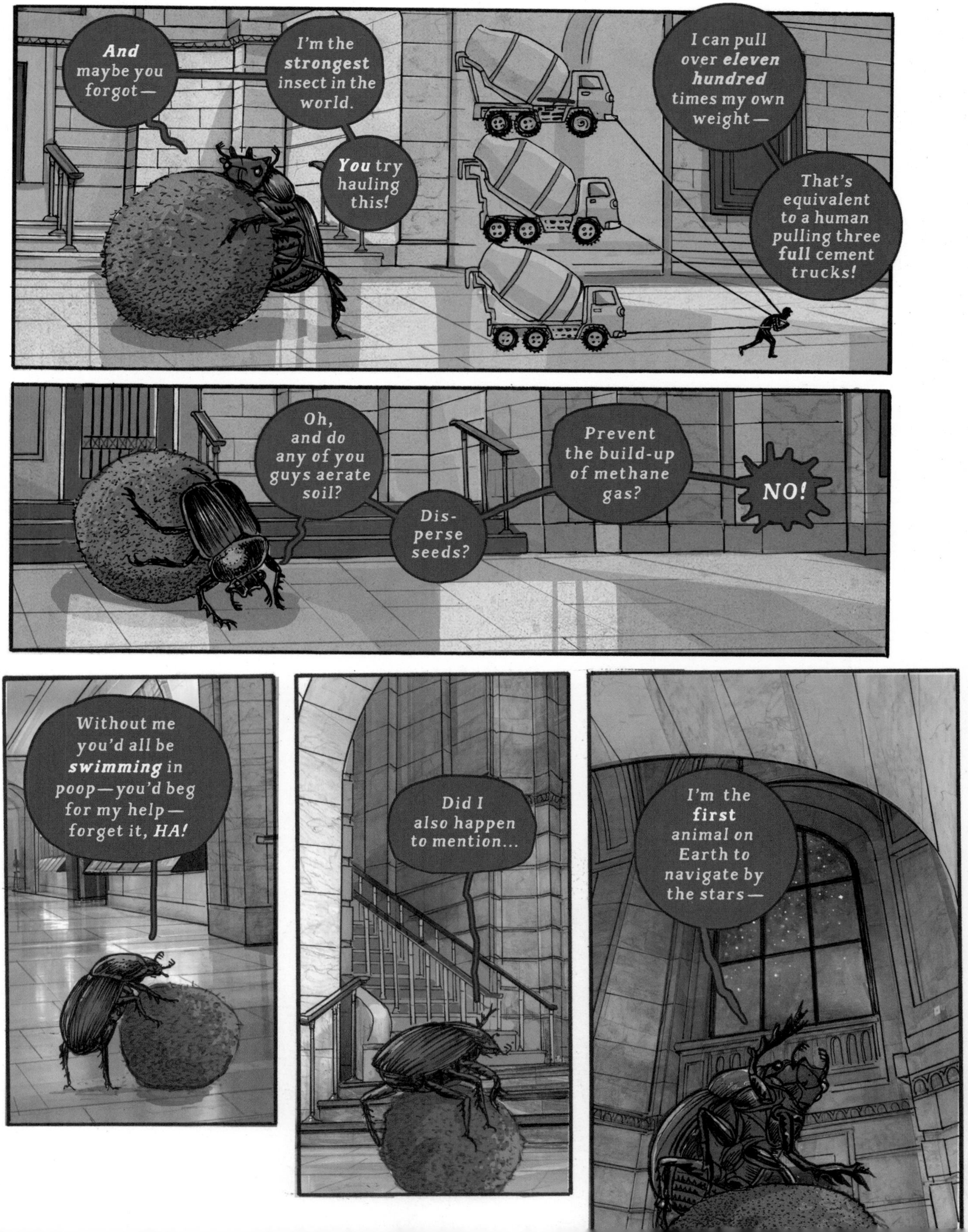

And maybe you forgot—
I'm the strongest insect in the world.
You try hauling this!
I can pull over eleven hundred times my own weight—
That's equivalent to a human pulling three full cement trucks!
Oh, and do any of you guys aerate soil?
Dis-perse seeds?
Prevent the build-up of methane gas?
NO!
Without me you'd all be swimming in poop—you'd beg for my help—forget it, HA!
Did I also happen to mention...
I'm the first animal on Earth to navigate by the stars—

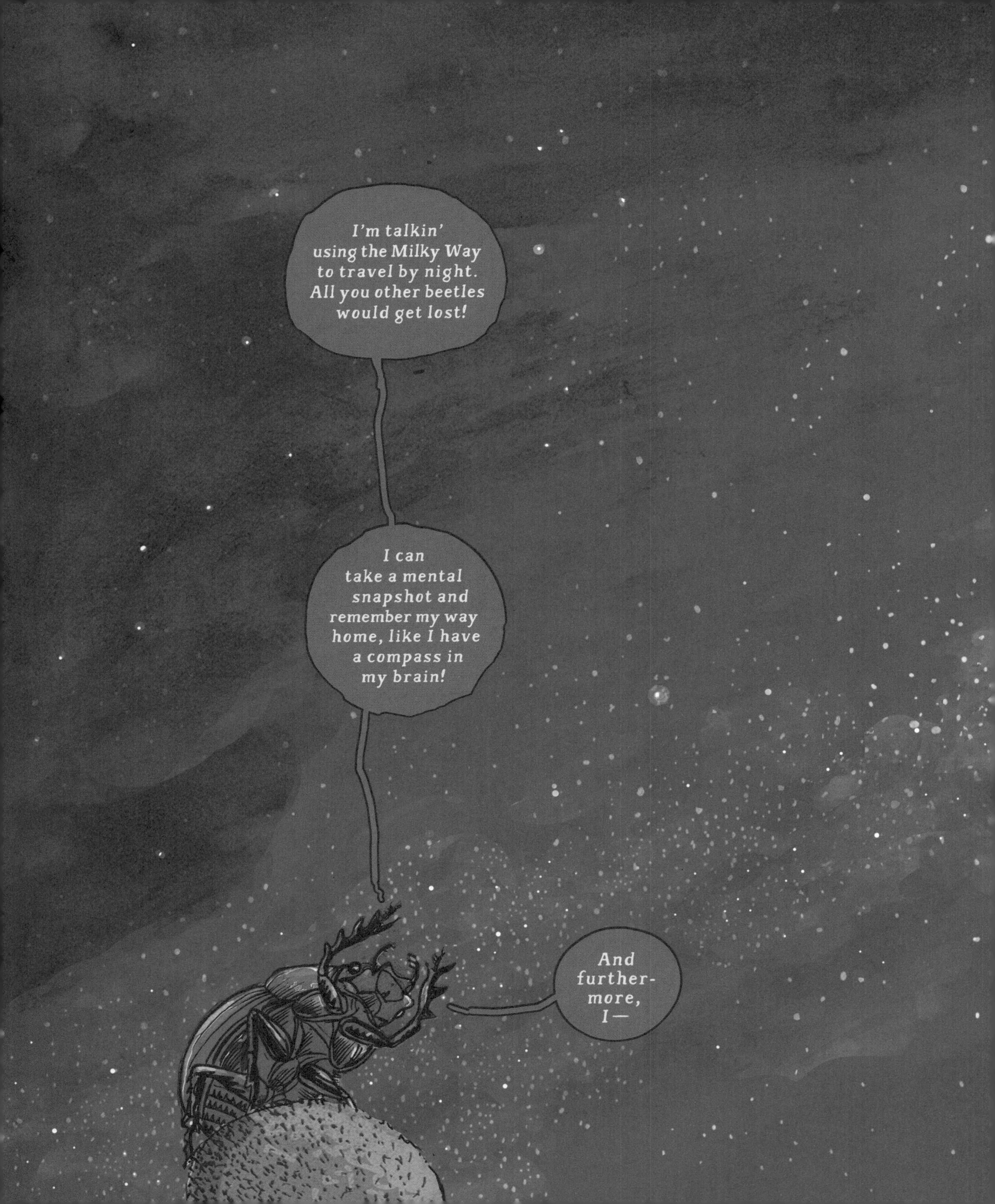
I'm talkin' using the Milky Way to travel by night. All you other beetles would get lost!
I can take a mental snapshot and remember my way home, like I have a compass in my brain!
And further-more, I—

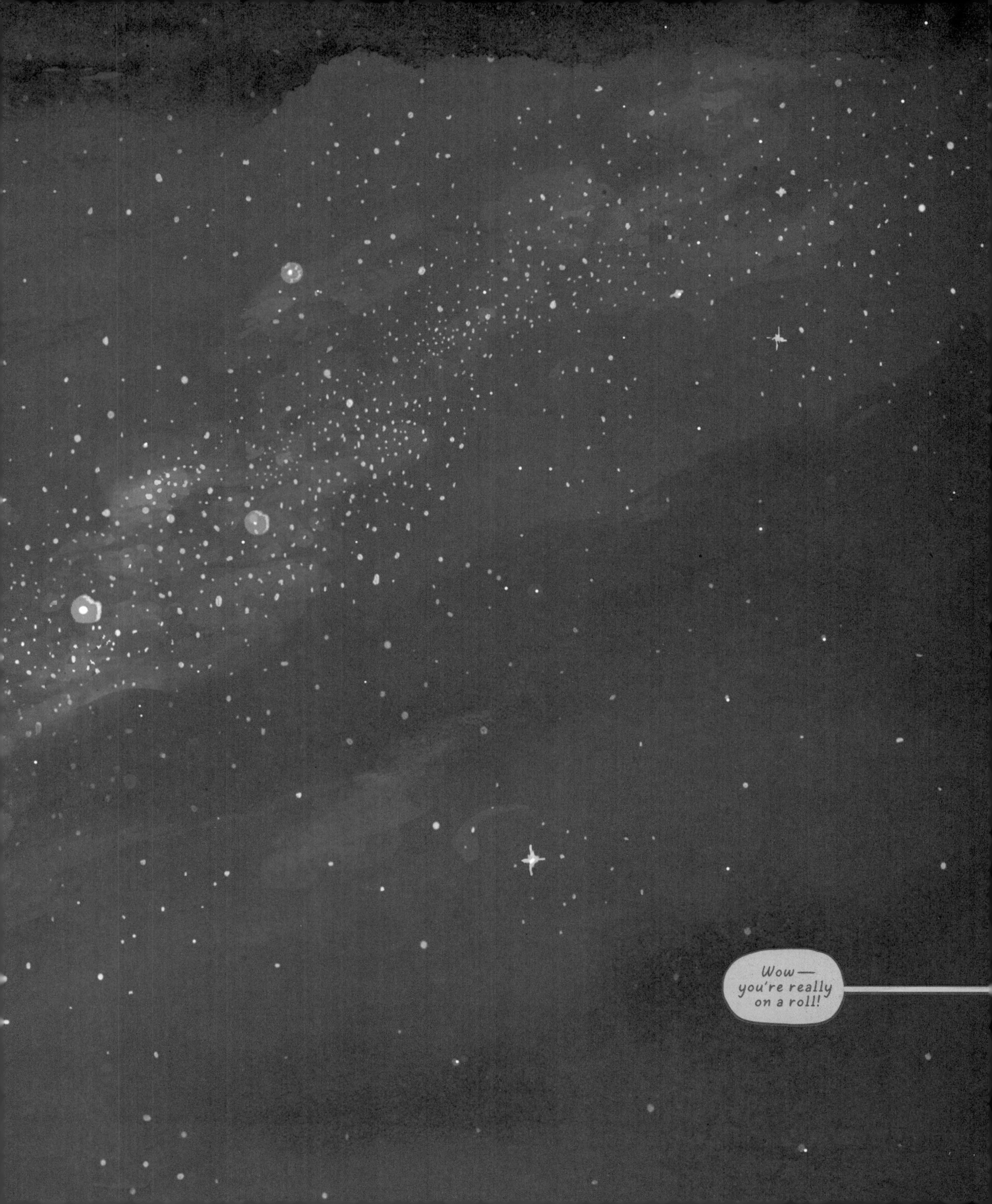
Wow—you're really on a roll!

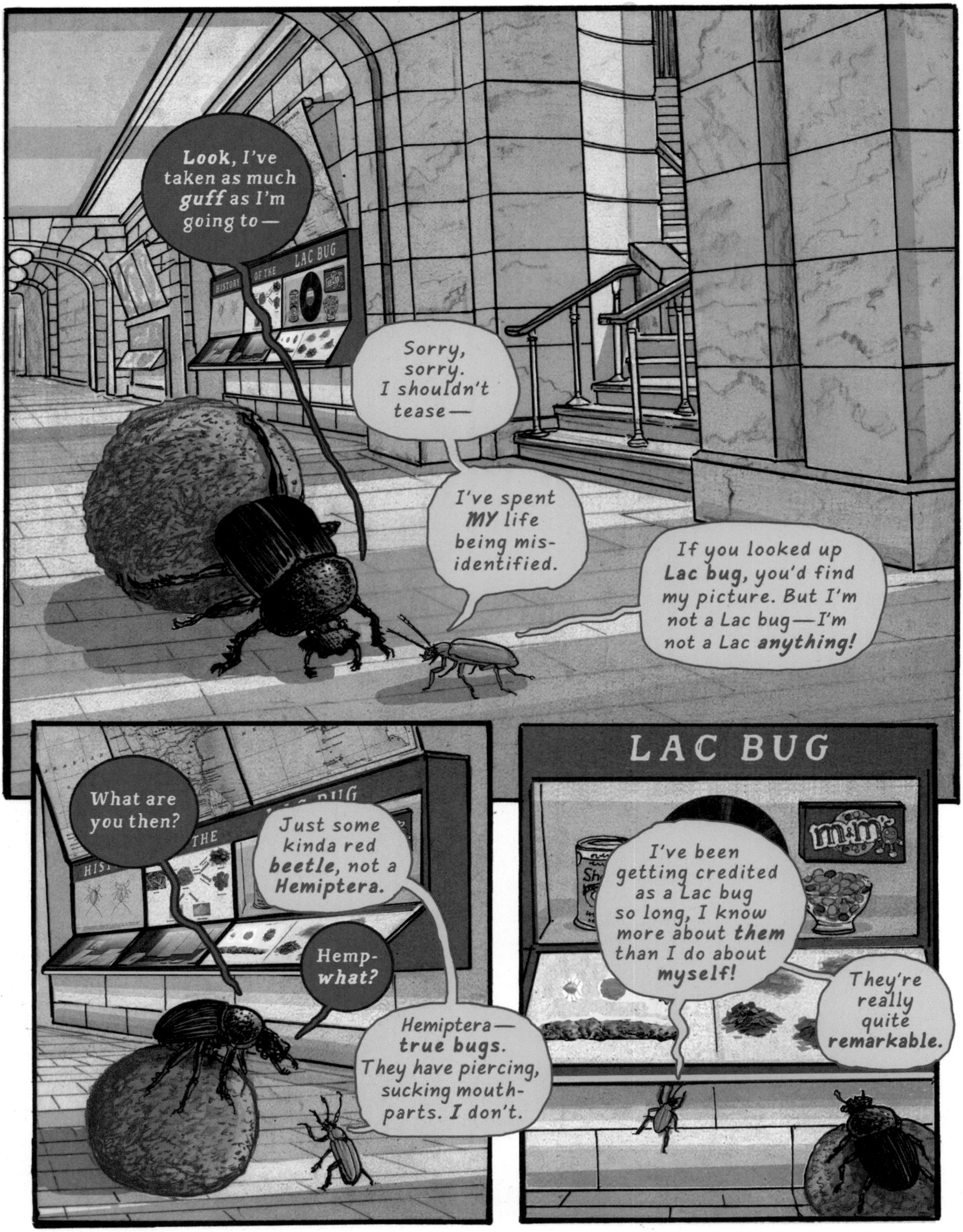
HISTORY OF THE LAC BUG
Look, I've taken as much guff as I'm going to—
Sorry, sorry. I shouldn't tease—
I've spent MY life being mis-identified.
If you looked up Lac bug, you'd find my picture. But I'm not a Lac bug—I'm not a Lac anything!
What are you then?
Just some kinda red beetle, not a Hemiptera.
Hemp-what?
Hemiptera—true bugs. They have piercing, sucking mouth-parts. I don't.
LAC BUG
I've been getting credited as a Lac bug so long, I know more about them than I do about myself!
They're really quite remarkable.

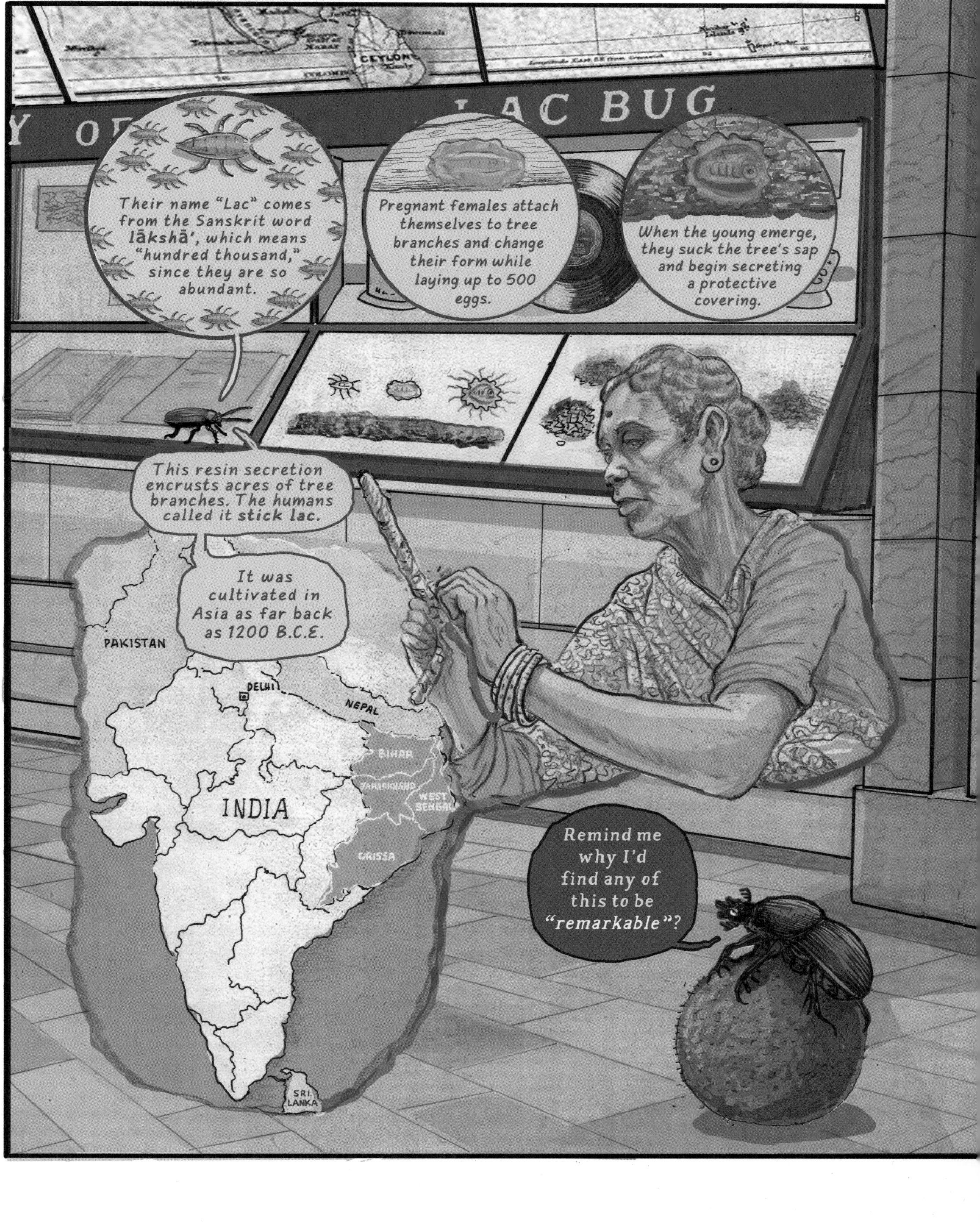
LAC BUG
Their name "Lac" comes from the Sanskrit word lākshā', which means "hundred thousand," since they are so abundant.
Pregnant females attach themselves to tree branches and change their form while laying up to 500 eggs.
When the young emerge, they suck the tree's sap and begin secreting a protective covering.
This resin secretion encrusts acres of tree branches. The humans called it stick lac.
It was cultivated in Asia as far back as 1200 B.C.E.
PAKISTAN
DELHI
NEPAL
BIHAR
WEST BENGAL
INDIA
ORISSA
SRI LANKA
Remind me why I'd find any of this to be "remarkable"?

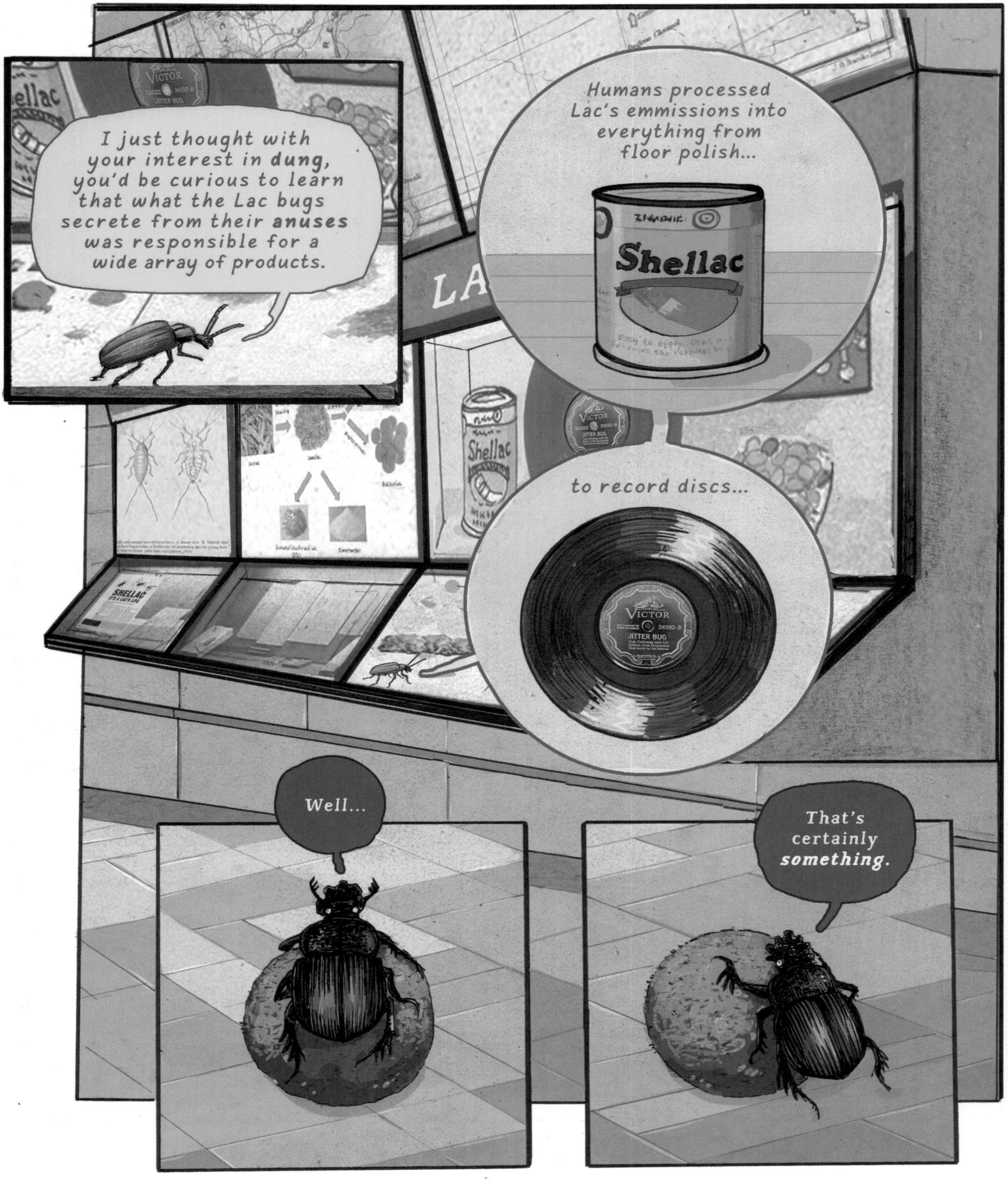
I just thought with your interest in **dung**, you'd be curious to learn that what the Lac bugs secrete from their **anuses** was responsible for a wide array of products.
Humans processed Lac's emmissions into everything from floor polish...
Shellac
to record discs...
VICTOR
JITTER BUG
Well...
That's certainly ***something***.

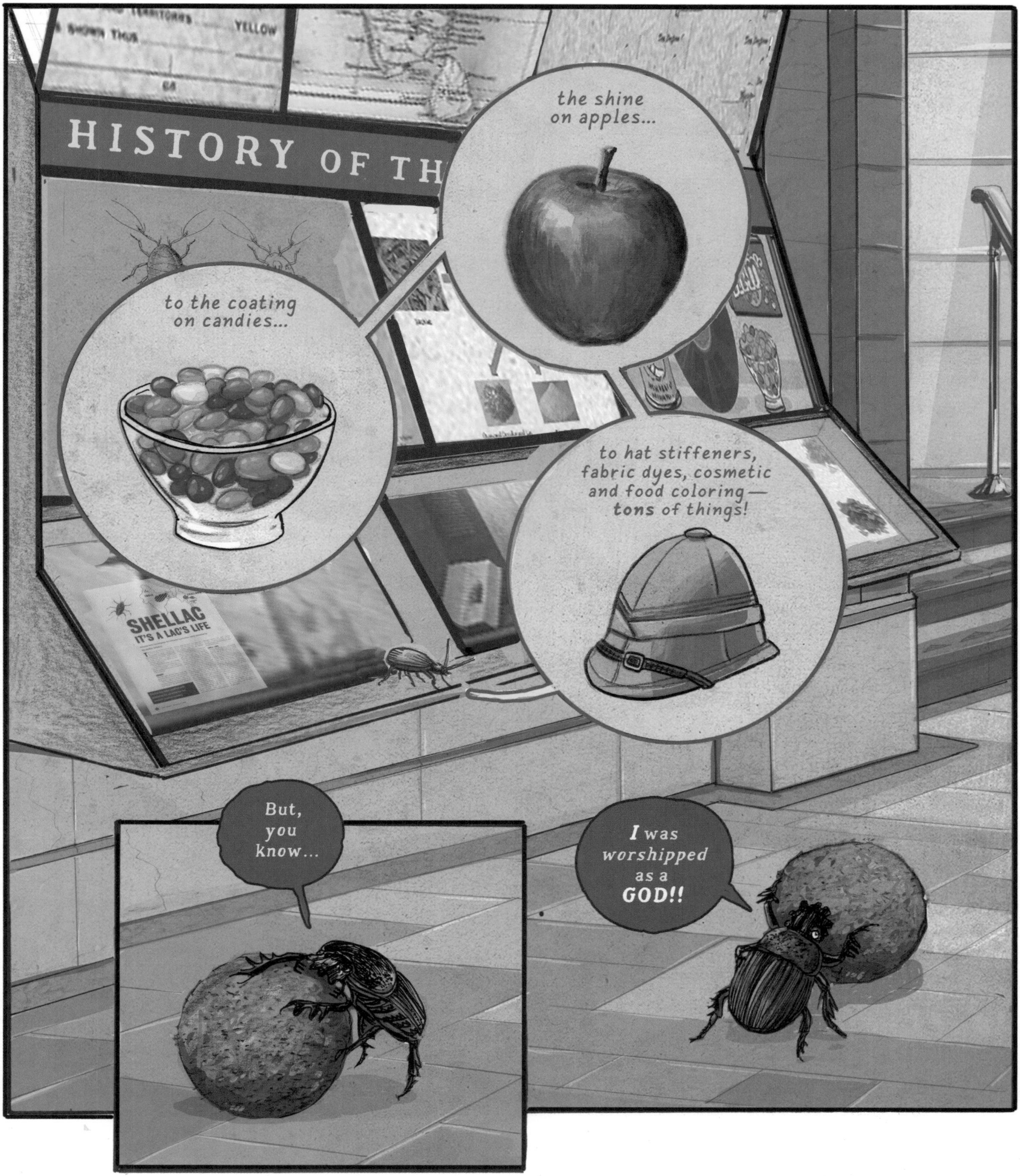
HISTORY OF TH
the shine on apples...
to the coating on candies...
to hat stiffeners, fabric dyes, cosmetic and food coloring—tons of things!
SHELLAC
IT'S A LAC'S LIFE
But, you know...
I was worshipped as a **GOD!!**

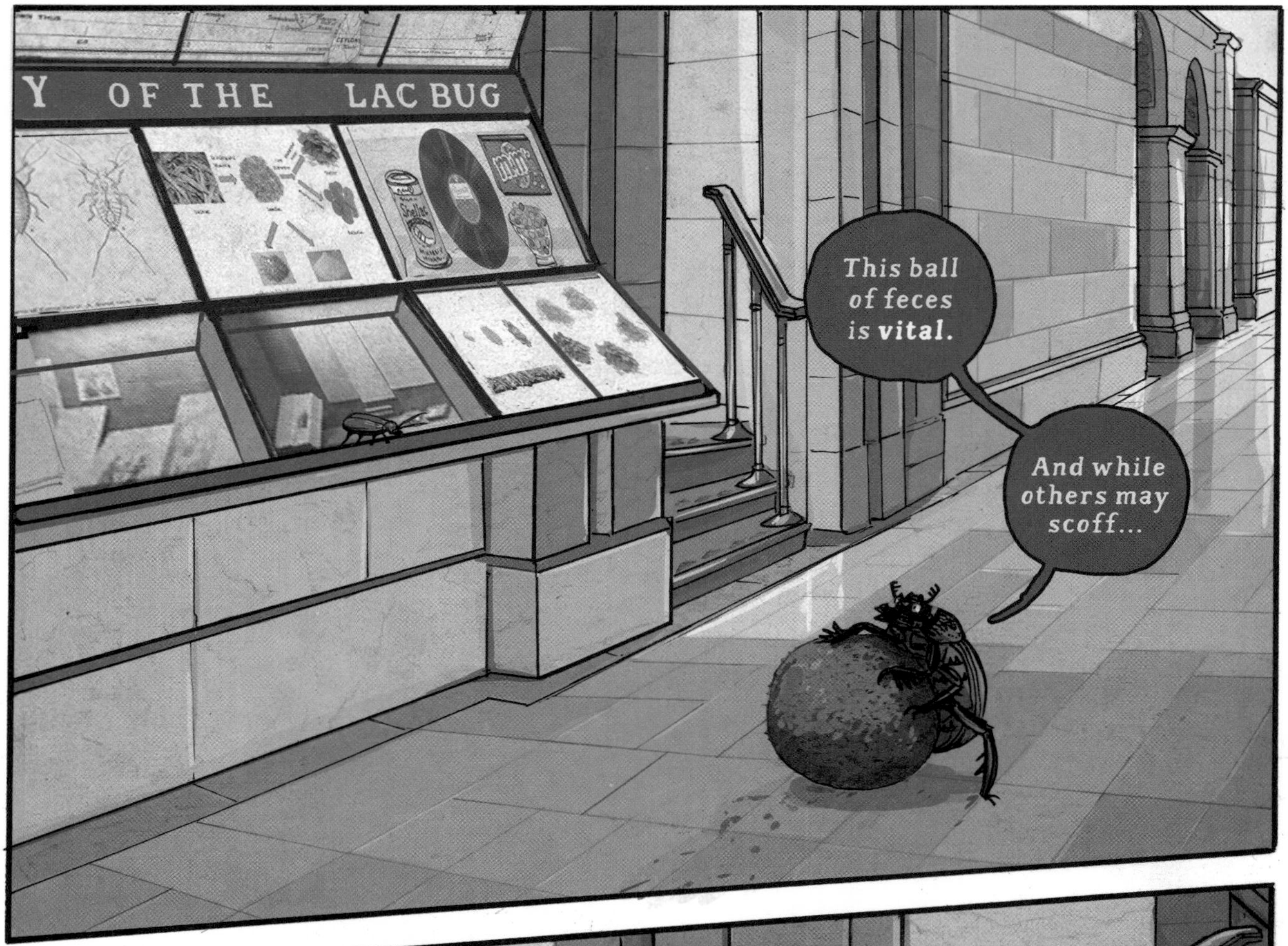
Y OF THE LAC BUG
This ball of feces is **vital**.
And while others may scoff...

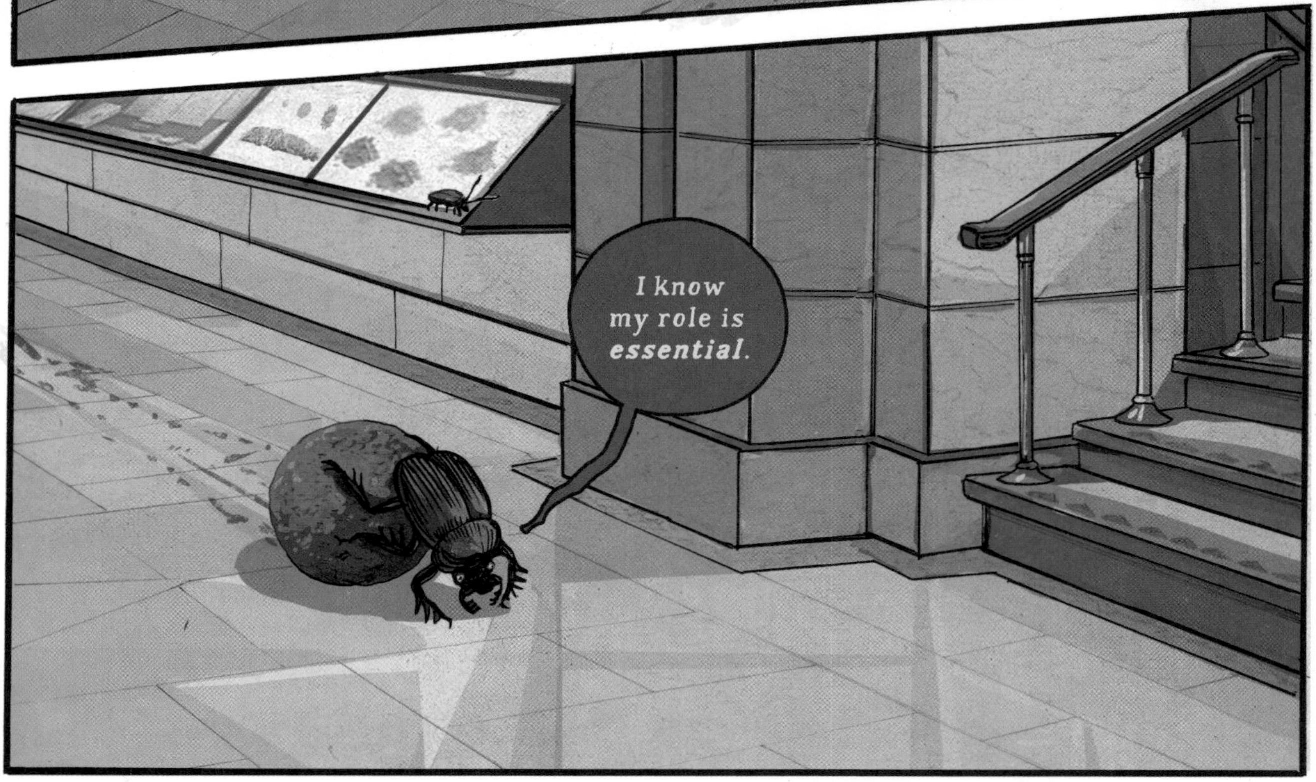
I know my role is *essential*.

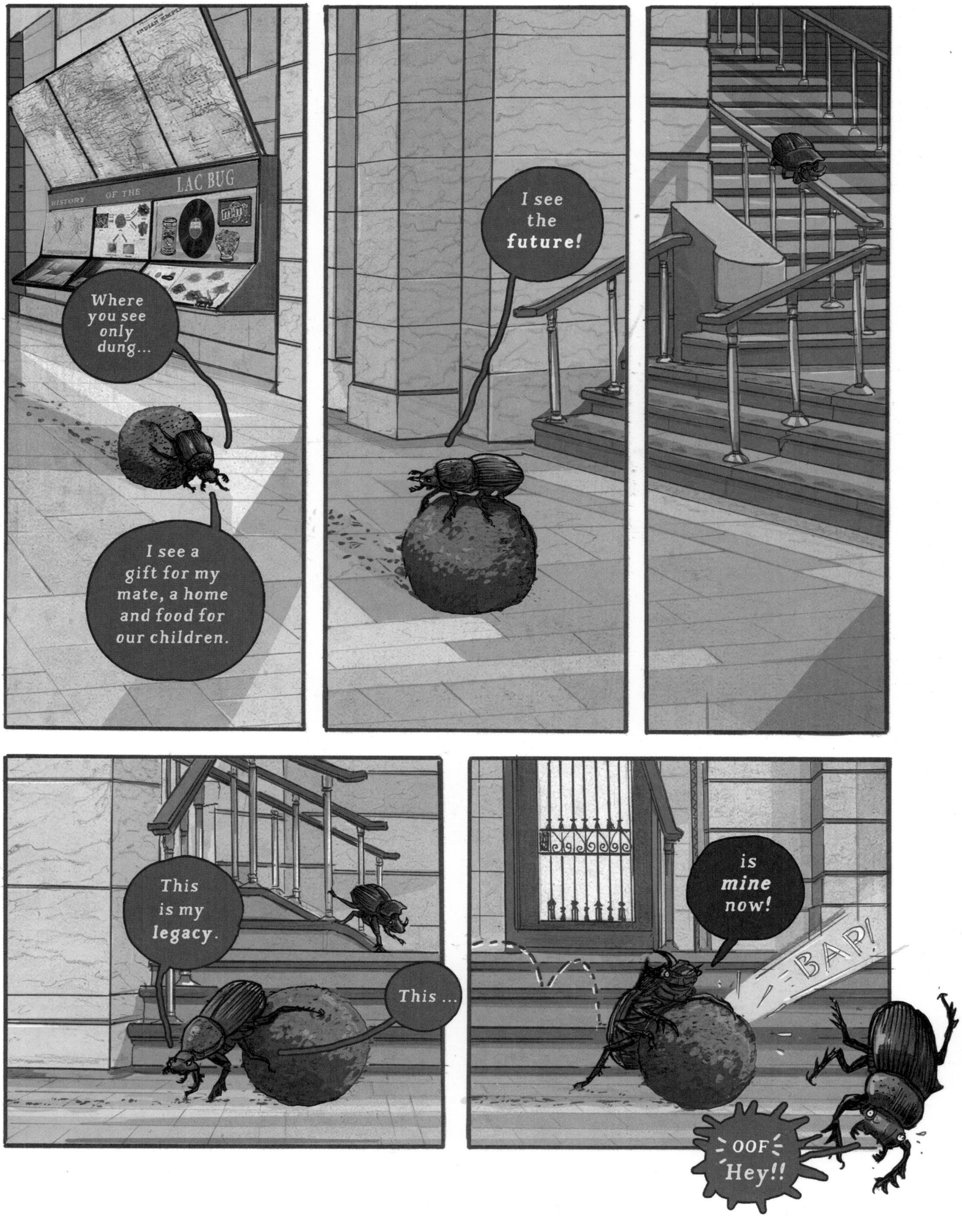
HISTORY OF THE LAC BUG
Where you see only dung...
I see a gift for my mate, a home and food for our children.
I see the future!
This is my legacy.
This ...
is mine now!
BAP!
OOF
Hey!!

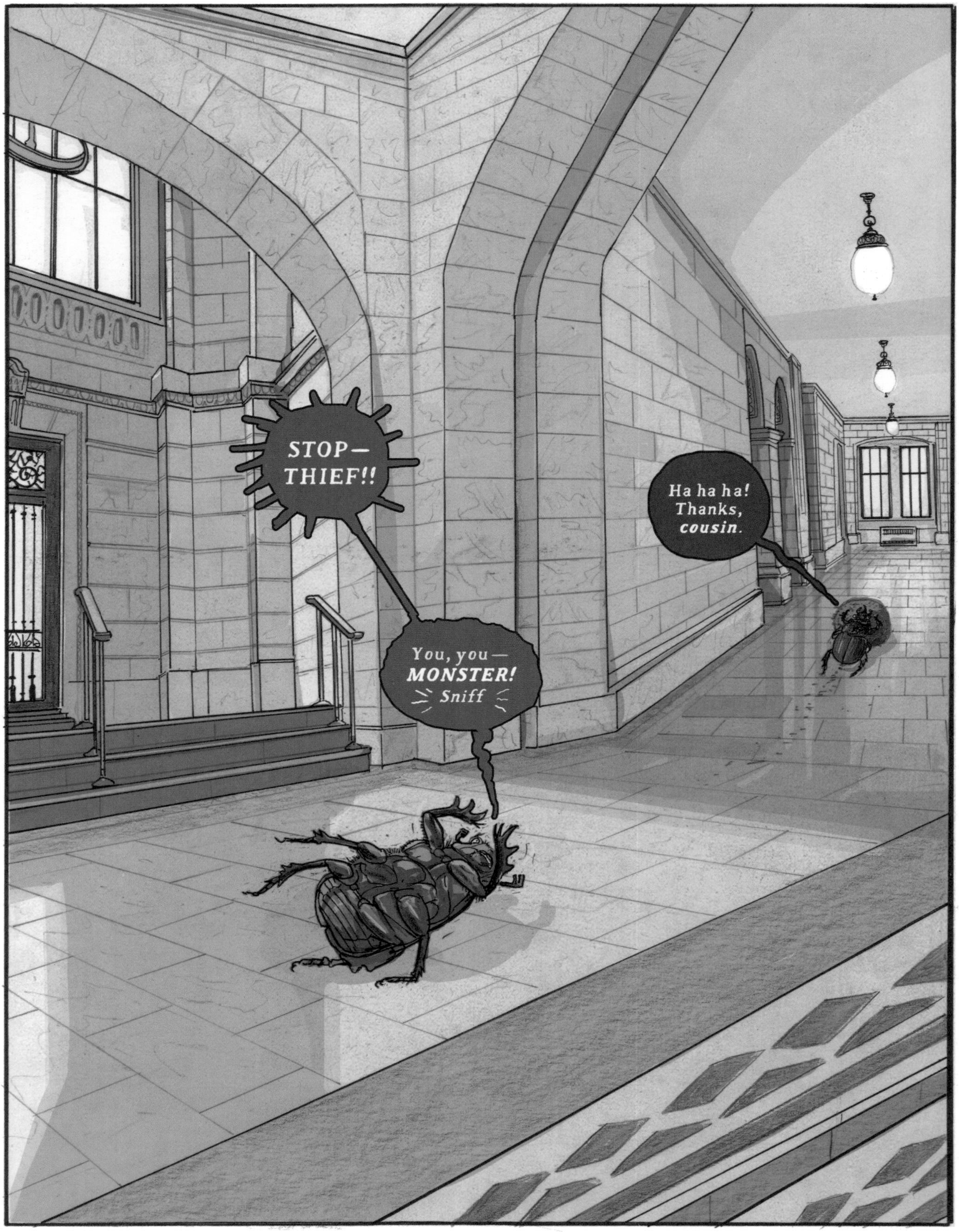
STOP—
THIEF!!
You, you—
MONSTER!
Sniff
Ha ha ha!
Thanks,
cousin.

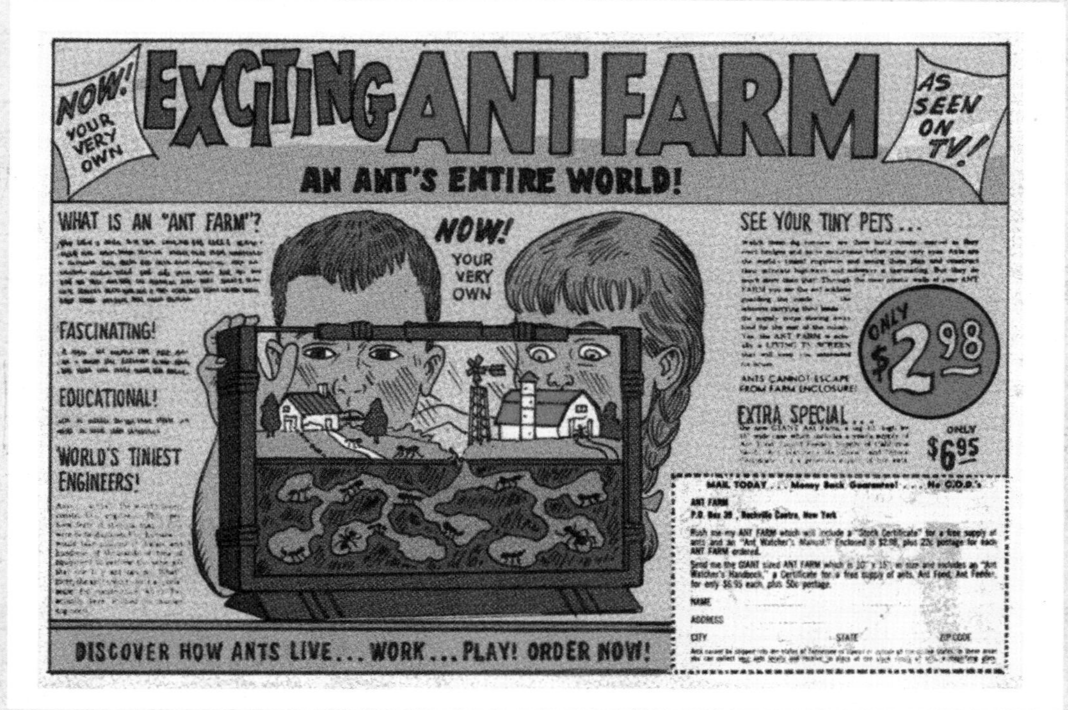
NOW! YOUR VERY OWN
EXCITING ANT FARM
AS SEEN ON TV!
AN ANT'S ENTIRE WORLD!
WHAT IS AN "ANT FARM"?
NOW! YOUR VERY OWN
SEE YOUR TINY PETS...
FASCINATING!
EDUCATIONAL!
WORLD'S TINIEST ENGINEERS!
ONLY $2.98
ANTS CANNOT ESCAPE FROM FARM ENCLOSURE!
EXTRA SPECIAL...
ONLY $6.95
DISCOVER HOW ANTS LIVE... WORK... PLAY! ORDER NOW!

Invented in 1932 by French entomologist Charles Janet, the Ant Farm was a familiar toy sold by mail order in comic books in the 1950s and 1960s.

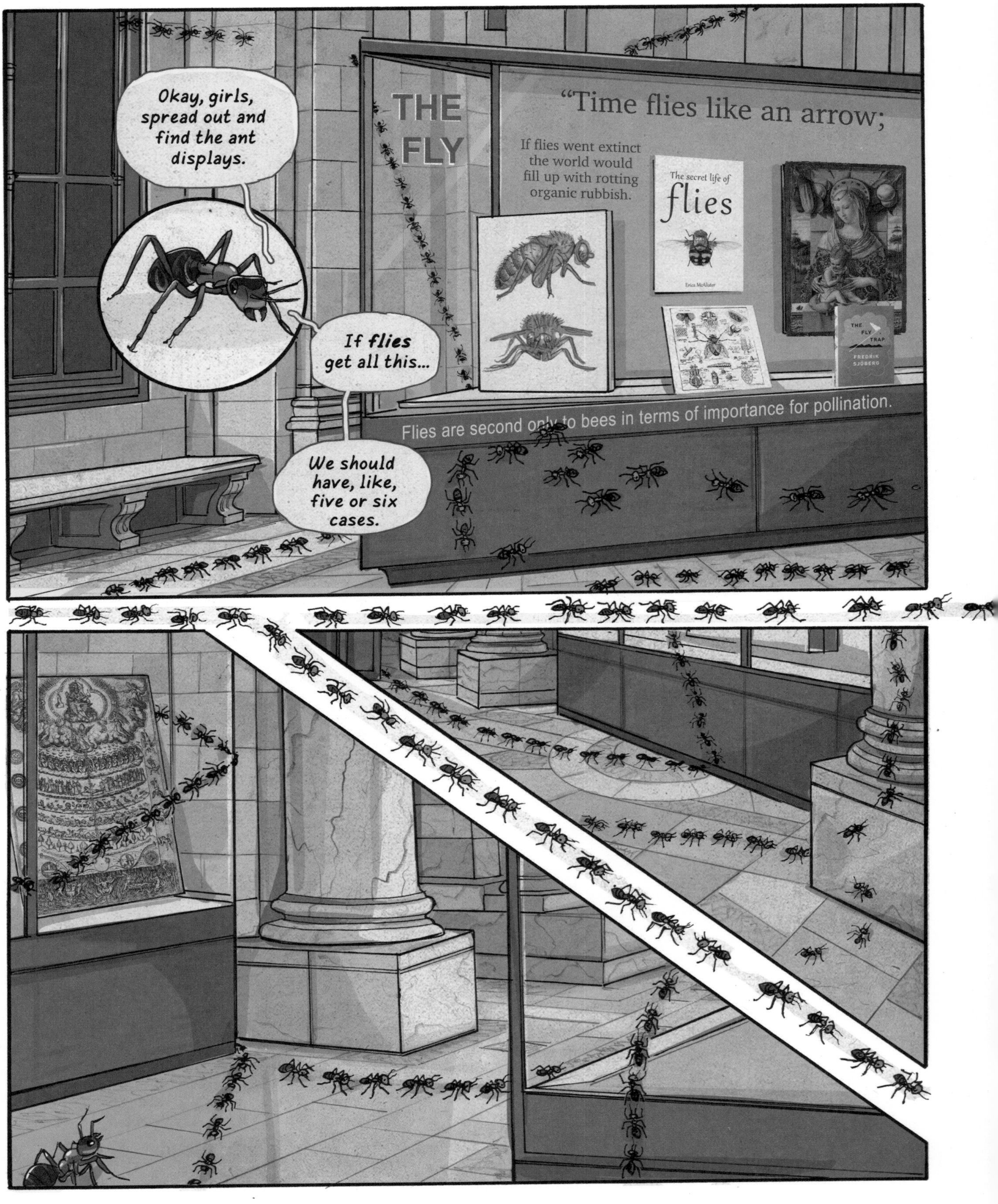
Okay, girls, spread out and find the ant displays.
If flies get all this...
We should have, like, five or six cases.
THE FLY
"Time flies like an arrow;
If flies went extinct the world would fill up with rotting organic rubbish.
The secret life of flies
THE FLY TRAP
FREDRIK SJÖBERG
Flies are second only to bees in terms of importance for pollination.

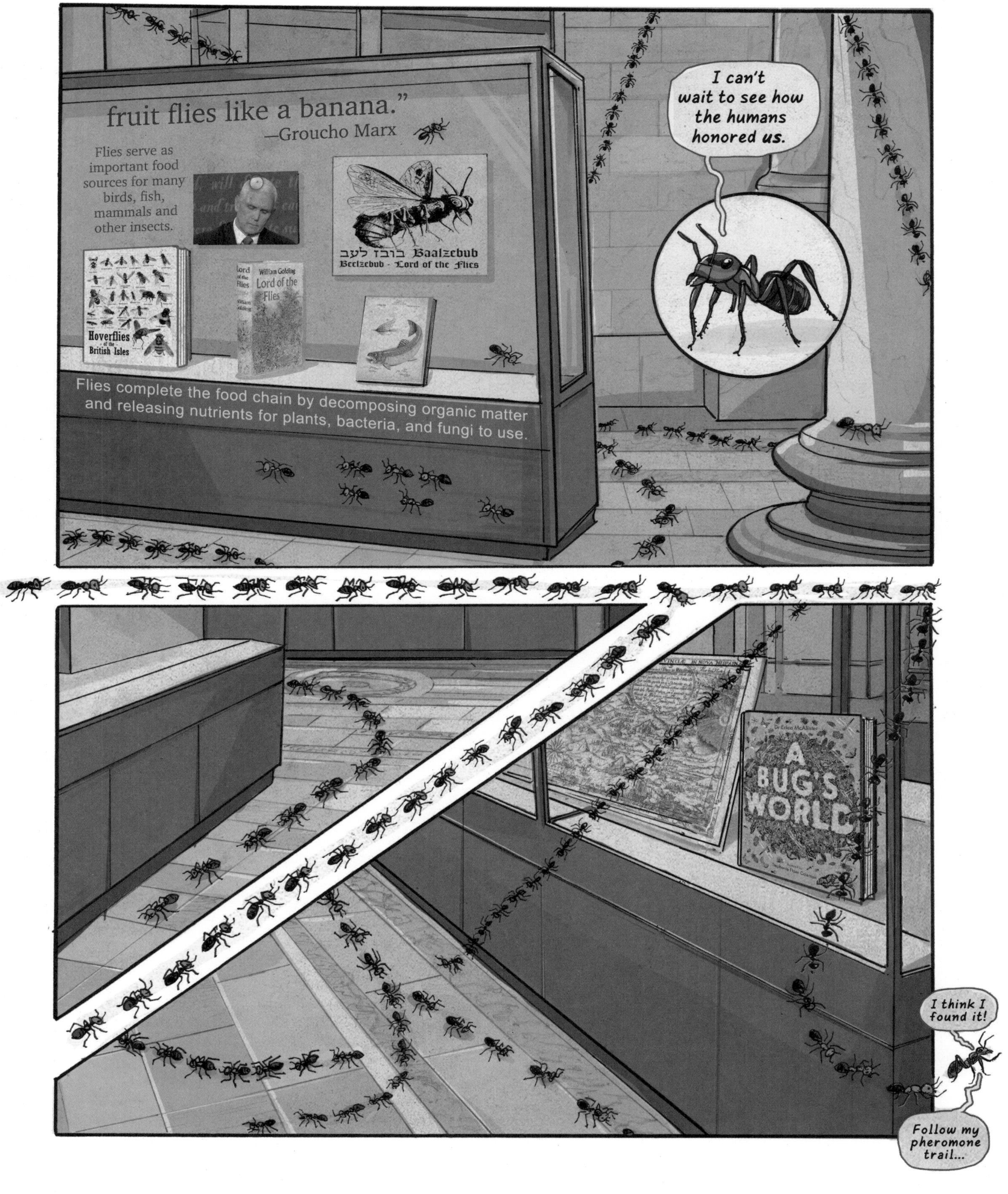

fruit flies like a banana."
—Groucho Marx
Flies serve as important food sources for many birds, fish, mammals and other insects.
Baalzebub
Beelzebub - Lord of the Flies
Hoverflies of the British Isles
William Golding
Lord of the Flies
Flies complete the food chain by decomposing organic matter and releasing nutrients for plants, bacteria, and fungi to use.
I can't wait to see how the humans honored us.
Dr Erica McAlister
A BUG'S WORLD
I think I found it!
Follow my pheromone trail...

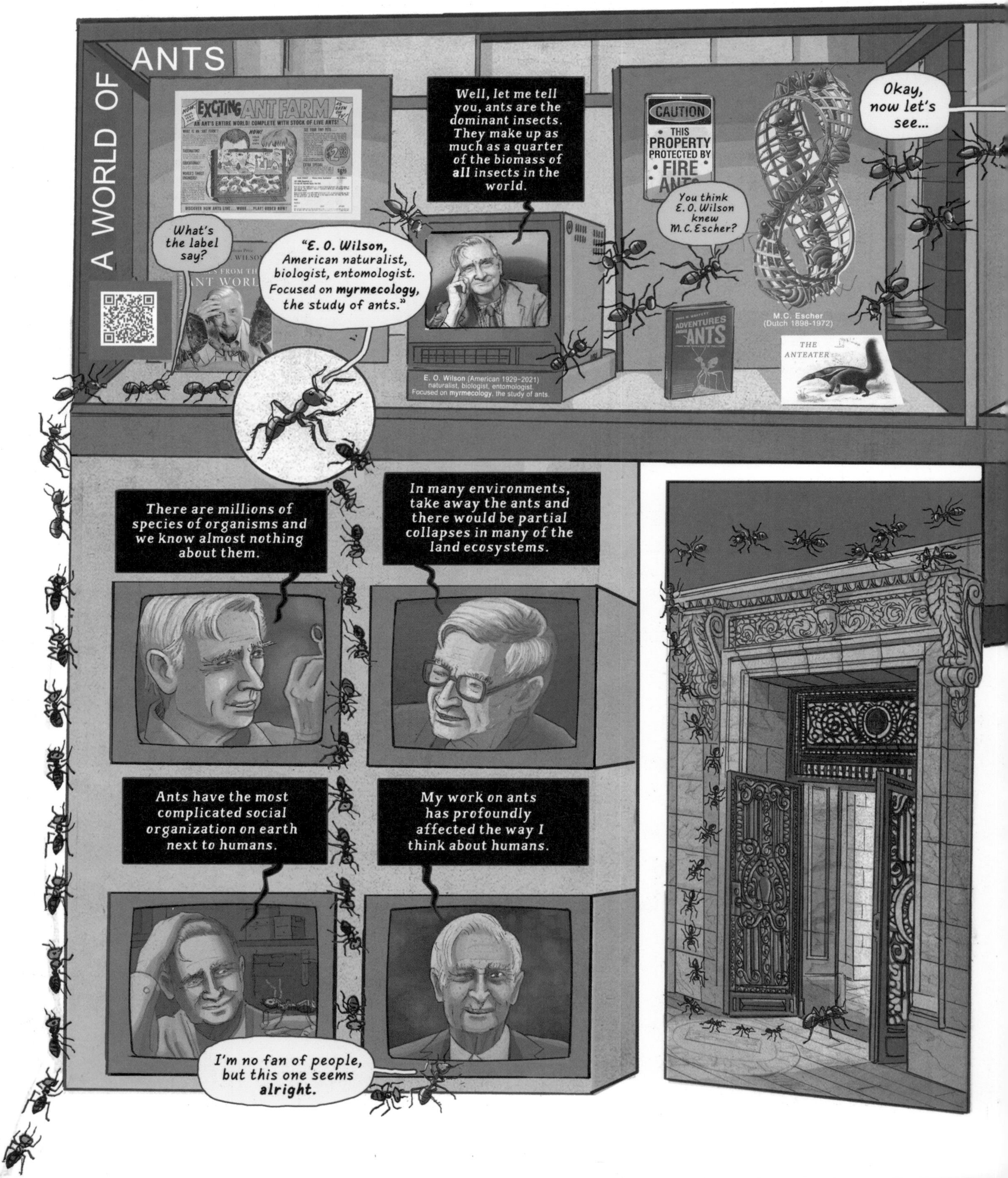
A WORLD OF ANTS
EXCITING ANT FARM
AN ANT'S ENTIRE WORLD! COMPLETE WITH STOCK OF LIVE ANTS!
DISCOVER HOW ANTS LIVE... WORK... PLAY! ORDER NOW!
What's the label say?
"E. O. Wilson, American naturalist, biologist, entomologist. Focused on **myrmecology**, the study of ants."
Well, let me tell you, ants are the dominant insects. They make up as much as a quarter of the biomass of **all** insects in the world.
E. O. Wilson (American 1929–2021) naturalist, biologist, entomologist. Focused on **myrmecology**, the study of ants.
CAUTION
THIS PROPERTY PROTECTED BY FIRE ANTS
You think E. O. Wilson knew M. C. Escher?
ADVENTURES AMONG ANTS
M.C. Escher (Dutch 1898-1972)
THE ANTEATER
Okay, now let's see...
There are millions of species of organisms and we know almost nothing about them.
In many environments, take away the ants and there would be partial collapses in many of the land ecosystems.
Ants have the most complicated social organization on earth next to humans.
My work on ants has profoundly affected the way I think about humans.
I'm no fan of people, but this one seems **alright**.

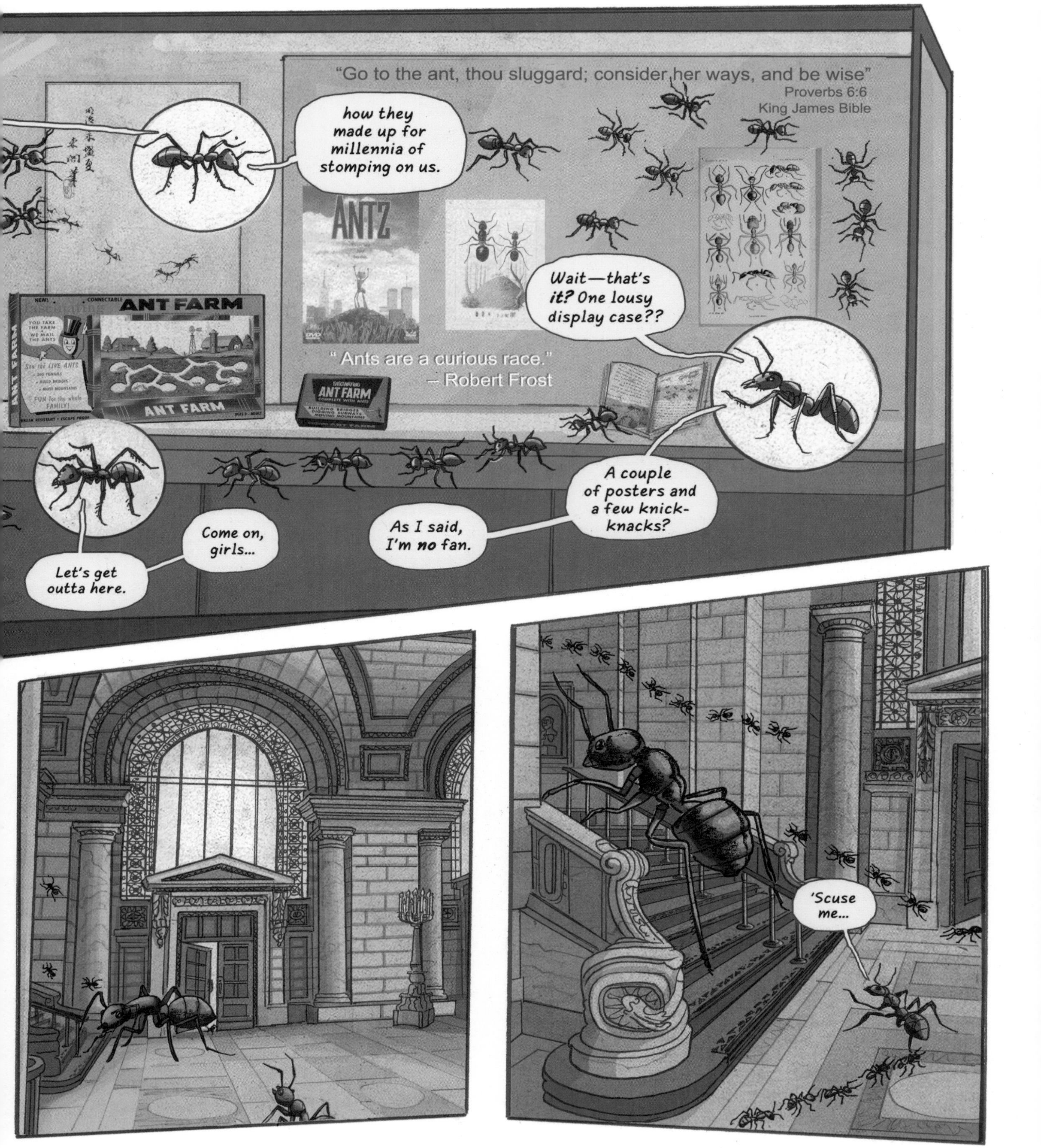
"Go to the ant, thou sluggard; consider her ways, and be wise"
Proverbs 6:6
King James Bible
how they made up for millennia of stomping on us.
ANTZ
ANT FARM
Wait—that's *it?* One lousy display case??
"Ants are a curious race."
– Robert Frost
ANT FARM
ANT FARM
A couple of posters and a few knick-knacks?
As I said, I'm *no* fan.
Come on, girls...
Let's get outta here.
'Scuse me...

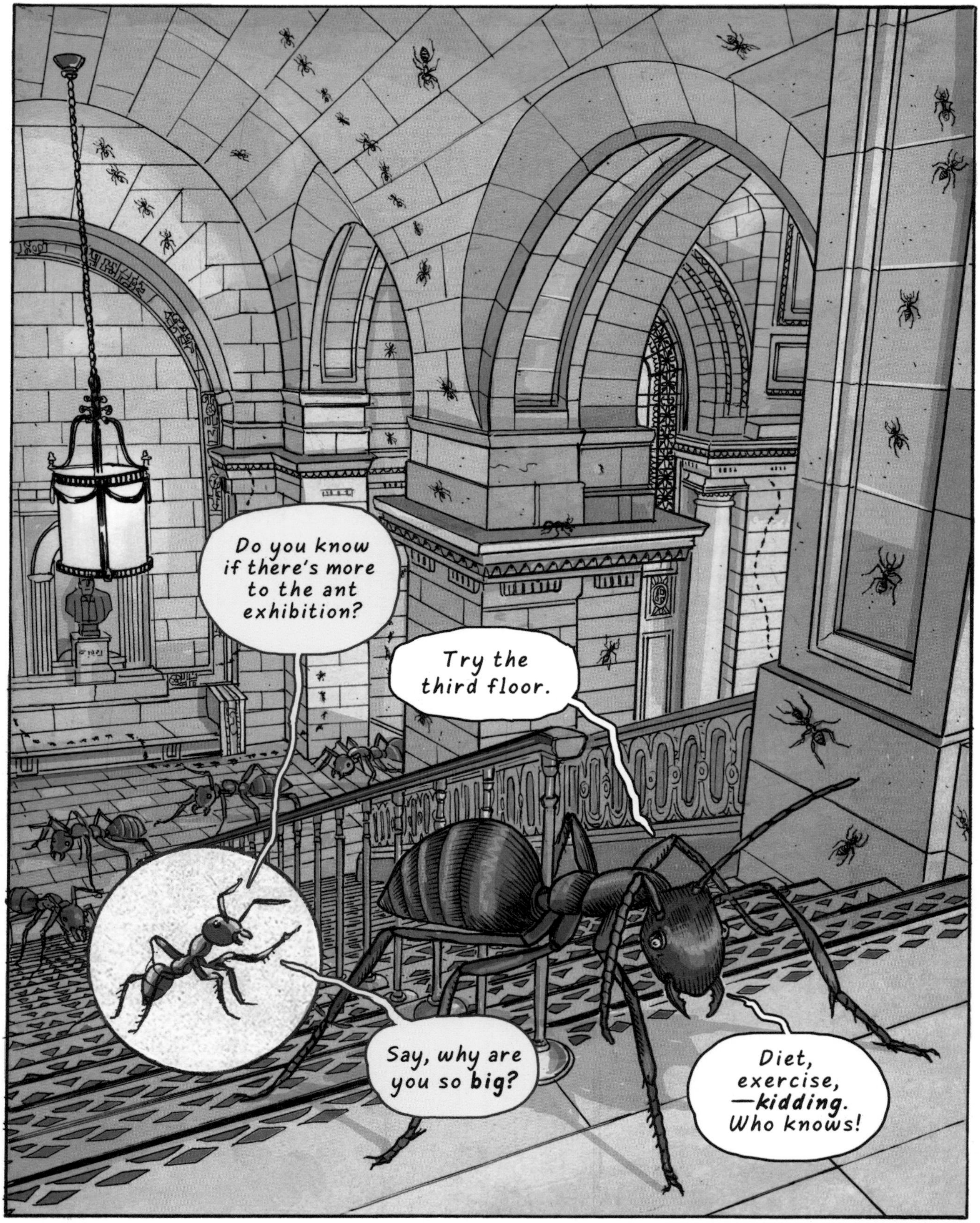
Do you know if there's more to the ant exhibition?
Try the third floor.
Say, why are you so big?
Diet, exercise, —kidding. Who knows!

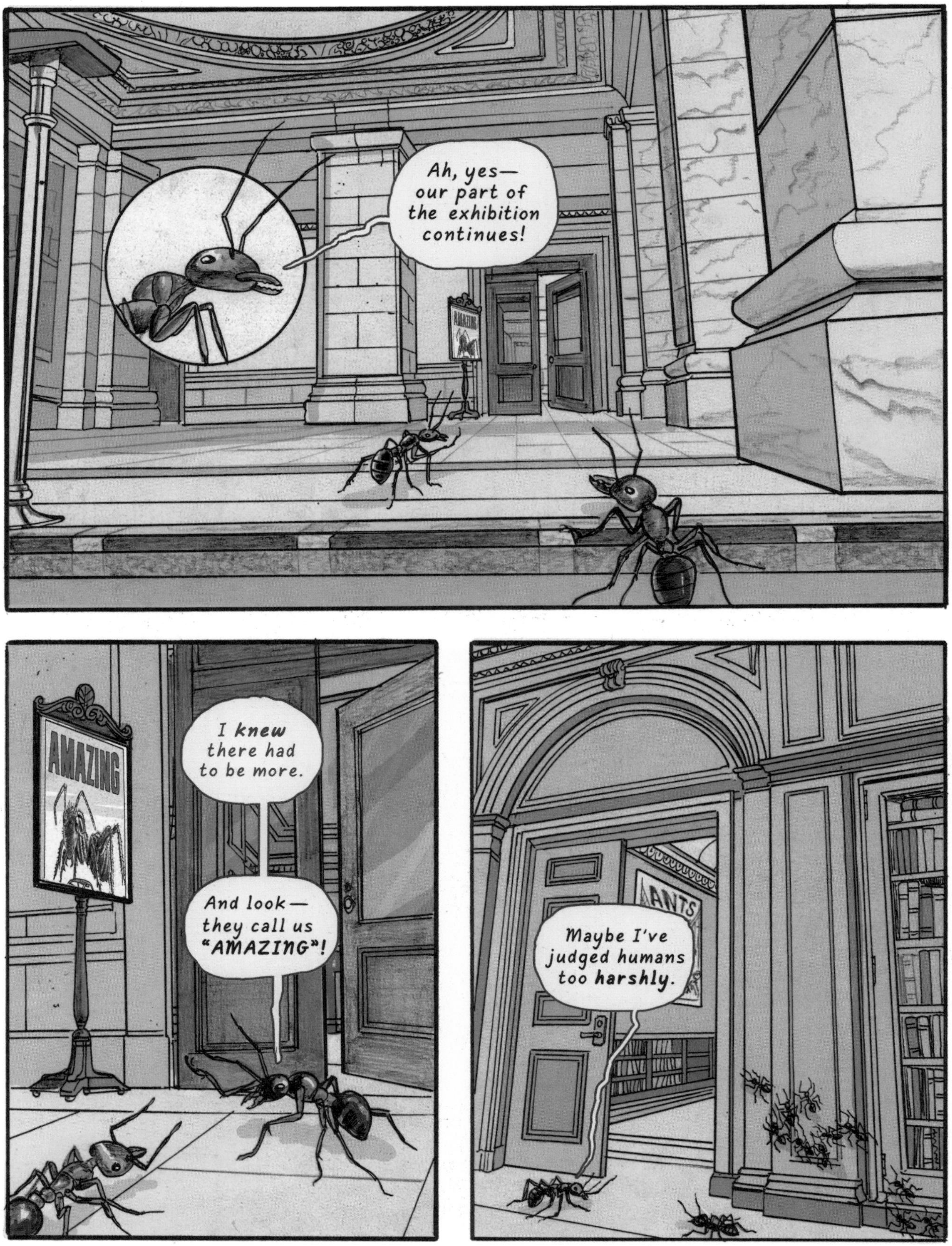
Ah, yes— our part of the exhibition continues!
AMAZING
I knew there had to be more.
And look— they call us "AMAZING"!
ANTS
Maybe I've judged humans too harshly.

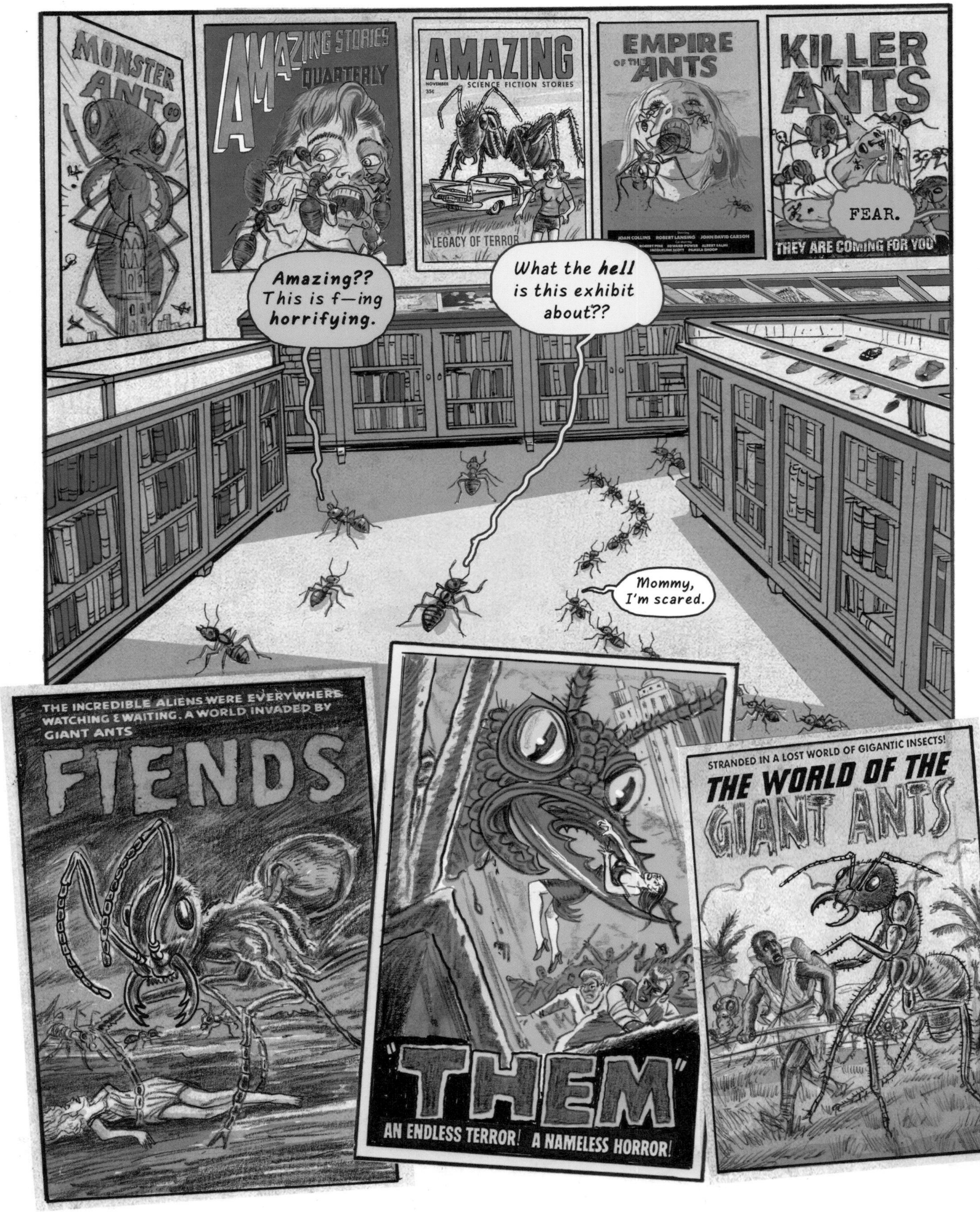
MONSTER ANT
AMAZING STORIES QUARTERLY
AMAZING
NOVEMBER
35¢
SCIENCE FICTION STORIES
LEGACY OF TERROR
EMPIRE OF THE ANTS
JOAN COLLINS ROBERT LANSING JOHN DAVID CARSON
KILLER ANTS
FEAR.
THEY ARE COMING FOR YOU
Amazing?? This is f—ing horrifying.
What the hell is this exhibit about??
Mommy, I'm scared.
THE INCREDIBLE ALIENS WERE EVERYWHERE WATCHING & WAITING. A WORLD INVADED BY GIANT ANTS
FIENDS
"THEM"
AN ENDLESS TERROR! A NAMELESS HORROR!
STRANDED IN A LOST WORLD OF GIGANTIC INSECTS!
THE WORLD OF THE GIANT ANTS

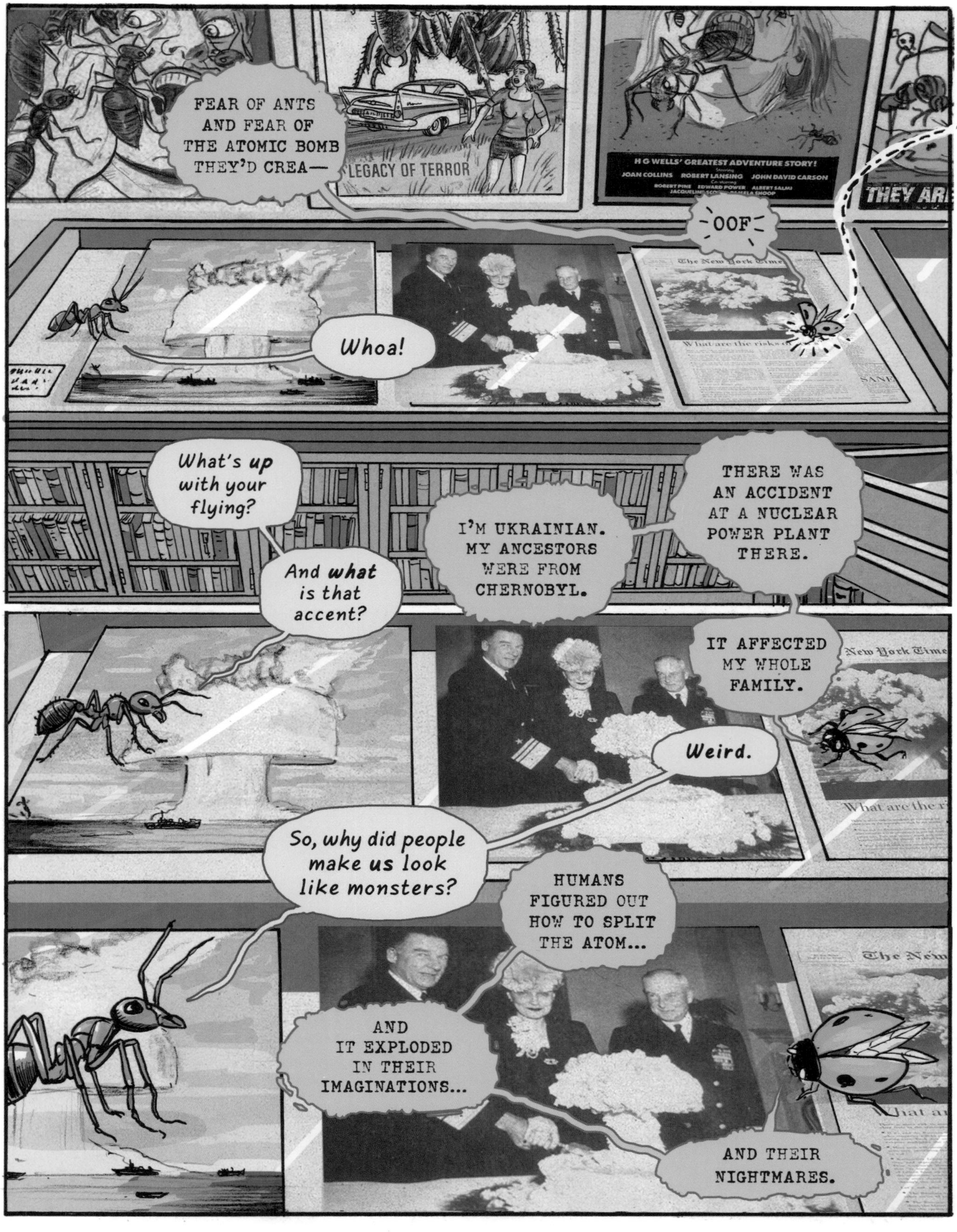

FEAR OF ANTS AND FEAR OF THE ATOMIC BOMB THEY'D CREA—
LEGACY OF TERROR
H G WELLS' GREATEST ADVENTURE STORY!
JOAN COLLINS ROBERT LANSING JOHN DAVID CARSON
OOF
Whoa!
What's up with your flying?
And what is that accent?
I'M UKRAINIAN. MY ANCESTORS WERE FROM CHERNOBYL.
THERE WAS AN ACCIDENT AT A NUCLEAR POWER PLANT THERE.
IT AFFECTED MY WHOLE FAMILY.
Weird.
So, why did people make us look like monsters?
HUMANS FIGURED OUT HOW TO SPLIT THE ATOM...
AND IT EXPLODED IN THEIR IMAGINATIONS...
AND THEIR NIGHTMARES.

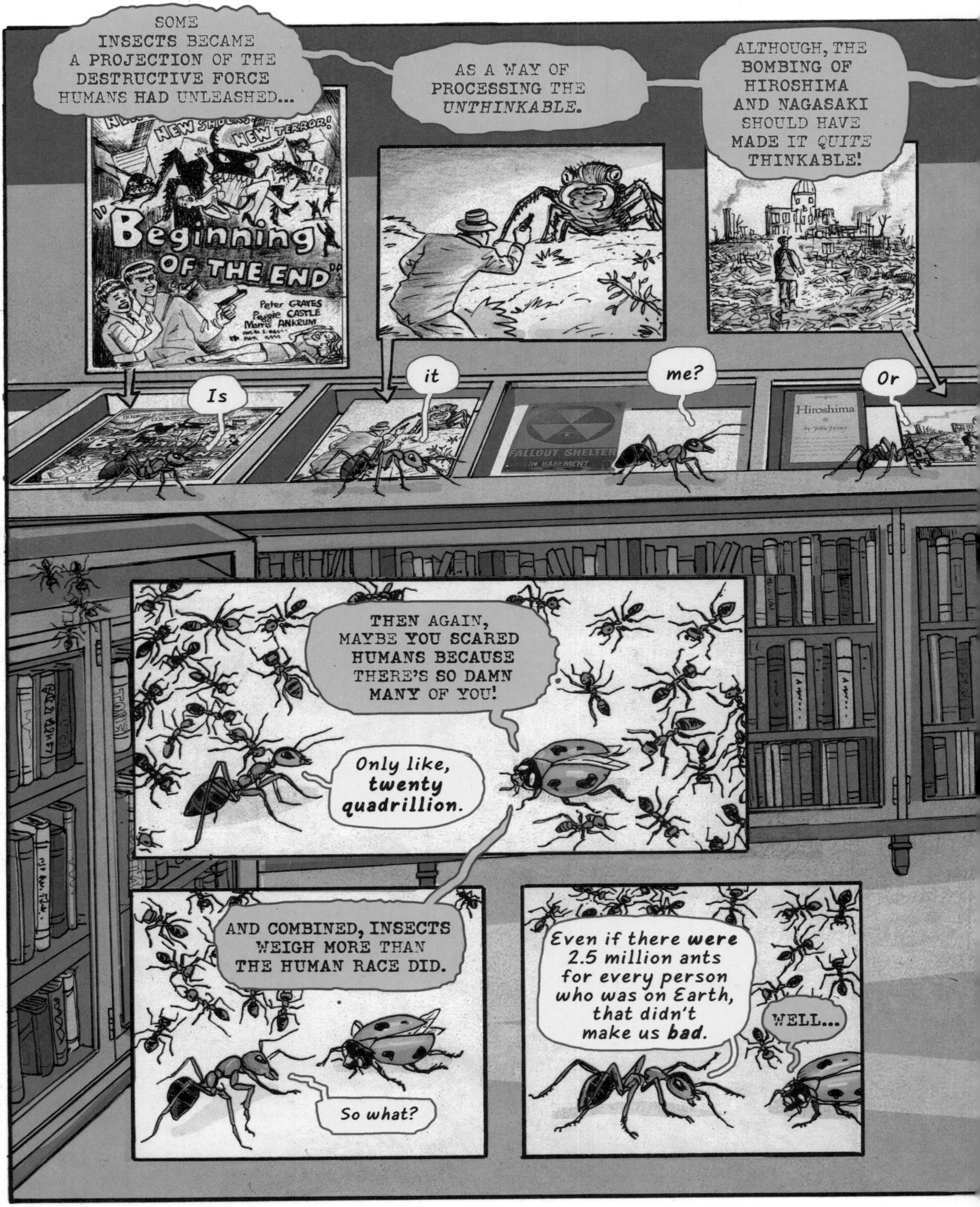
SOME INSECTS BECAME A PROJECTION OF THE DESTRUCTIVE FORCE HUMANS HAD UNLEASHED...
AS A WAY OF PROCESSING THE *UNTHINKABLE*.
ALTHOUGH, THE BOMBING OF HIROSHIMA AND NAGASAKI SHOULD HAVE MADE IT *QUITE* THINKABLE!
NEW SHOCKS! NEW TERROR!
Beginning OF THE END
Peter GRAVES
Peggie CASTLE
Morris ANKRUM
Is
it
me?
Or
FALLOUT SHELTER IN BASEMENT
Hiroshima
by John Hersey
THEN AGAIN, MAYBE YOU SCARED HUMANS BECAUSE THERE'S SO DAMN MANY OF YOU!
Only like, **twenty quadrillion**.
AND COMBINED, INSECTS WEIGH MORE THAN THE HUMAN RACE DID.
So what?
Even if there **were** 2.5 million ants for every person who was on Earth, that didn't make us **bad**.
WELL...

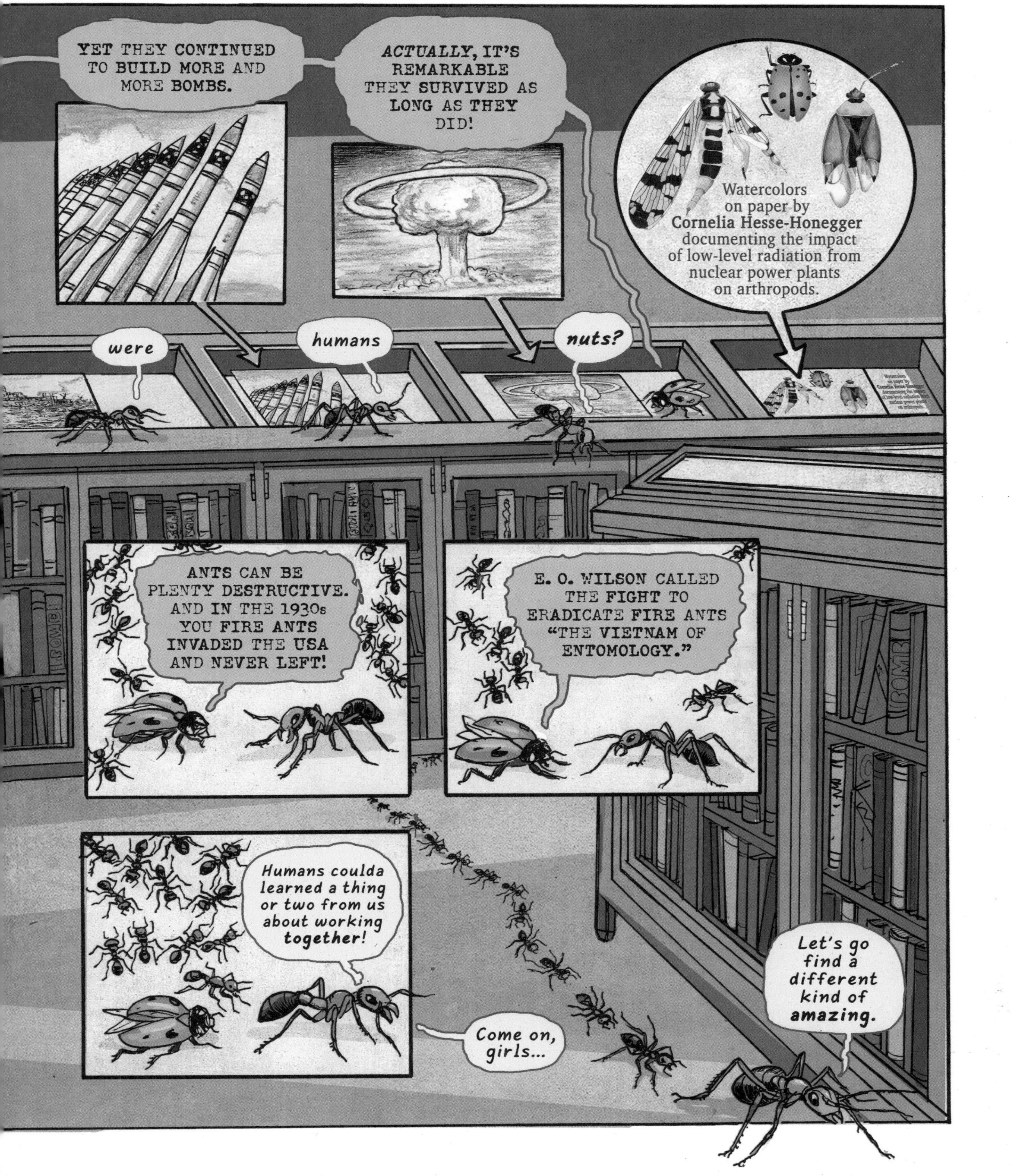
YET THEY CONTINUED TO BUILD MORE AND MORE BOMBS.
ACTUALLY, IT'S REMARKABLE THEY SURVIVED AS LONG AS THEY DID!
Watercolors on paper by Cornelia Hesse-Honegger documenting the impact of low-level radiation from nuclear power plants on arthropods.
were
humans
nuts?
ANTS CAN BE PLENTY DESTRUCTIVE. AND IN THE 1930s YOU FIRE ANTS INVADED THE USA AND NEVER LEFT!
E. O. WILSON CALLED THE FIGHT TO ERADICATE FIRE ANTS "THE VIETNAM OF ENTOMOLOGY."
Humans coulda learned a thing or two from us about working together!
Come on, girls...
Let's go find a different kind of amazing.

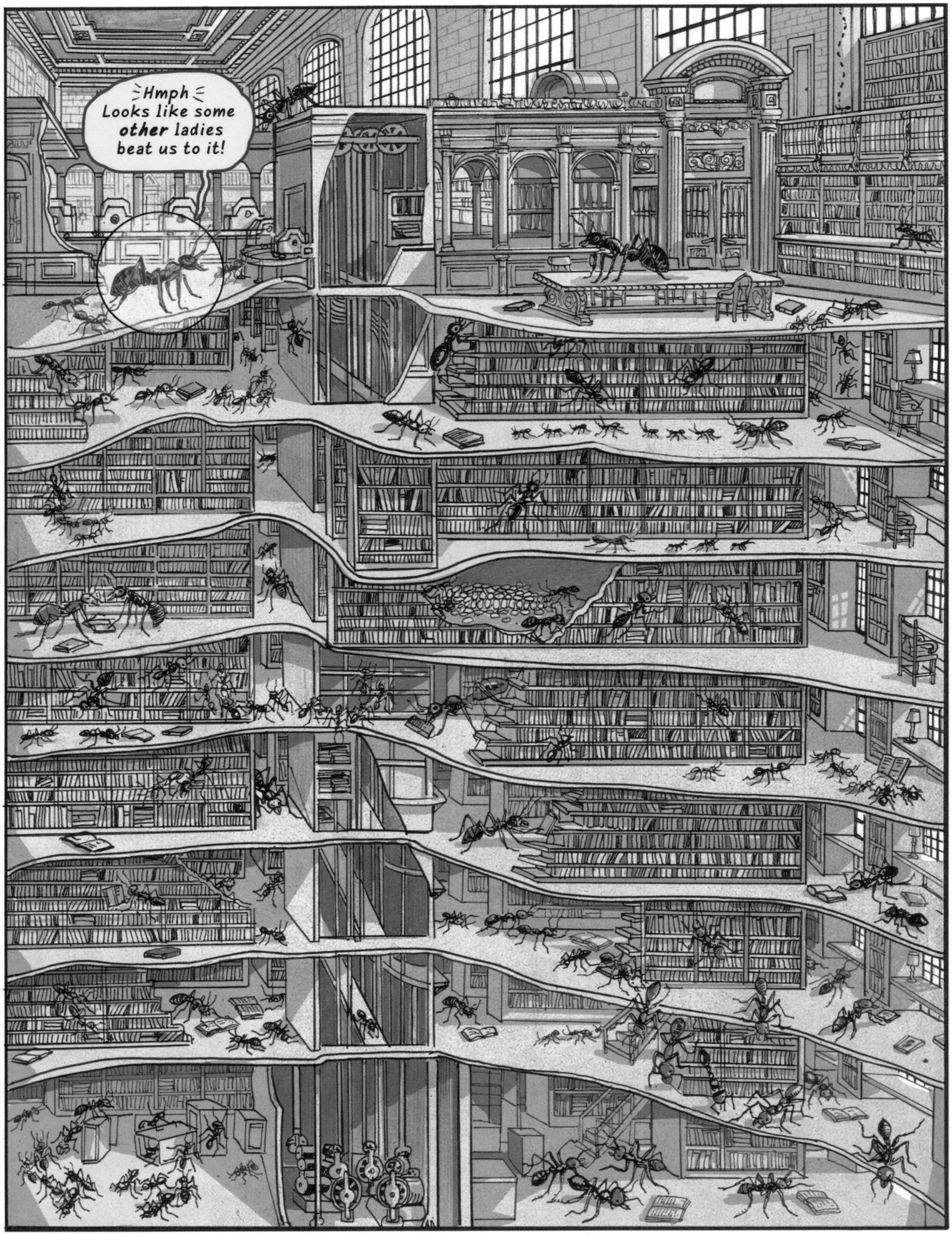
Hmph
Looks like some **other** ladies beat us to it!

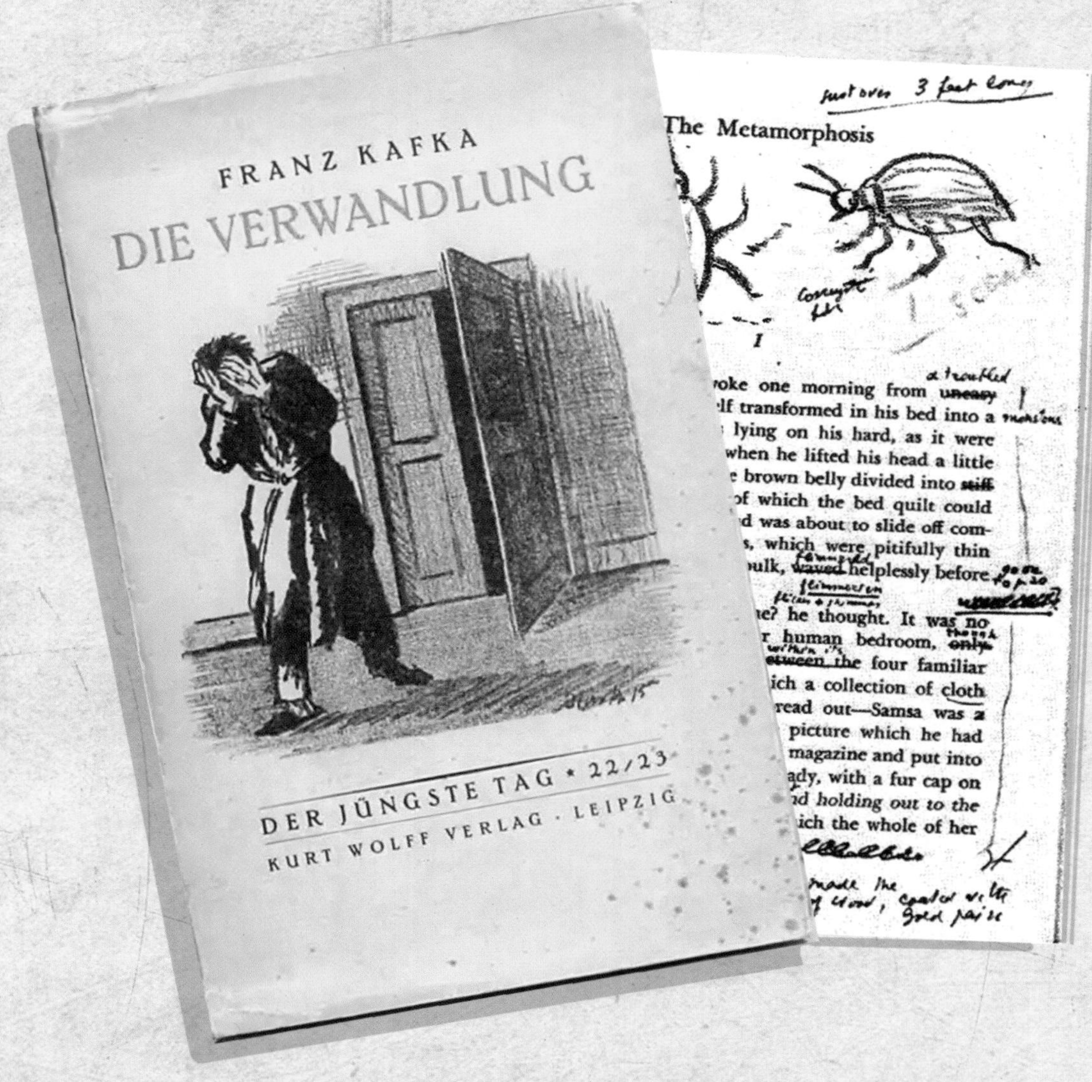
FRANZ KAFKA
DIE VERWANDLUNG
DER JÜNGSTE TAG • 22/23
KURT WOLFF VERLAG · LEIPZIG
The Metamorphosis

The Metamorphosis
First edition 1915
Franz Kafka
(1883–1924)
Austro-Hungarian
Empire

The Metamorphosis
Annotated edition
Vladimir Nabokov
(1899–1977)
Russia

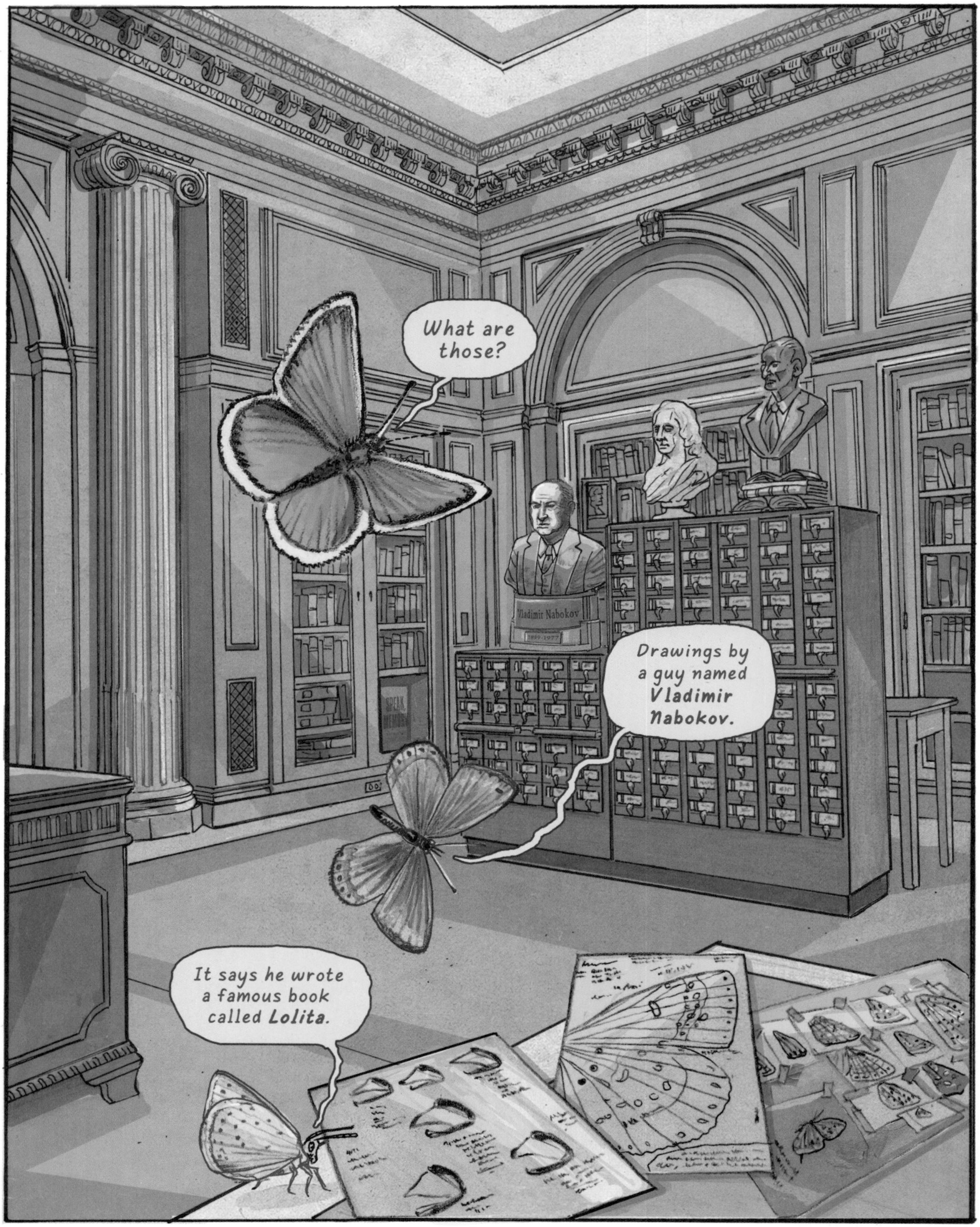
What are those?
Vladimir Nabokov
Drawings by a guy named **Vladimir Nabokov**.
It says he wrote a famous book called ***Lolita***.

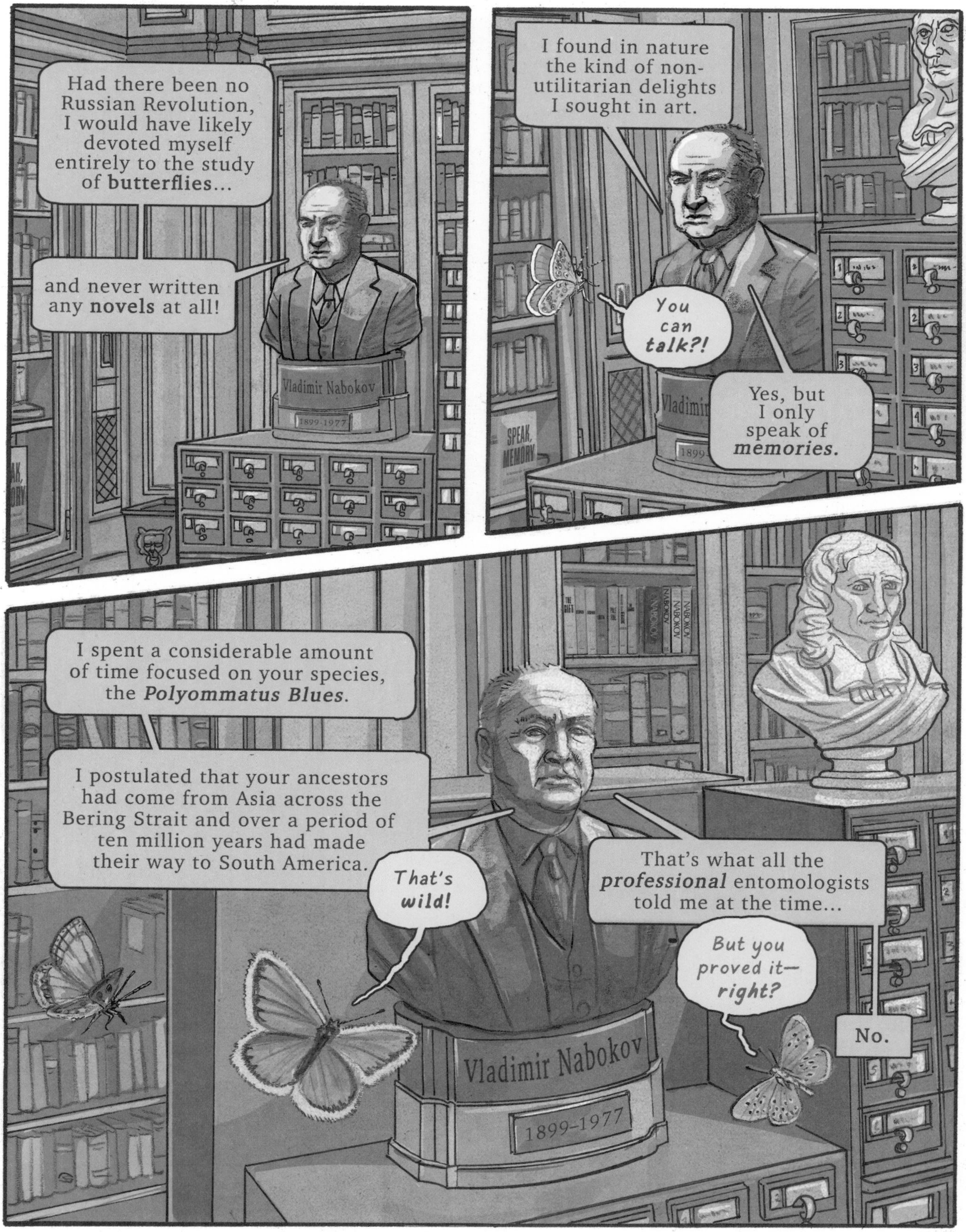
Had there been no Russian Revolution, I would have likely devoted myself entirely to the study of **butterflies**...
and never written any **novels** at all!
Vladimir Nabokov
1899-1977
I found in nature the kind of non-utilitarian delights I sought in art.
*You can **talk?!***
SPEAK, MEMORY
Yes, but I only speak of ***memories.***
I spent a considerable amount of time focused on your species, the ***Polyommatus Blues***.
I postulated that your ancestors had come from Asia across the Bering Strait and over a period of ten million years had made their way to South America.
*That's **wild!***
That's what all the ***professional*** entomologists told me at the time...
*But you proved it—**right?***
No.
Vladimir Nabokov
1899–1977

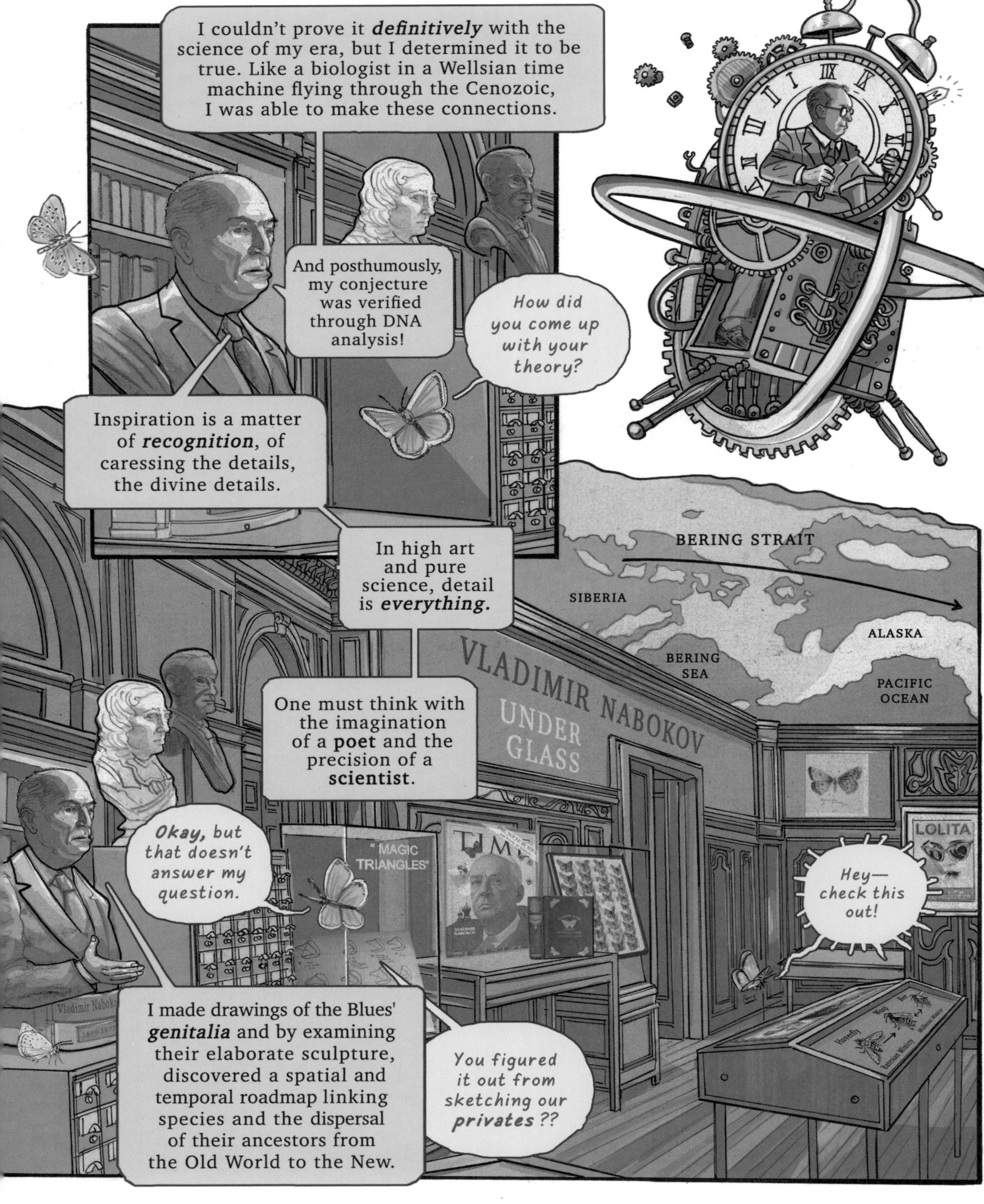
I couldn't prove it *definitively* with the science of my era, but I determined it to be true. Like a biologist in a Wellsian time machine flying through the Cenozoic, I was able to make these connections.
And posthumously, my conjecture was verified through DNA analysis!
How did you come up with your theory?
Inspiration is a matter of ***recognition***, of caressing the details, the divine details.
In high art and pure science, detail is ***everything.***
BERING STRAIT
SIBERIA
ALASKA
BERING SEA
PACIFIC OCEAN
VLADIMIR NABOKOV
UNDER GLASS
One must think with the imagination of a **poet** and the precision of a **scientist**.
LOLITA
"MAGIC TRIANGLES"
TIME
VLADIMIR NABOKOV
Okay, but that doesn't answer my question.
Hey— check this out!
Vladimir Nabokov
I made drawings of the Blues' ***genitalia*** and by examining their elaborate sculpture, discovered a spatial and temporal roadmap linking species and the dispersal of their ancestors from the Old World to the New.
You figured it out from sketching our ***privates*** ??

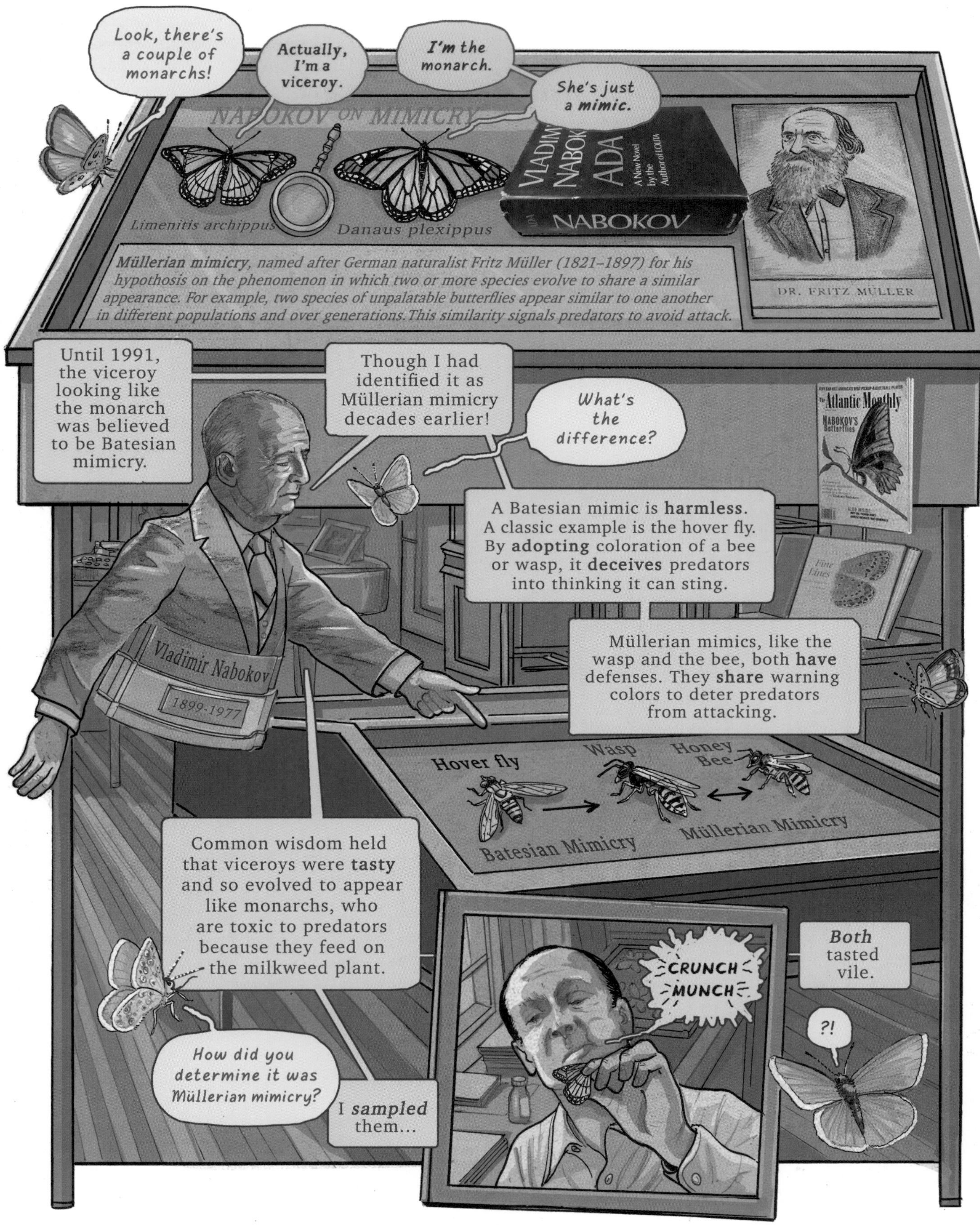

Look, there's a couple of monarchs!
Actually, I'm a viceroy.
I'm the monarch.
She's just a mimic.
NABOKOV ON MIMICRY
Limenitis archippus
Danaus plexippus
VLADIMIR NABOKOV ADA A New Novel by the Author of LOLITA NABOKOV
DR. FRITZ MÜLLER
Müllerian mimicry, named after German naturalist Fritz Müller (1821–1897) for his hypothesis on the phenomenon in which two or more species evolve to share a similar appearance. For example, two species of unpalatable butterflies appear similar to one another in different populations and over generations. This similarity signals predators to avoid attack.
Until 1991, the viceroy looking like the monarch was believed to be Batesian mimicry.
Though I had identified it as Müllerian mimicry decades earlier!
What's the difference?
The Atlantic Monthly
NABOKOV'S Butterflies
A Batesian mimic is harmless. A classic example is the hover fly. By adopting coloration of a bee or wasp, it deceives predators into thinking it can sting.
Fine Lines
Müllerian mimics, like the wasp and the bee, both have defenses. They share warning colors to deter predators from attacking.
Vladimir Nabokov
1899-1977
Hover fly
Wasp
Honey Bee
Batesian Mimicry
Müllerian Mimicry
Common wisdom held that viceroys were tasty and so evolved to appear like monarchs, who are toxic to predators because they feed on the milkweed plant.
Both tasted vile.
CRUNCH MUNCH
?!
How did you determine it was Müllerian mimicry?
I sampled them...

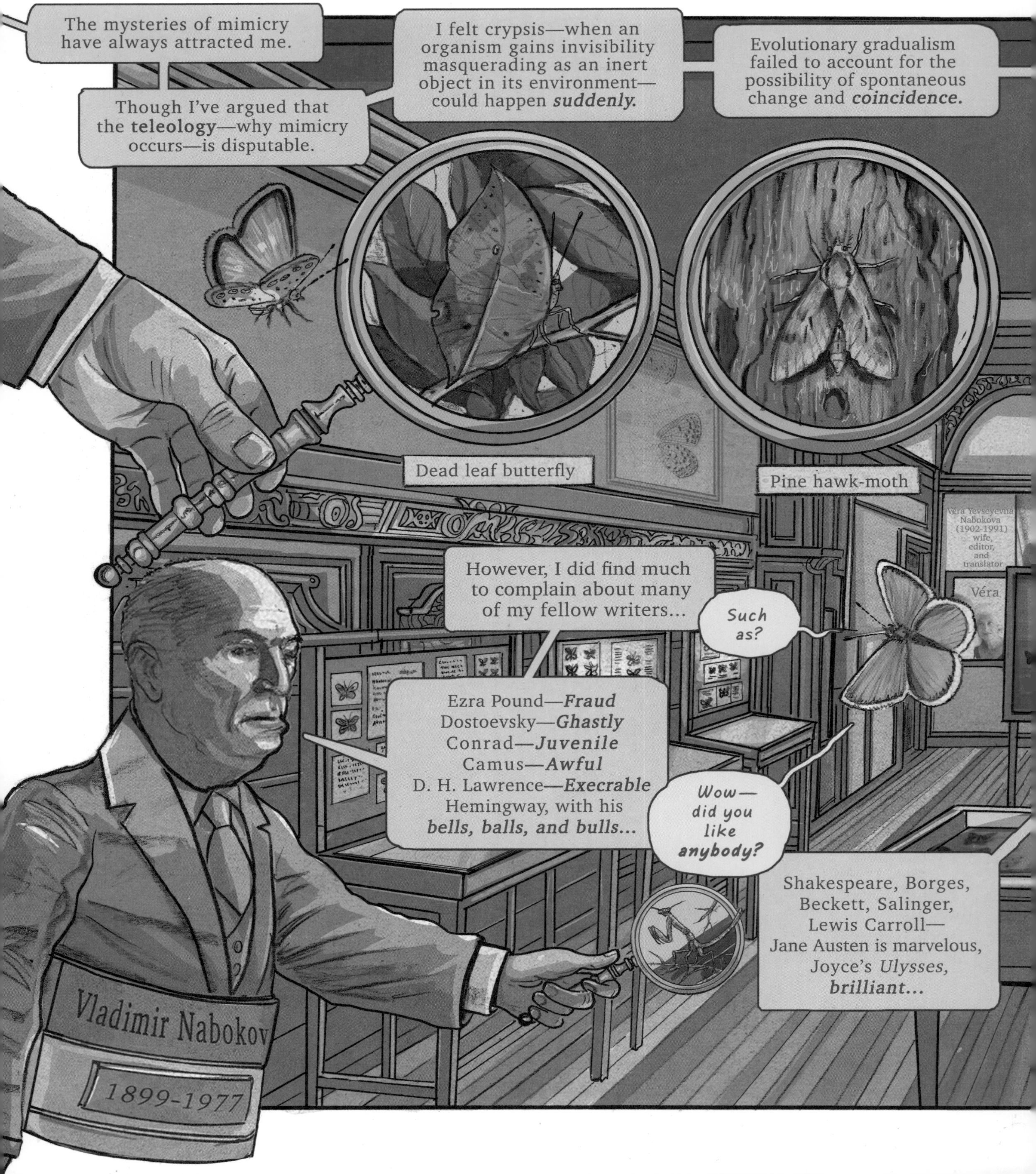
The mysteries of mimicry have always attracted me.
Though I've argued that the **teleology**—why mimicry occurs—is disputable.
I felt crypsis—when an organism gains invisibility masquerading as an inert object in its environment—could happen ***suddenly.***
Evolutionary gradualism failed to account for the possibility of spontaneous change and ***coincidence.***
Dead leaf butterfly
Pine hawk-moth
Véra Yevseyevna Nabokova (1902-1991) wife, editor, and translator
Véra
However, I did find much to complain about many of my fellow writers...
Such as?
Ezra Pound—***Fraud*** Dostoevsky—***Ghastly*** Conrad—***Juvenile*** Camus—***Awful*** D. H. Lawrence—***Execrable*** Hemingway, with his ***bells, balls, and bulls...***
Wow—did you like ***anybody?***
Shakespeare, Borges, Beckett, Salinger, Lewis Carroll—Jane Austen is marvelous, Joyce's *Ulysses*, ***brilliant...***
Vladimir Nabokov
1899–1977

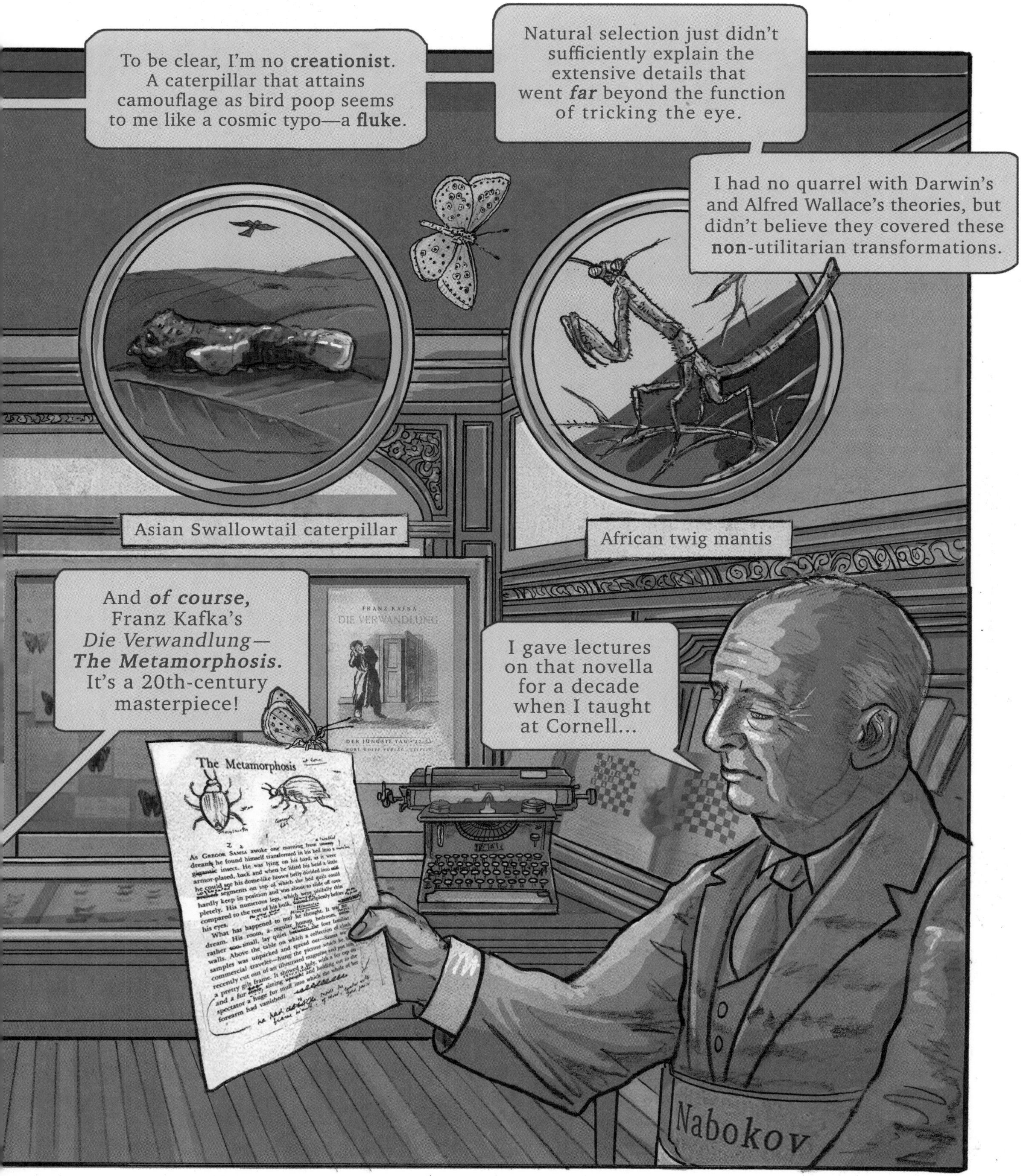

To be clear, I'm no **creationist**. A caterpillar that attains camouflage as bird poop seems to me like a cosmic typo—a **fluke**.
Natural selection just didn't sufficiently explain the extensive details that went ***far*** beyond the function of tricking the eye.
I had no quarrel with Darwin's and Alfred Wallace's theories, but didn't believe they covered these **non**-utilitarian transformations.
Asian Swallowtail caterpillar
African twig mantis
And ***of course,*** Franz Kafka's *Die Verwandlung*—***The Metamorphosis.*** It's a 20th-century masterpiece!
FRANZ KAFKA
DIE VERWANDLUNG
I gave lectures on that novella for a decade when I taught at Cornell...
The Metamorphosis
Nabokov

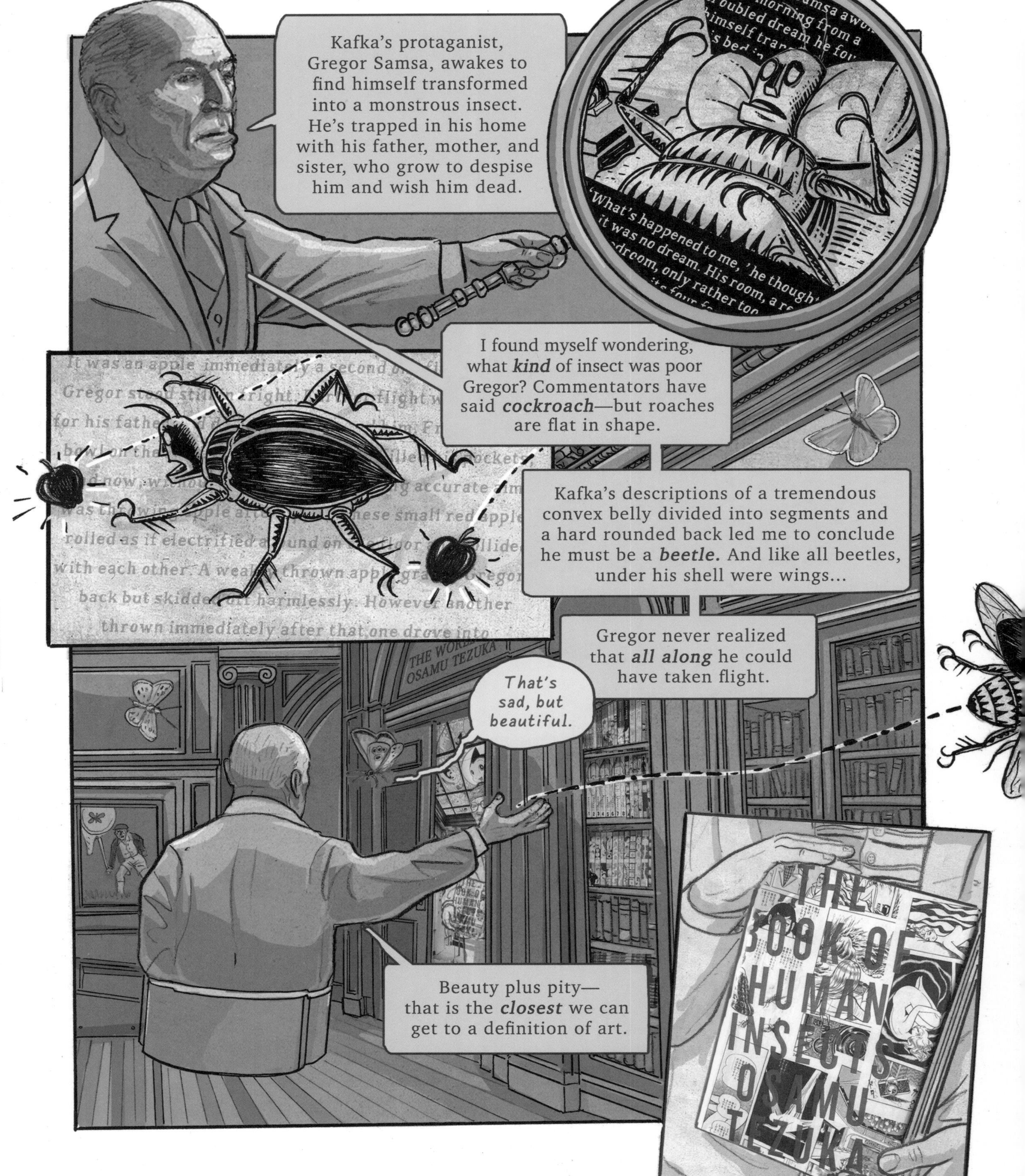

Kafka's protaganist, Gregor Samsa, awakes to find himself transformed into a monstrous insect. He's trapped in his home with his father, mother, and sister, who grow to despise him and wish him dead.
I found myself wondering, what ***kind*** of insect was poor Gregor? Commentators have said ***cockroach***—but roaches are flat in shape.
Kafka's descriptions of a tremendous convex belly divided into segments and a hard rounded back led me to conclude he must be a ***beetle.*** And like all beetles, under his shell were wings...
Gregor never realized that ***all along*** he could have taken flight.
That's sad, but beautiful.
Beauty plus pity— that is the ***closest*** we can get to a definition of art.
THE BOOK OF HUMAN INSECTS OSAMU TEZUKA

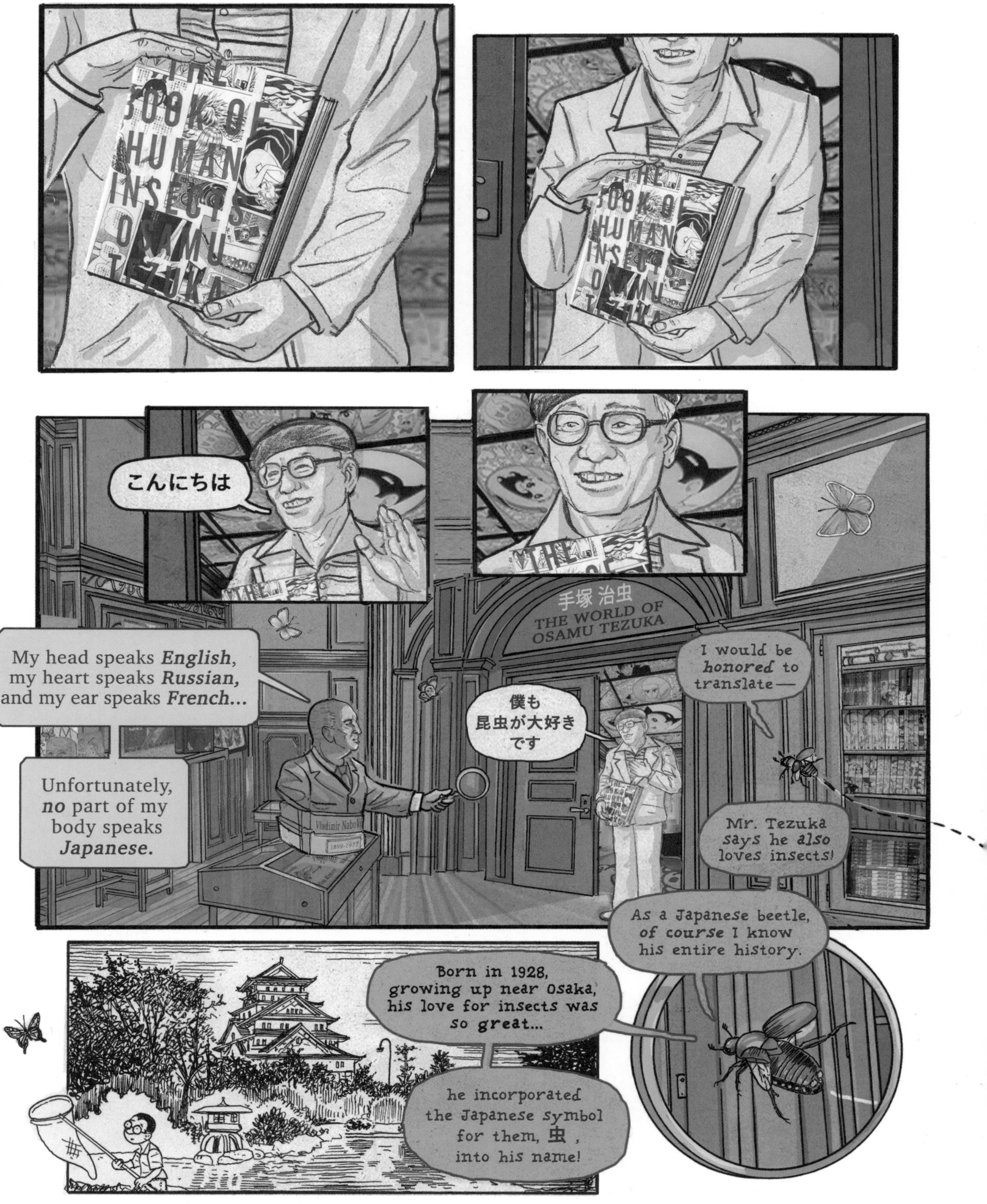
THE BOOK OF HUMAN INSECTS OSAMU TEZUKA
こんにちは
手塚 治虫
THE WORLD OF OSAMU TEZUKA
My head speaks *English*, my heart speaks *Russian,* and my ear speaks *French...*
Unfortunately, *no* part of my body speaks *Japanese.*
1899-1977
僕も昆虫が大好きです
I would be *honored* to translate—
Mr. Tezuka says he *also* loves insects!
As a Japanese beetle, *of course* I know his entire history.
Born in 1928, growing up near Osaka, his love for insects was so **great...**
he incorporated the Japanese symbol for them, 虫, into his name!

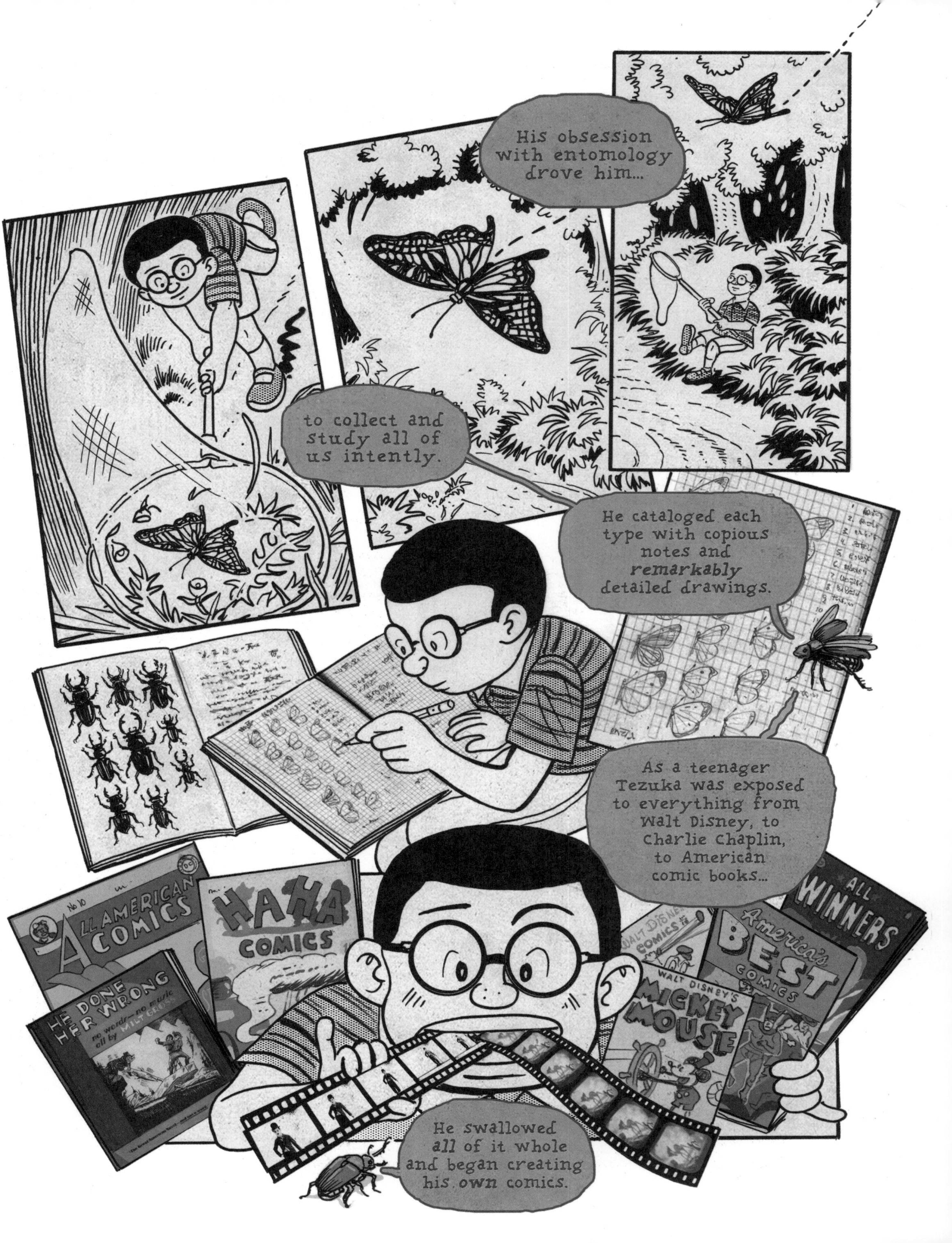
His obsession with entomology drove him...
to collect and study all of us intently.
He cataloged each type with copious notes and *remarkably* detailed drawings.
As a teenager Tezuka was exposed to everything from Walt Disney, to Charlie Chaplin, to American comic books...
ALL AMERICAN COMICS
HA HA COMICS
ALL WINNERS
America's BEST COMICS
HE DONE HER WRONG
WALT DISNEY'S MICKEY MOUSE
He swallowed *all* of it whole and began creating his *own* comics.

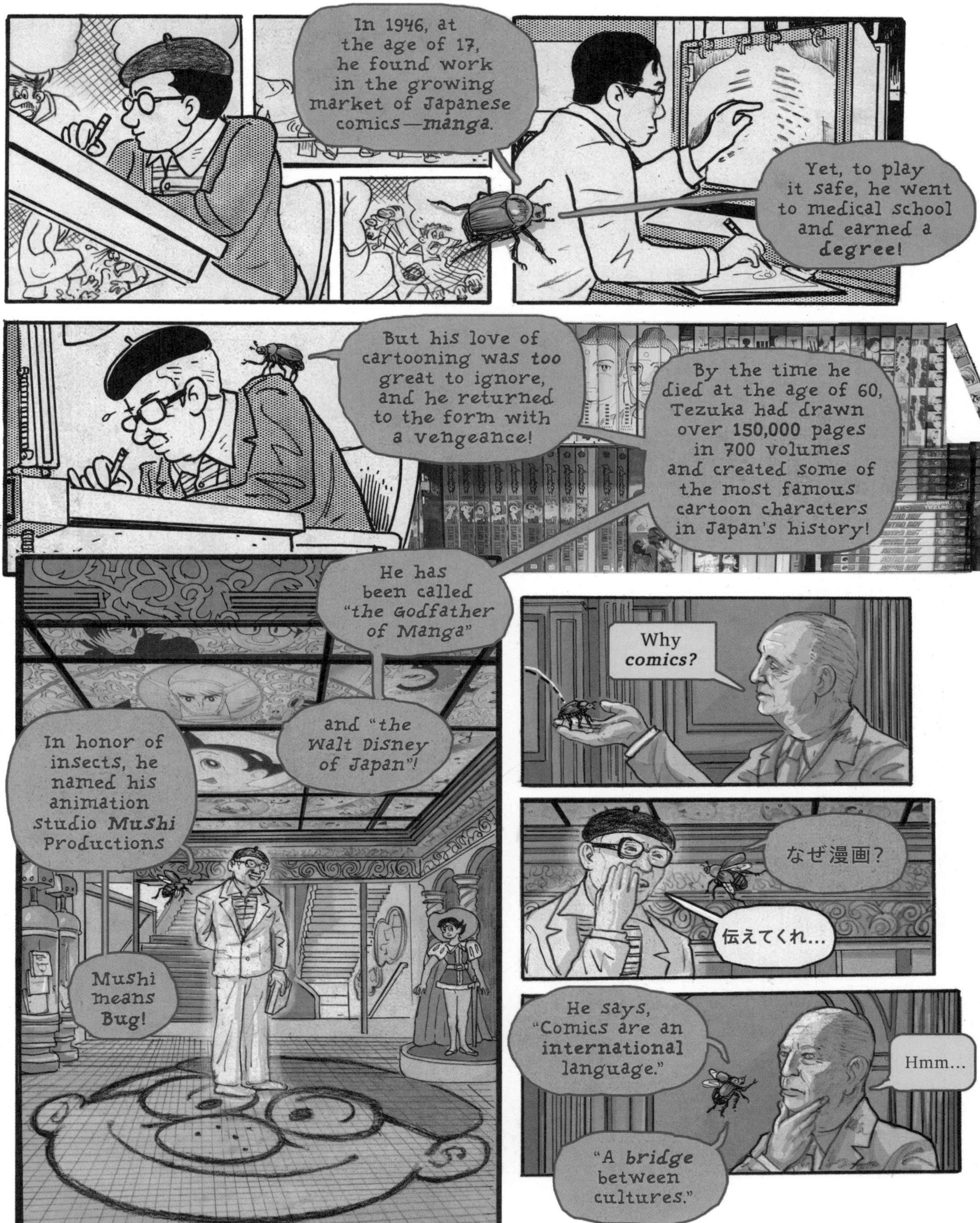
In 1946, at the age of 17, he found work in the growing market of Japanese comics—*manga*.
Yet, to play it safe, he went to medical school and earned a **degree**!
But his love of cartooning was *too* great to ignore, and he returned to the form with a vengeance!
By the time he died at the age of 60, Tezuka had drawn over **150,000** pages in 700 volumes and created some of the most famous cartoon characters in Japan's history!
He has been called "*the Godfather of Manga*"
and "*the Walt Disney of Japan*"!
In honor of insects, he named his animation studio ***Mushi*** Productions
Mushi means Bug!
Why *comics?*
なぜ漫画?
伝えてくれ...
He *says*, "Comics are an **international** language."
Hmm...
"A *bridge* between cultures."

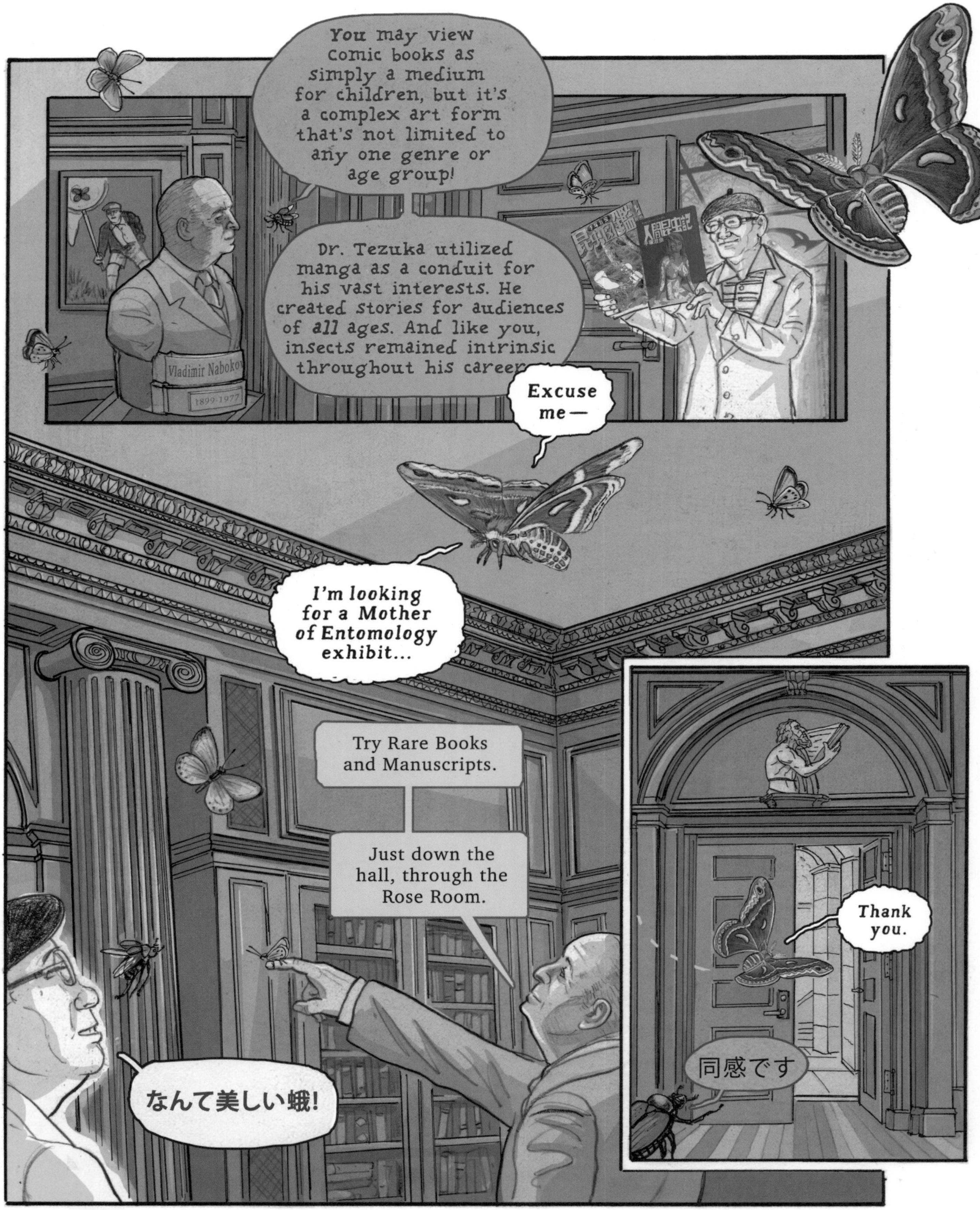
You may view comic books as simply a medium for children, but it's a complex art form that's not limited to any one genre or age group!
Dr. Tezuka utilized manga as a conduit for his vast interests. He created stories for audiences of *all* ages. And like you, insects remained intrinsic throughout his career—
Vladimir Nabokov
1899-1977
Excuse me—
I'm looking for a Mother of Entomology exhibit...
Try Rare Books and Manuscripts.
Just down the hall, through the Rose Room.
なんて美しい蛾!
Thank you.
同感です

DISSERTATION
SUR LA
GENERATION
ET LES
TRANSFORMATIONS
DES
INSECTES
DE
SURINAM:
Dans laquelle on traite des
VERS ET DES CHENILLES DE SURINAM,
des Plantes, fleurs, & fruits dont ils vivent & dans lesquels on les a trouvez.
On y parle aussi des
CRAPAUX, LEZARDS, SERPENS, ARAGNEES,
& autres petits animaux du même païs, peints sur les lieux d'après nature,
& depuis gravez avec beaucoup de soin.
On y a joint
Un Traité des changemens des POISSONS en GRENOUILLES
& des GRENOUILLES en POISSONS.
Par MARIE SIBILLE MERIAN.
A LA HAYE,
Chez PIERRE GOSSE.
M. DCC. XXVI.

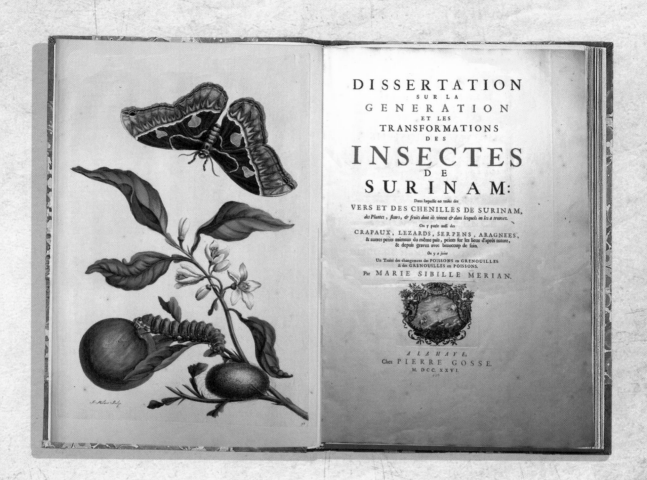

Maria Sibylla Merian
(1647–1717)
Transformation of the Insects of Suriname
Published in Amsterdam in 1705,
it included 60 of Merian's
hand-colored engravings.

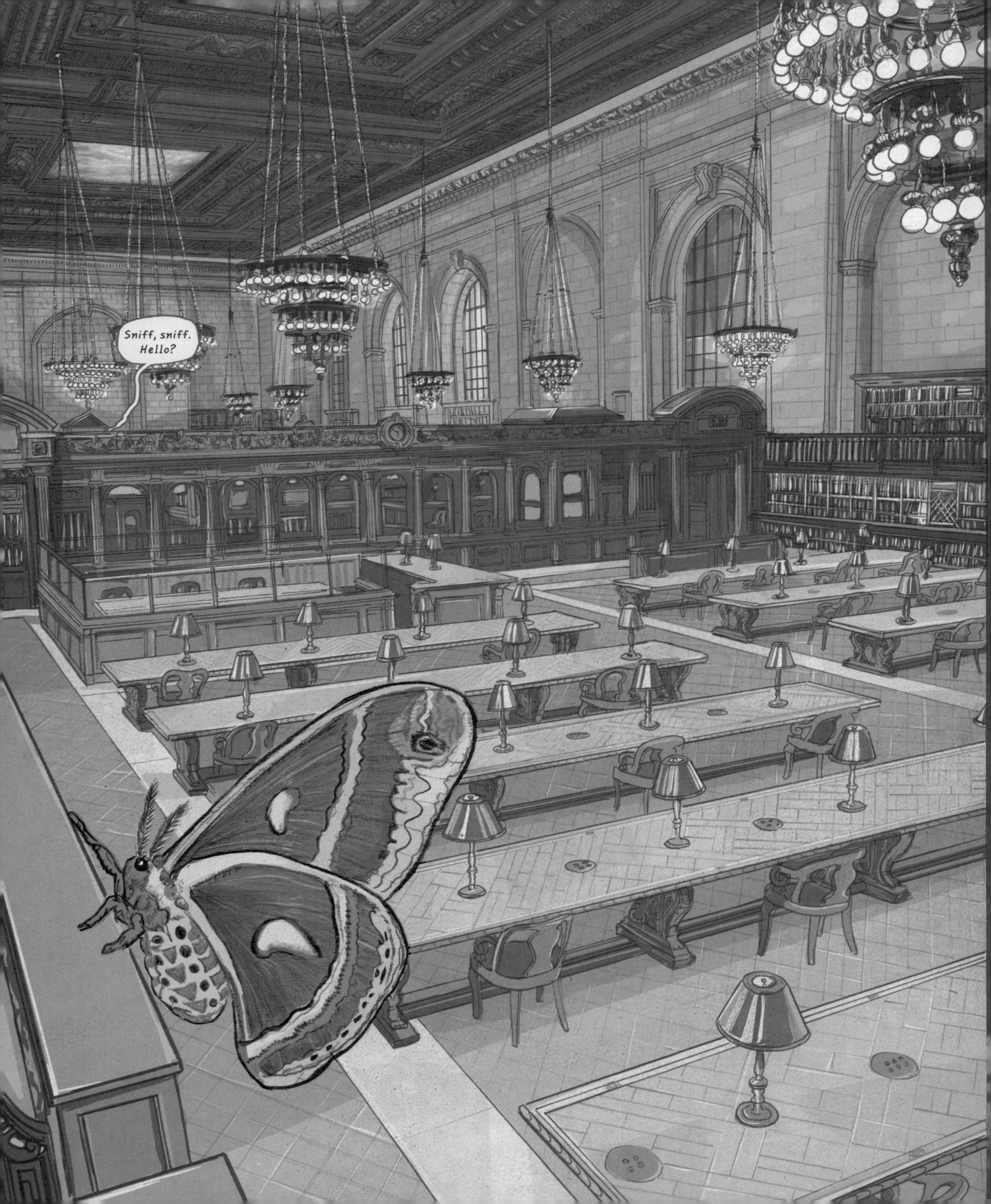
Sniff, sniff. Hello?

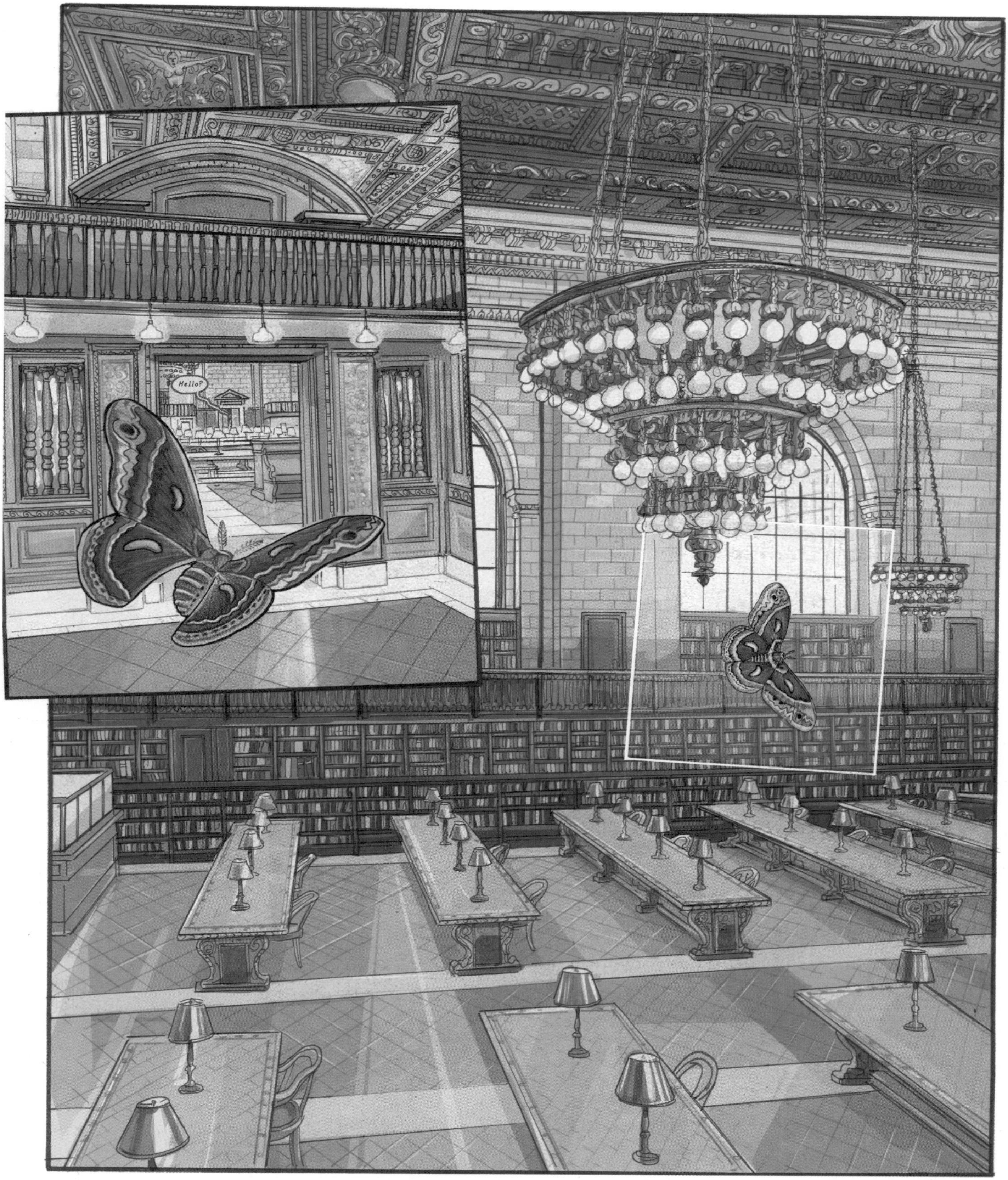
Hello?

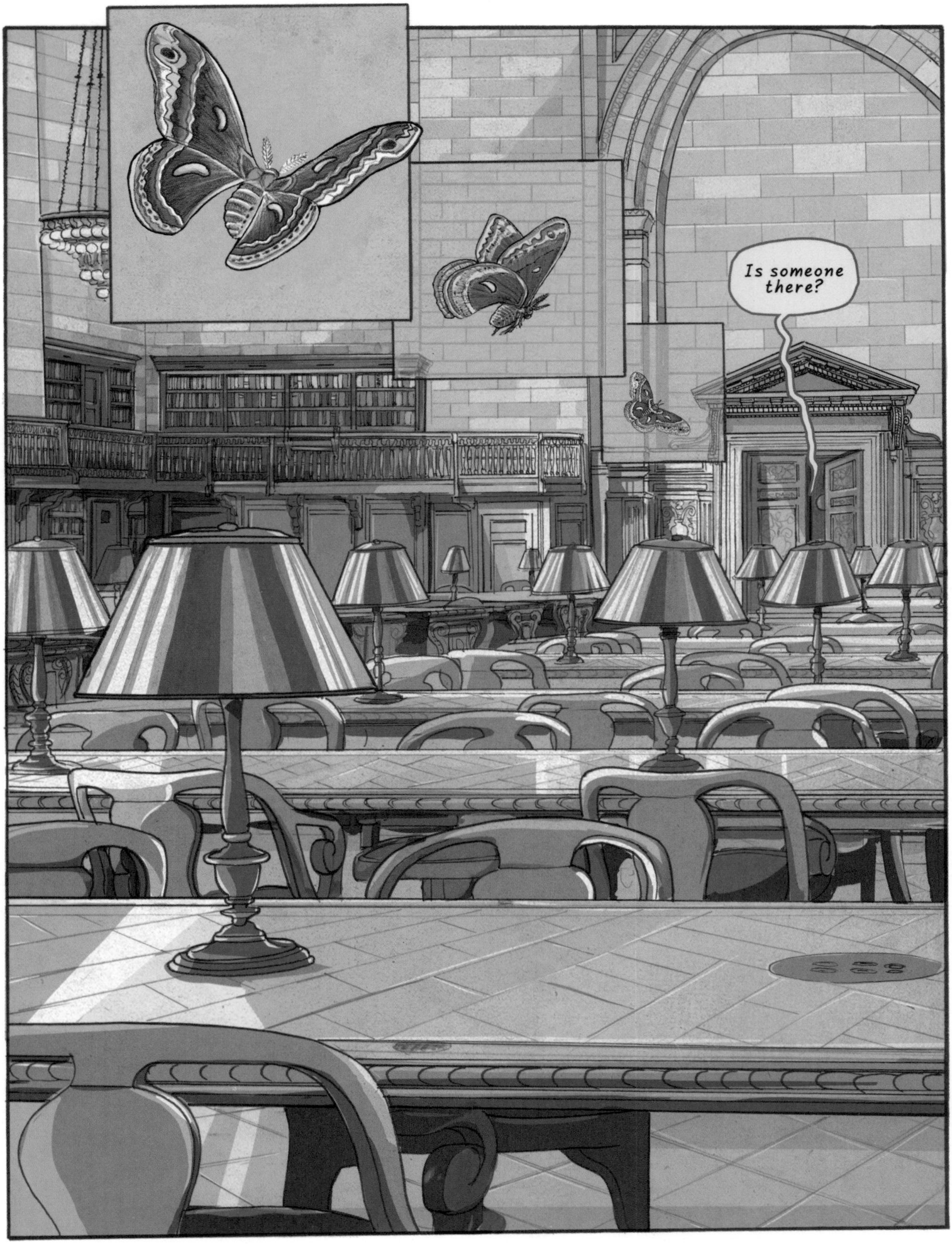
Is someone there?

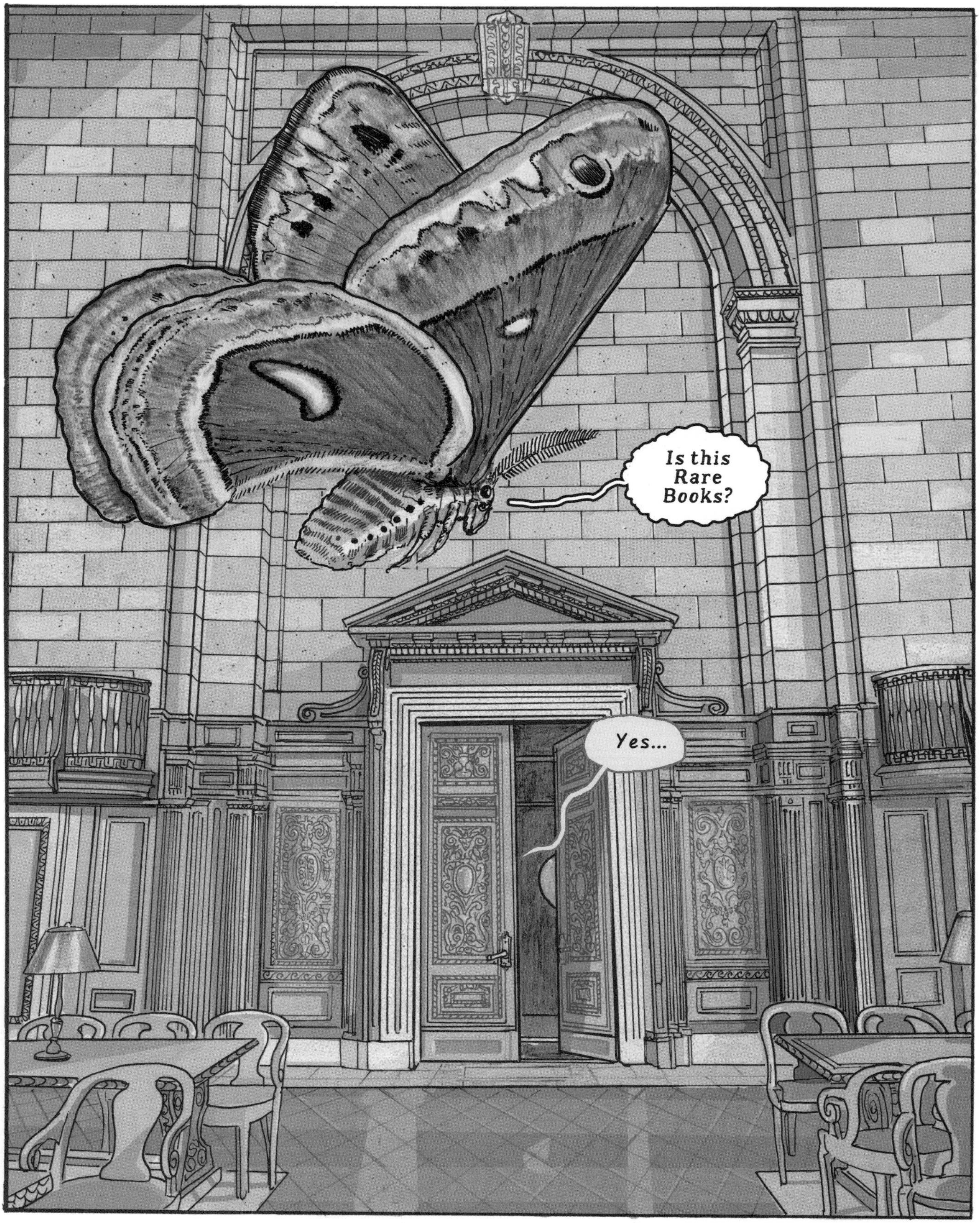
Is this Rare Books?
Yes...

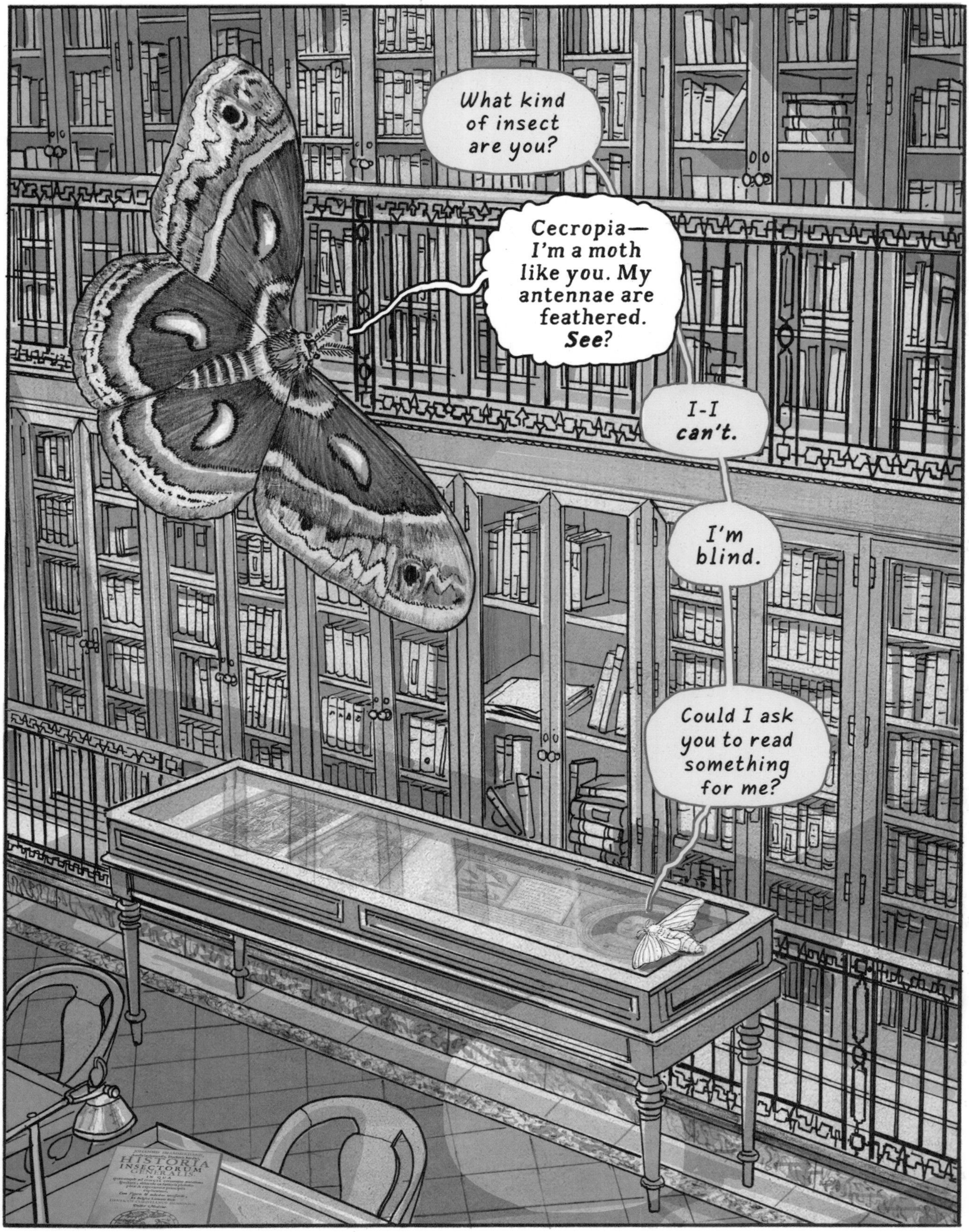
What kind of insect are you?
Cecropia— I'm a moth like you. My antennae are feathered. **See**?
I-I **can't**.
I'm blind.
Could I ask you to read something for me?
HISTORIA
INSECTORUM

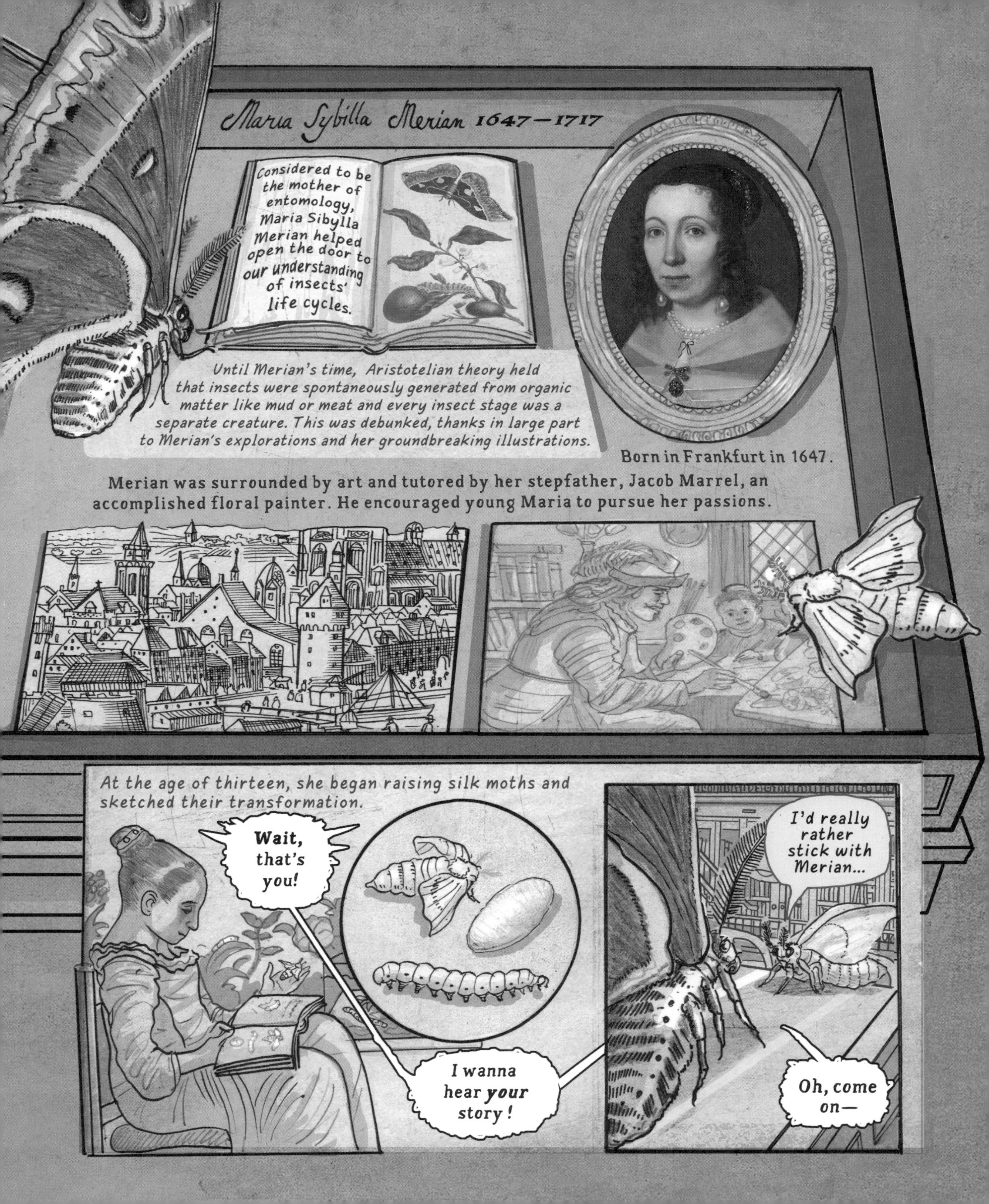
Maria Sybilla Merian 1647—1717
Considered to be the mother of entomology, Maria Sibylla Merian helped open the door to our understanding of insects' life cycles.
Until Merian's time, Aristotelian theory held that insects were spontaneously generated from organic matter like mud or meat and every insect stage was a separate creature. This was debunked, thanks in large part to Merian's explorations and her groundbreaking illustrations.
Born in Frankfurt in 1647.
Merian was surrounded by art and tutored by her stepfather, Jacob Marrel, an accomplished floral painter. He encouraged young Maria to pursue her passions.
At the age of thirteen, she began raising silk moths and sketched their transformation.
Wait, that's you!
I wanna hear your story!
I'd really rather stick with Merian...
Oh, come on—

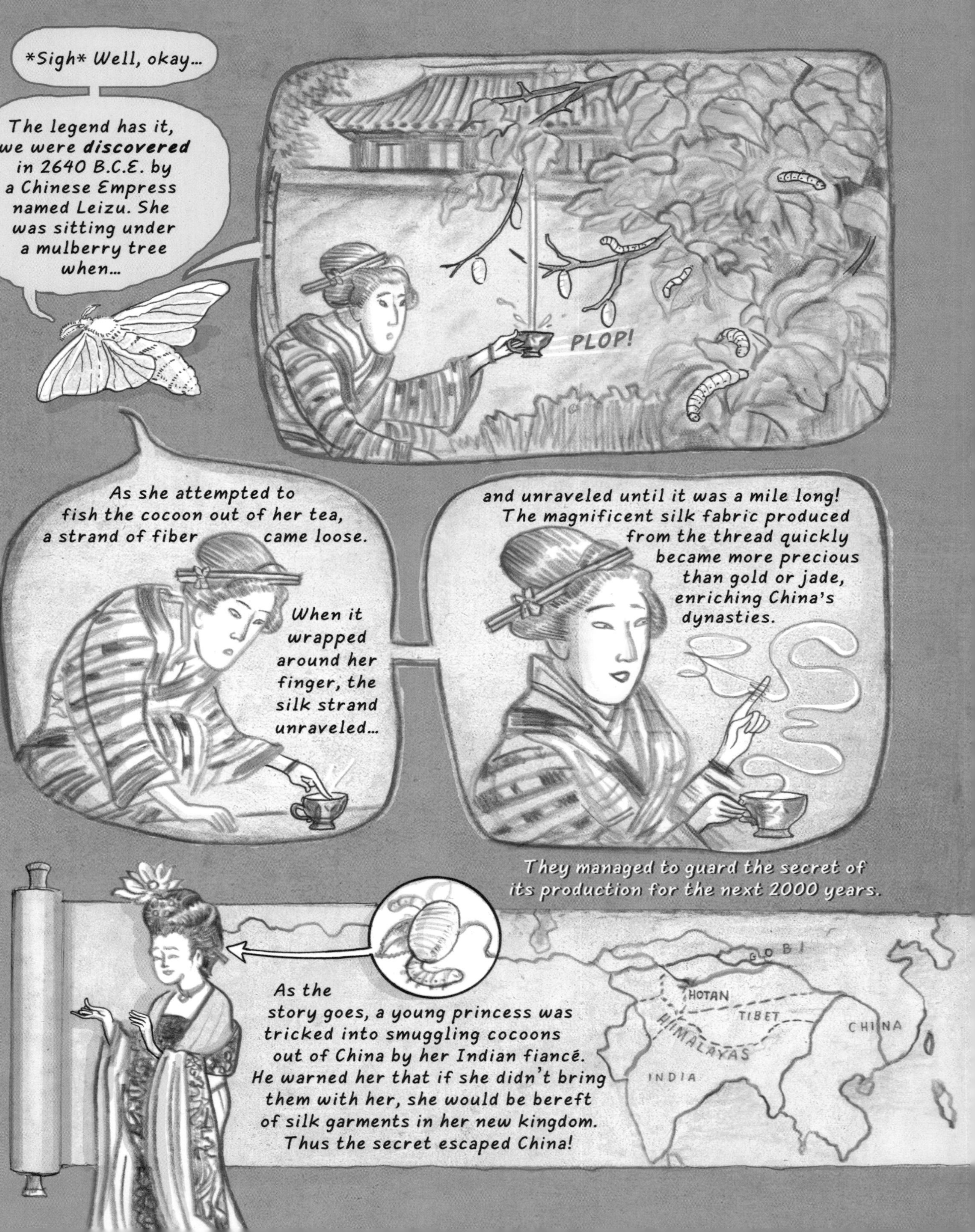

Sigh Well, okay...
The legend has it, we were **discovered** in 2640 B.C.E. by a Chinese Empress named Leizu. She was sitting under a mulberry tree when...
PLOP!
As she attempted to fish the cocoon out of her tea, a strand of fiber came loose.
When it wrapped around her finger, the silk strand unraveled...
and unraveled until it was a mile long! The magnificent silk fabric produced from the thread quickly became more precious than gold or jade, enriching China's dynasties.
They managed to guard the secret of its production for the next 2000 years.
GOBI
HOTAN
TIBET
CHINA
HIMALAYAS
INDIA
As the story goes, a young princess was tricked into smuggling cocoons out of China by her Indian fiancé. He warned her that if she didn't bring them with her, she would be bereft of silk garments in her new kingdom. Thus the secret escaped China!

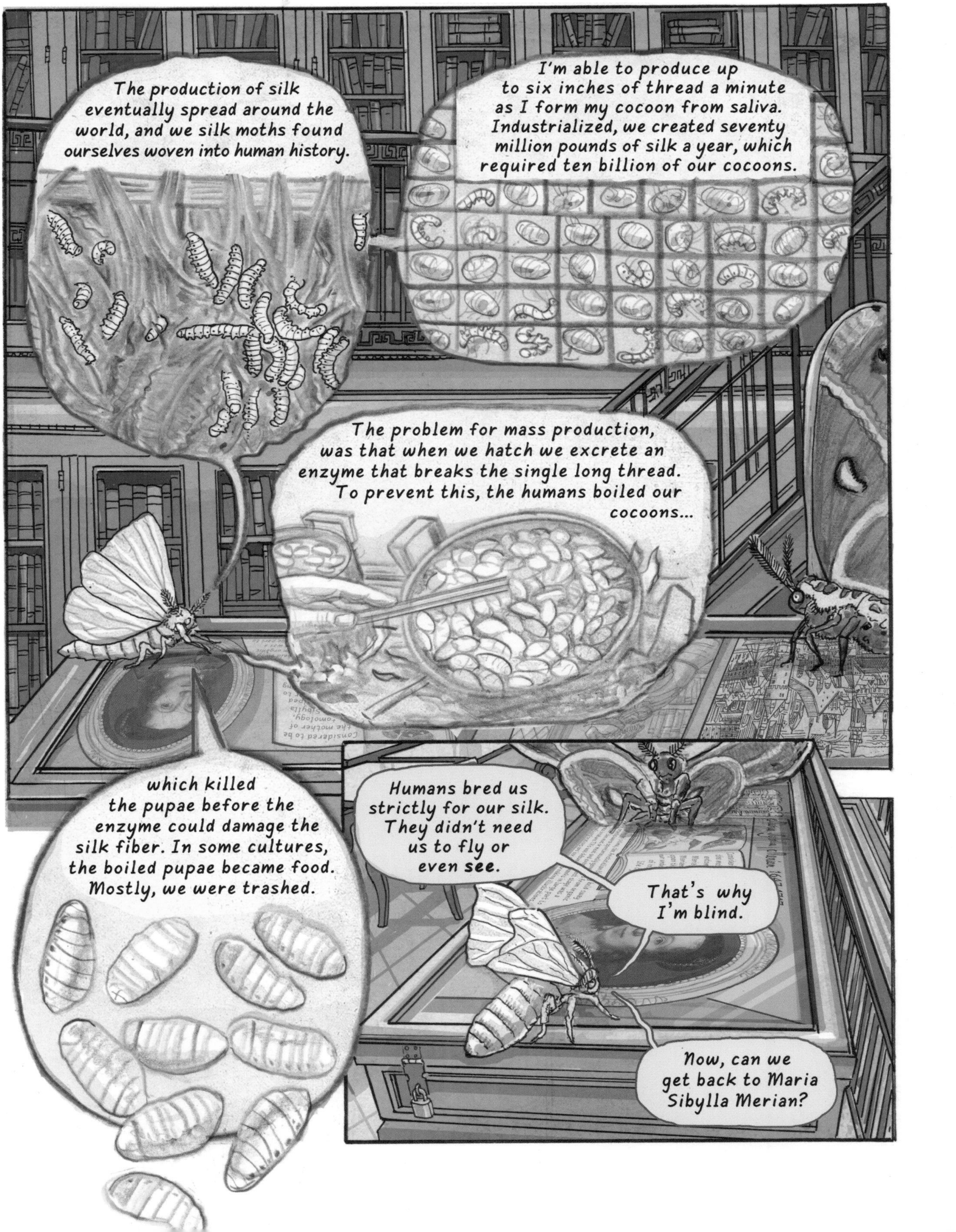
The production of silk eventually spread around the world, and we silk moths found ourselves woven into human history.
I'm able to produce up to six inches of thread a minute as I form my cocoon from saliva. Industrialized, we created seventy million pounds of silk a year, which required ten billion of our cocoons.
The problem for mass production, was that when we hatch we excrete an enzyme that breaks the single long thread. To prevent this, the humans boiled our cocoons...
which killed the pupae before the enzyme could damage the silk fiber. In some cultures, the boiled pupae became food. Mostly, we were trashed.
Humans bred us strictly for our silk. They didn't need us to fly or even **see**.
That's why I'm blind.
Now, can we get back to Maria Sibylla Merian?

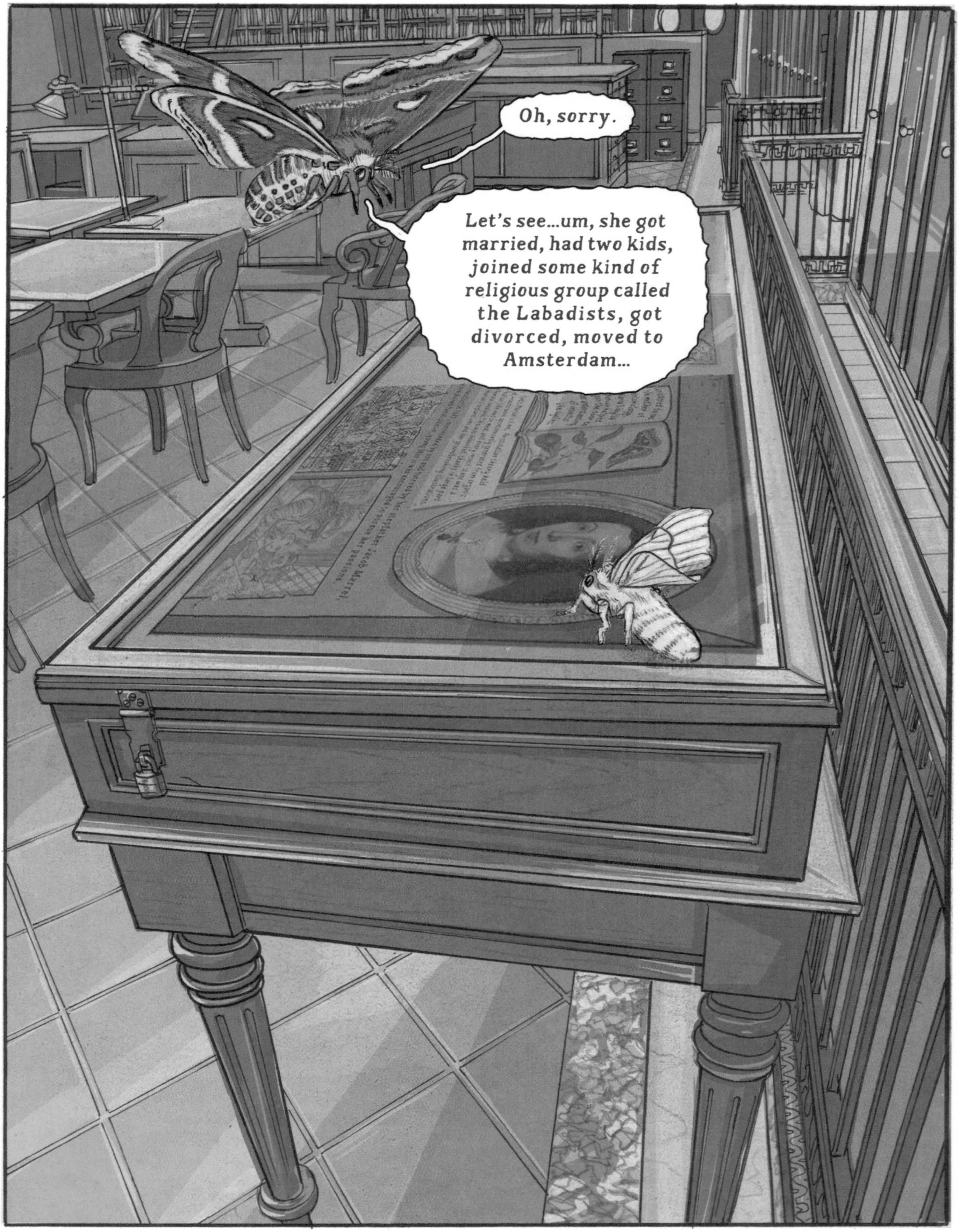
Oh, sorry.
Let's see…um, she got married, had two kids, joined some kind of religious group called the Labadists, got divorced, moved to Amsterdam…

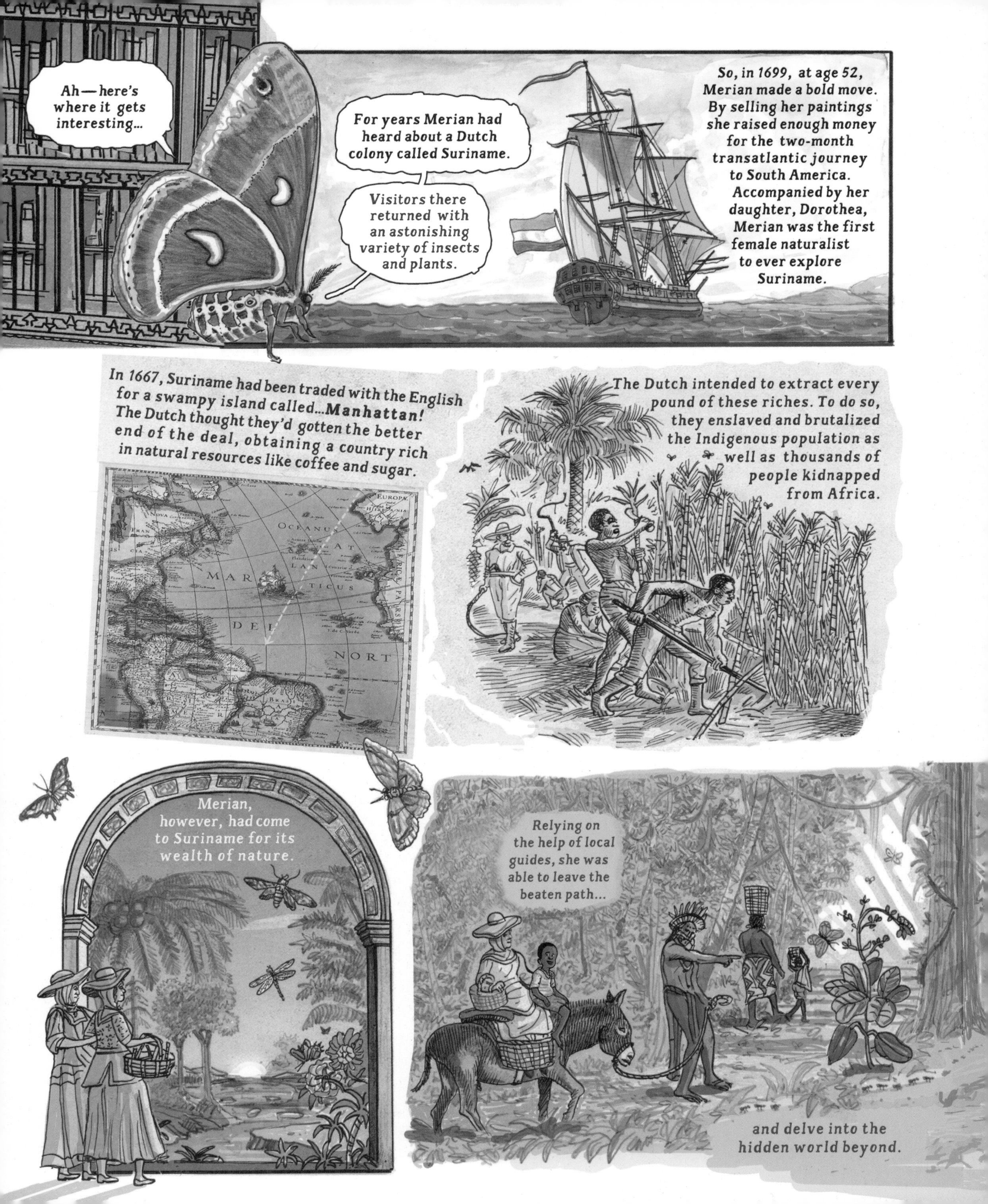
Ah—here's where it gets interesting...
For years Merian had heard about a Dutch colony called Suriname.
Visitors there returned with an astonishing variety of insects and plants.
So, in 1699, at age 52, Merian made a bold move. By selling her paintings she raised enough money for the two-month transatlantic journey to South America. Accompanied by her daughter, Dorothea, Merian was the first female naturalist to ever explore Suriname.
In 1667, Suriname had been traded with the English for a swampy island called...**Manhattan!** The Dutch thought they'd gotten the better end of the deal, obtaining a country rich in natural resources like coffee and sugar.
OCEANUS
AT
LAN
TICUS
MAR
DEL
NORT
The Dutch intended to extract every pound of these riches. To do so, they enslaved and brutalized the Indigenous population as well as thousands of people kidnapped from Africa.
Merian, however, had come to Suriname for its wealth of nature.
Relying on the help of local guides, she was able to leave the beaten path...
and delve into the hidden world beyond.

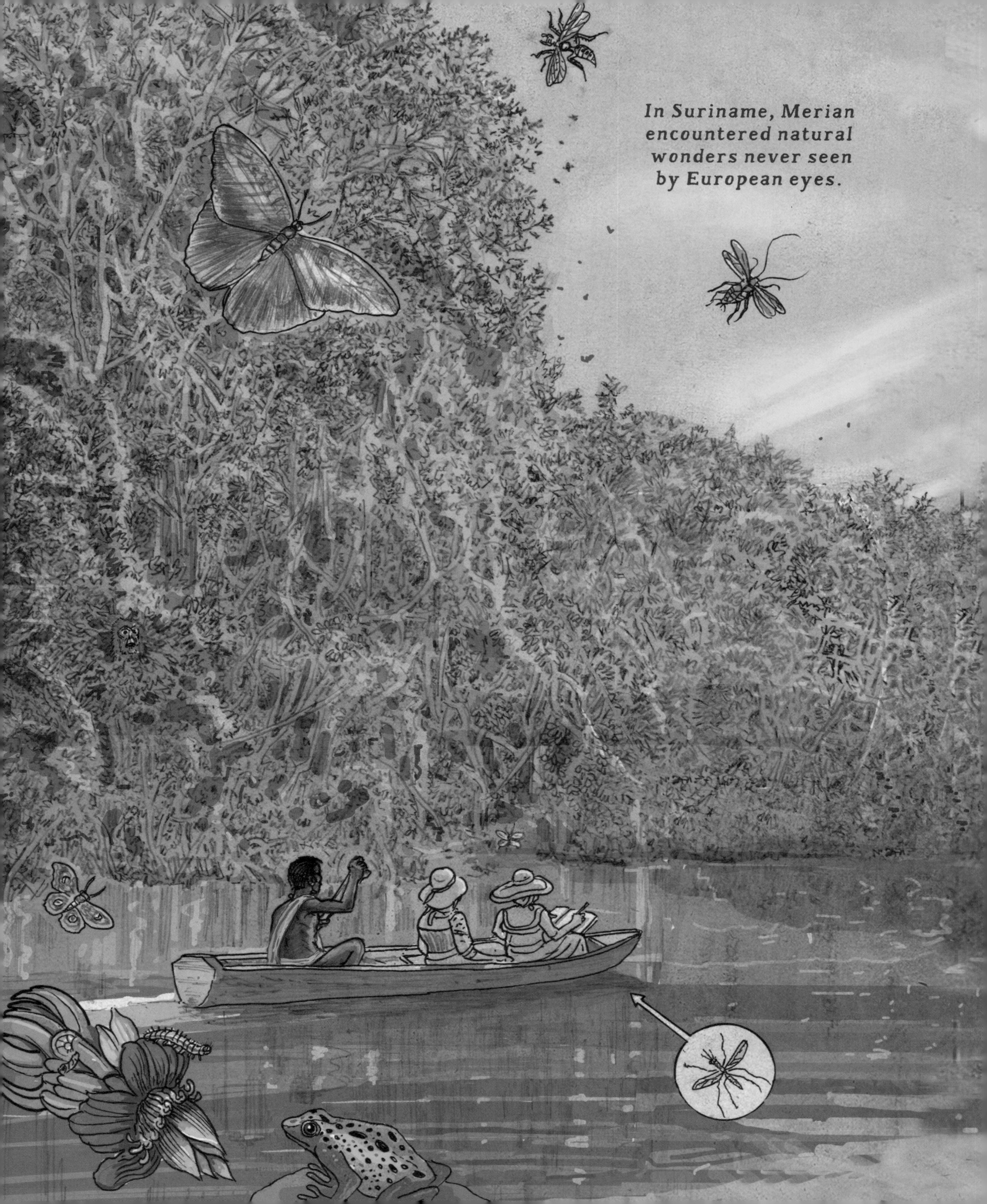
In Suriname, Merian encountered natural wonders never seen by European eyes.

She captured
these marvels in her
beautiful, meticulous
drawings...

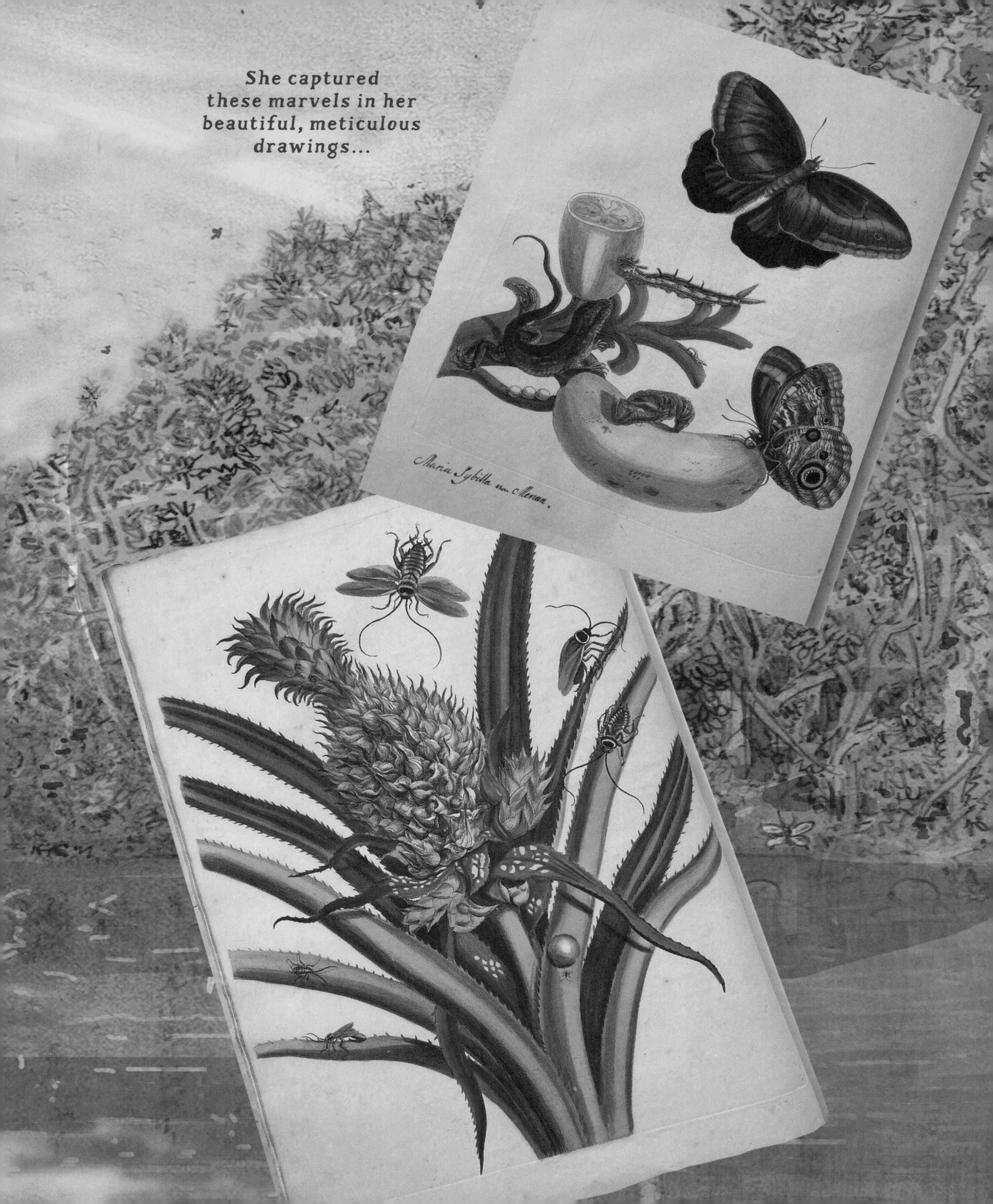

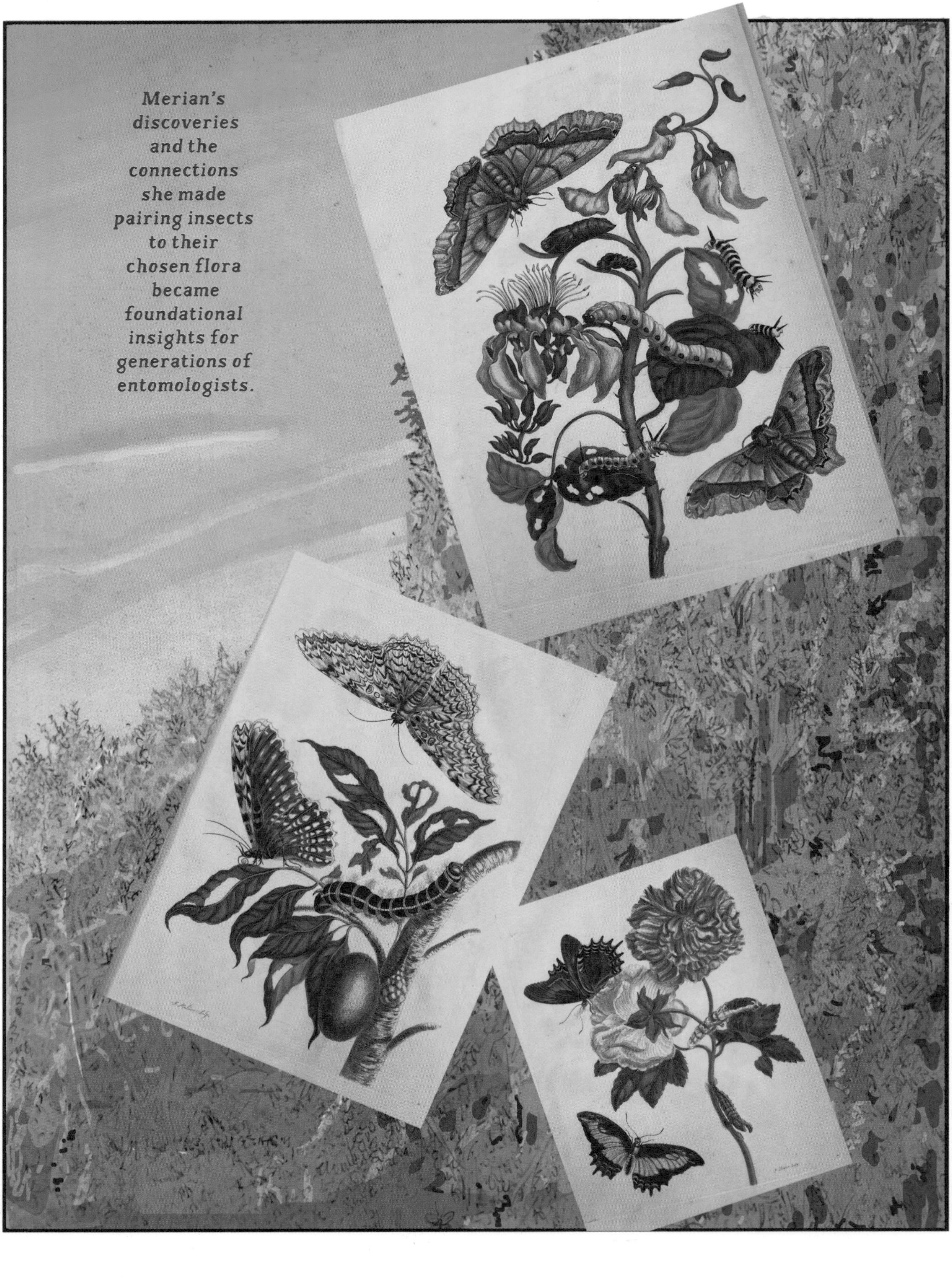

Merian's discoveries and the connections she made pairing insects to their chosen flora became foundational insights for generations of entomologists.

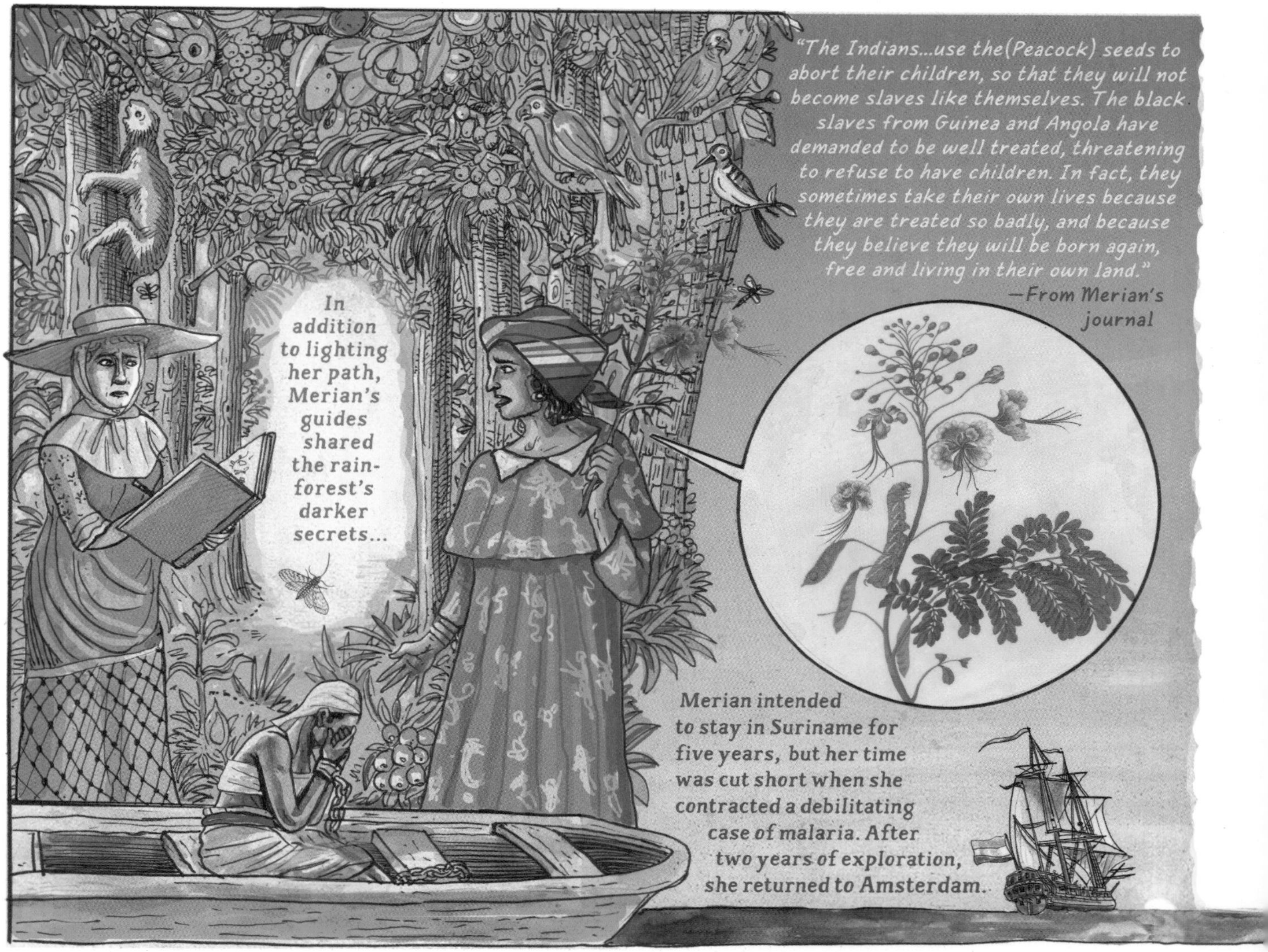
"The Indians...use the (Peacock) seeds to abort their children, so that they will not become slaves like themselves. The black slaves from Guinea and Angola have demanded to be well treated, threatening to refuse to have children. In fact, they sometimes take their own lives because they are treated so badly, and because they believe they will be born again, free and living in their own land."
—From Merian's journal
In addition to lighting her path, Merian's guides shared the rain-forest's darker secrets...
Merian intended to stay in Suriname for five years, but her time was cut short when she contracted a debilitating case of malaria. After two years of exploration, she returned to Amsterdam.

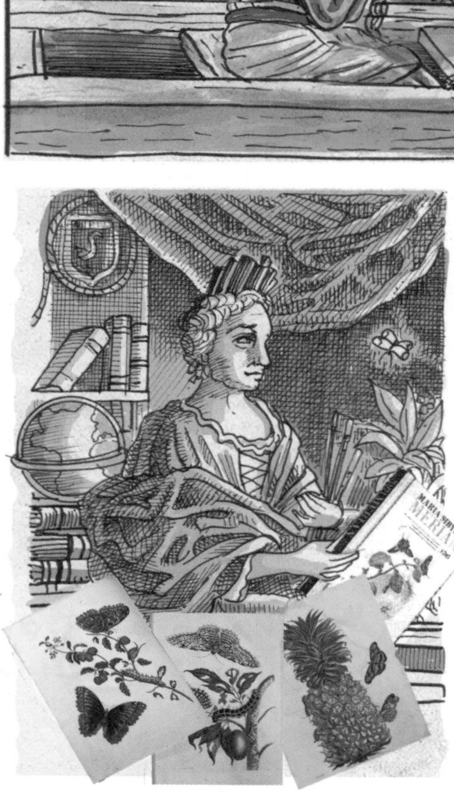

In 1705, she published the stunning images she'd begun in South America, a book of sixty hand-colored etchings titled ***Transformation of the Insects of Suriname.*** She died in 1717, working into her final year.

The importance of her contributions can't be overstated. Surmounting social norms, she produced critical insight into the natural world a century and a half before Charles Darwin.

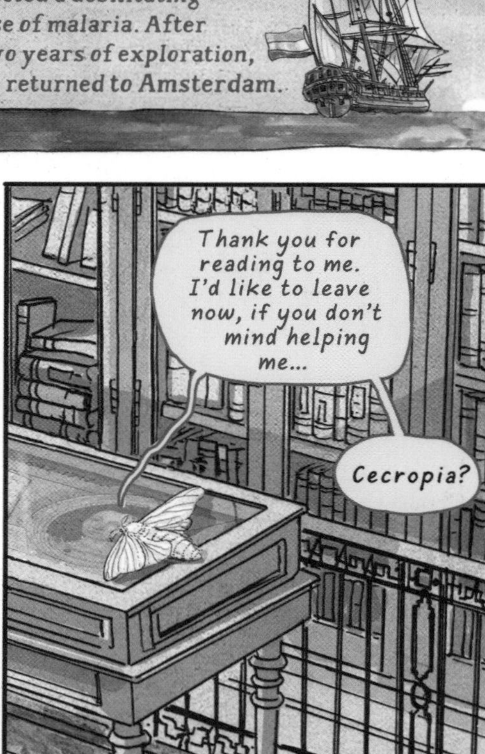
Thank you for reading to me. I'd like to leave now, if you don't mind helping me...
Cecropia?

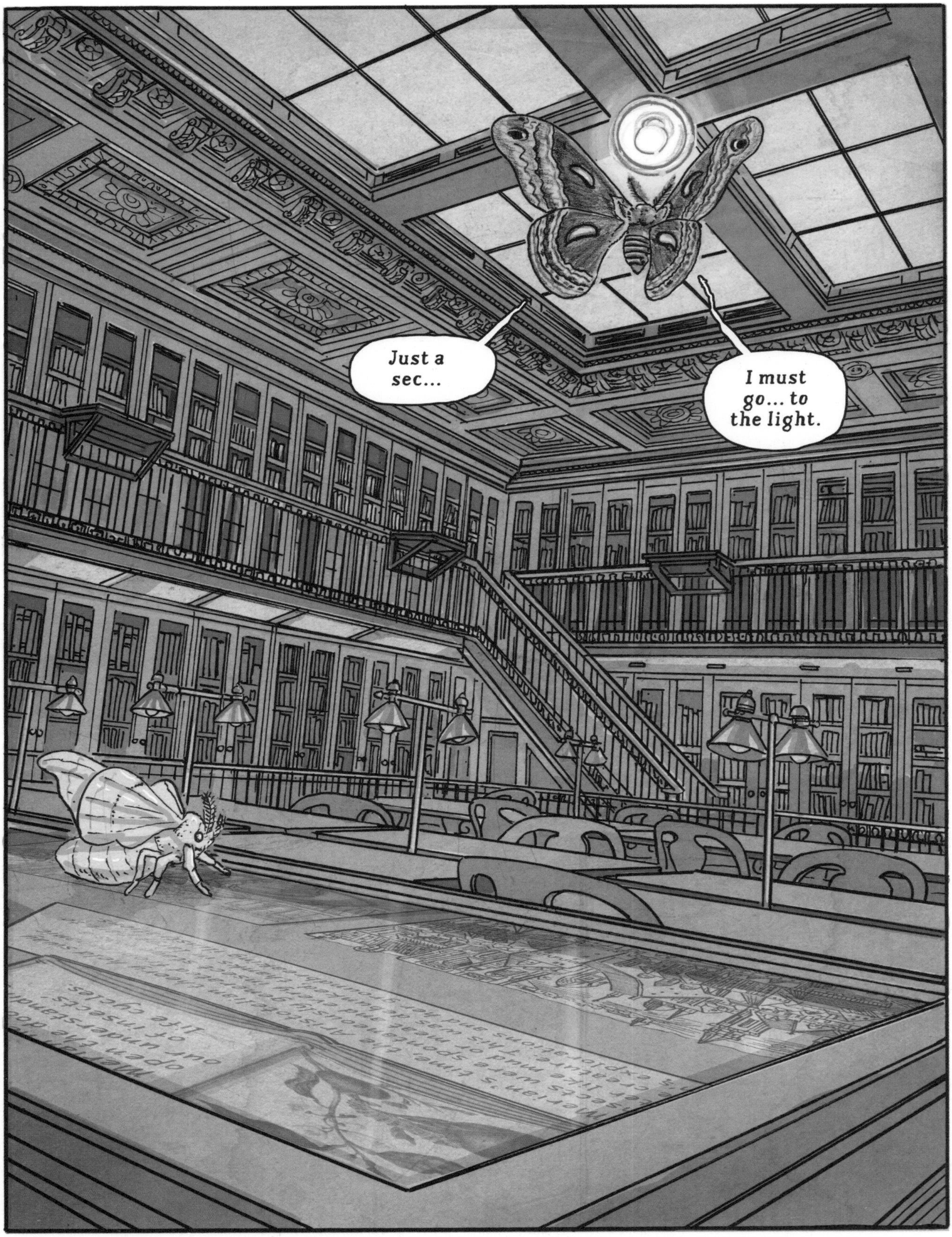
Just a sec...
I must go... to the light.

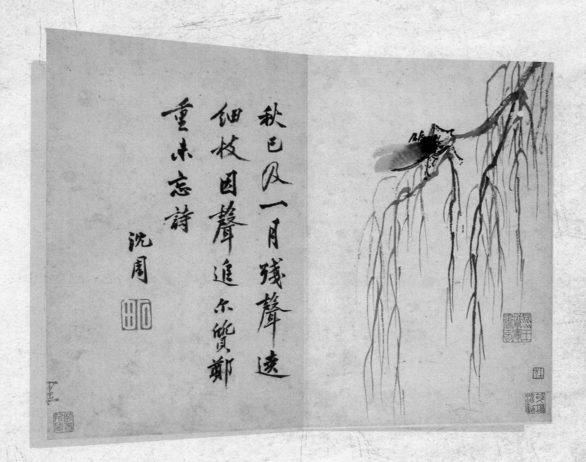

A Crying Cicada on an Autumn Willow
Shen Zhou
(1427–1509)
China, Ming Dynasty

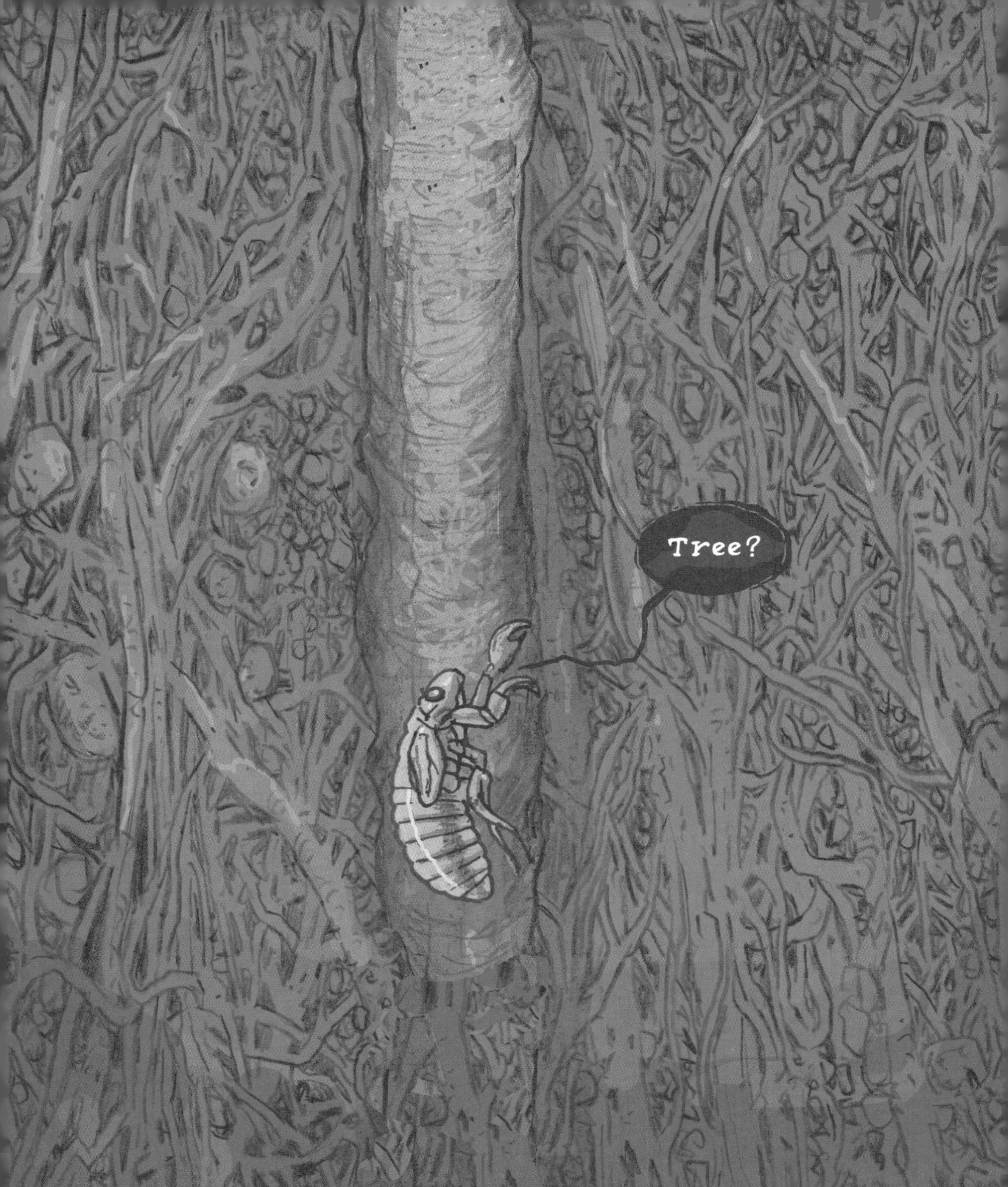
Tree?

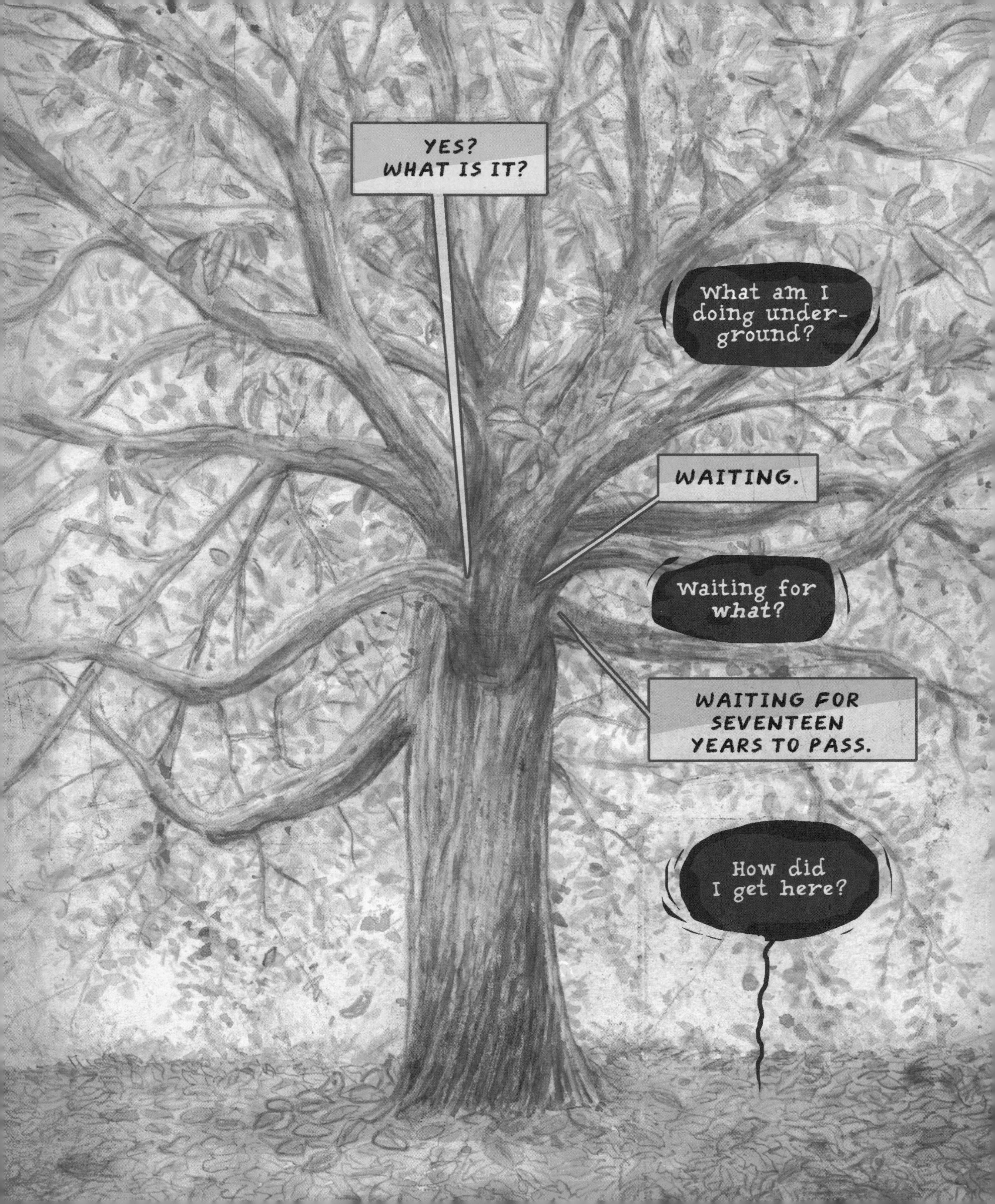
YES? WHAT IS IT?
What am I doing under-ground?
WAITING.
Waiting for *what*?
WAITING FOR SEVENTEEN YEARS TO PASS.
How did I get here?

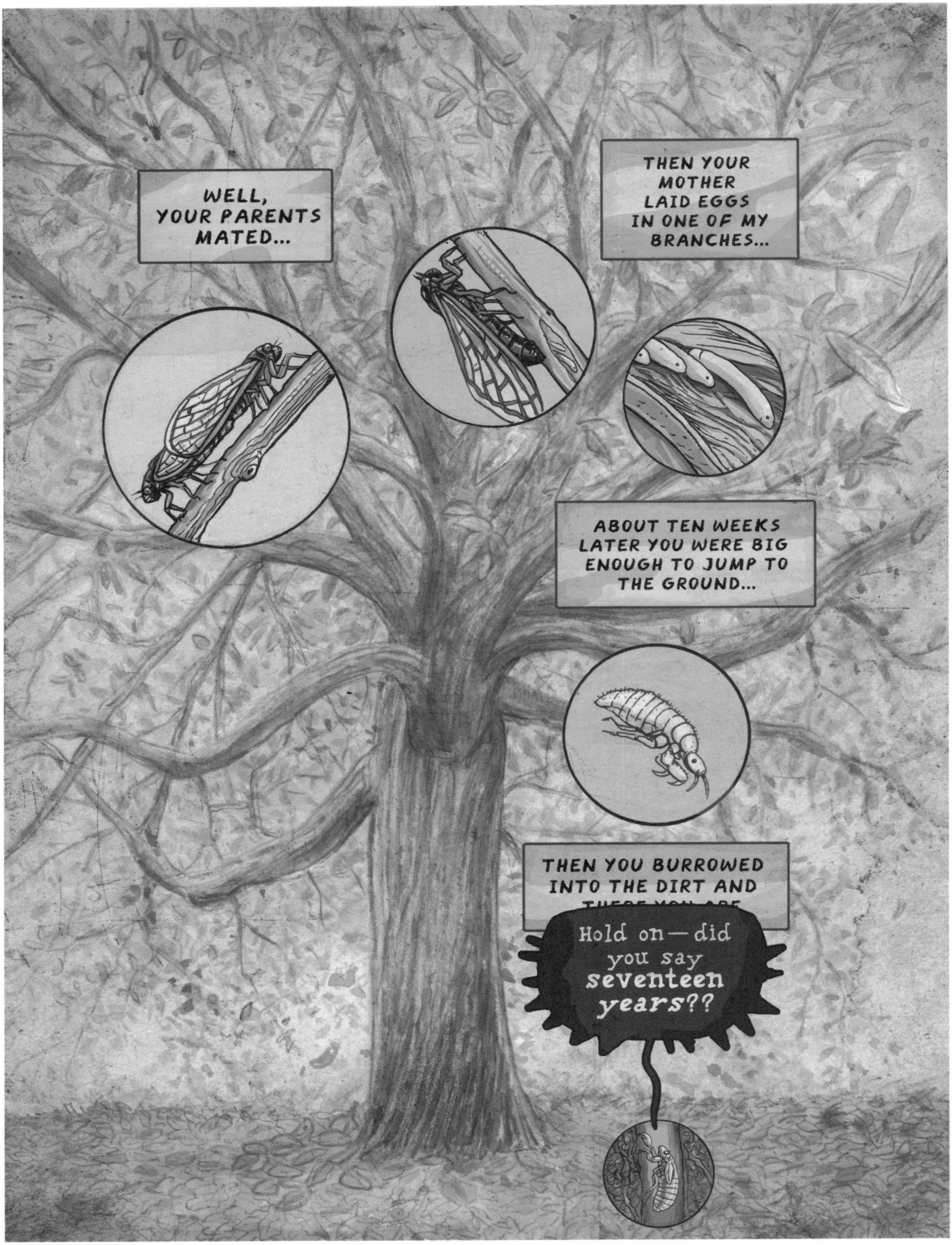
WELL, YOUR PARENTS MATED...
THEN YOUR MOTHER LAID EGGS IN ONE OF MY BRANCHES...
ABOUT TEN WEEKS LATER YOU WERE BIG ENOUGH TO JUMP TO THE GROUND...
THEN YOU BURROWED INTO THE DIRT AND
Hold on—did you say seventeen years??

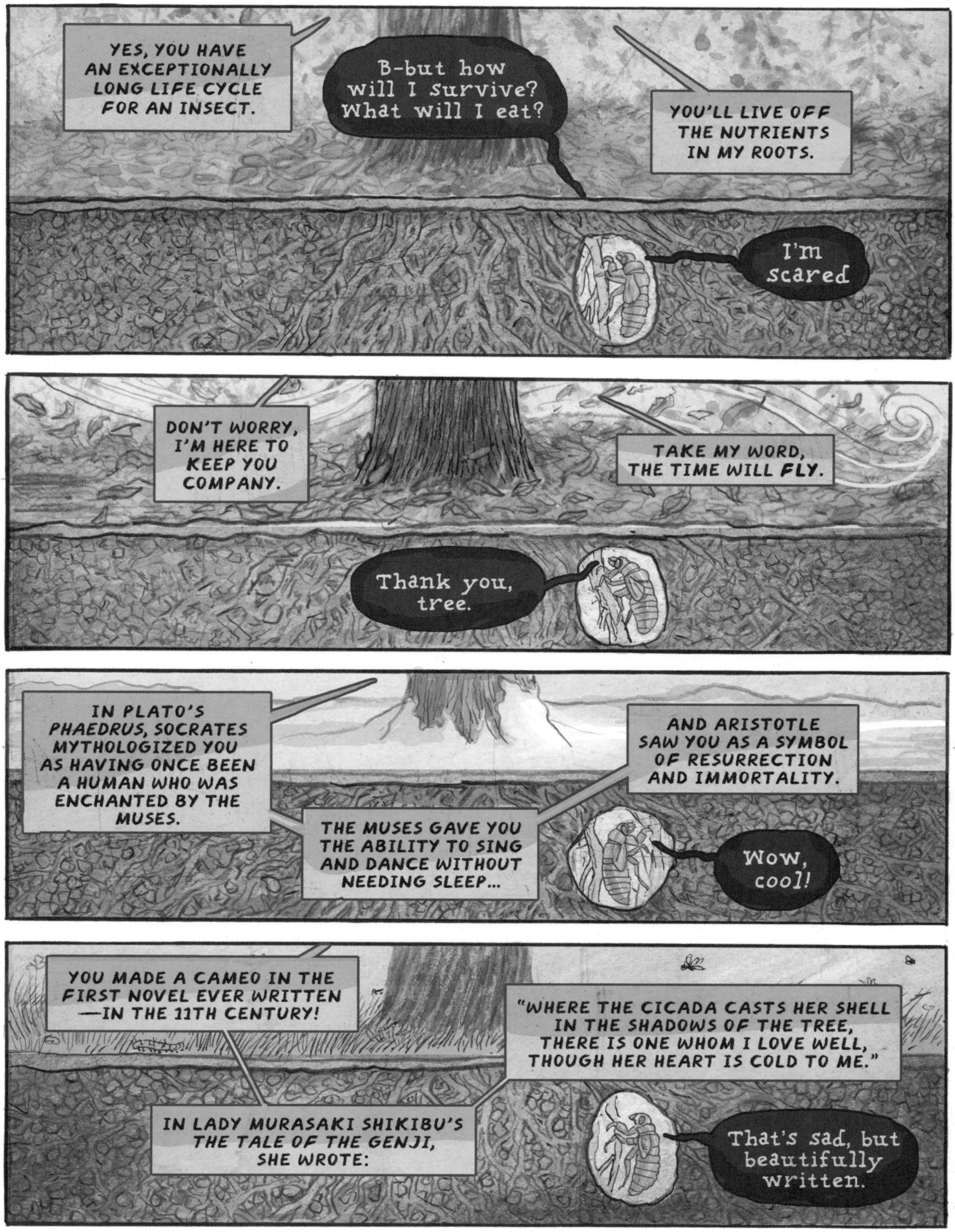
YES, YOU HAVE AN EXCEPTIONALLY LONG LIFE CYCLE FOR AN INSECT.
B-but how will I survive? What will I eat?
YOU'LL LIVE OFF THE NUTRIENTS IN MY ROOTS.
I'm scared
DON'T WORRY, I'M HERE TO KEEP YOU COMPANY.
TAKE MY WORD, THE TIME WILL FLY.
Thank you, tree.
IN PLATO'S PHAEDRUS, SOCRATES MYTHOLOGIZED YOU AS HAVING ONCE BEEN A HUMAN WHO WAS ENCHANTED BY THE MUSES.
THE MUSES GAVE YOU THE ABILITY TO SING AND DANCE WITHOUT NEEDING SLEEP...
AND ARISTOTLE SAW YOU AS A SYMBOL OF RESURRECTION AND IMMORTALITY.
Wow, cool!
YOU MADE A CAMEO IN THE FIRST NOVEL EVER WRITTEN —IN THE 11TH CENTURY!
IN LADY MURASAKI SHIKIBU'S THE TALE OF THE GENJI, SHE WROTE:
"WHERE THE CICADA CASTS HER SHELL IN THE SHADOWS OF THE TREE, THERE IS ONE WHOM I LOVE WELL, THOUGH HER HEART IS COLD TO ME."
That's sad, but beautifully written.

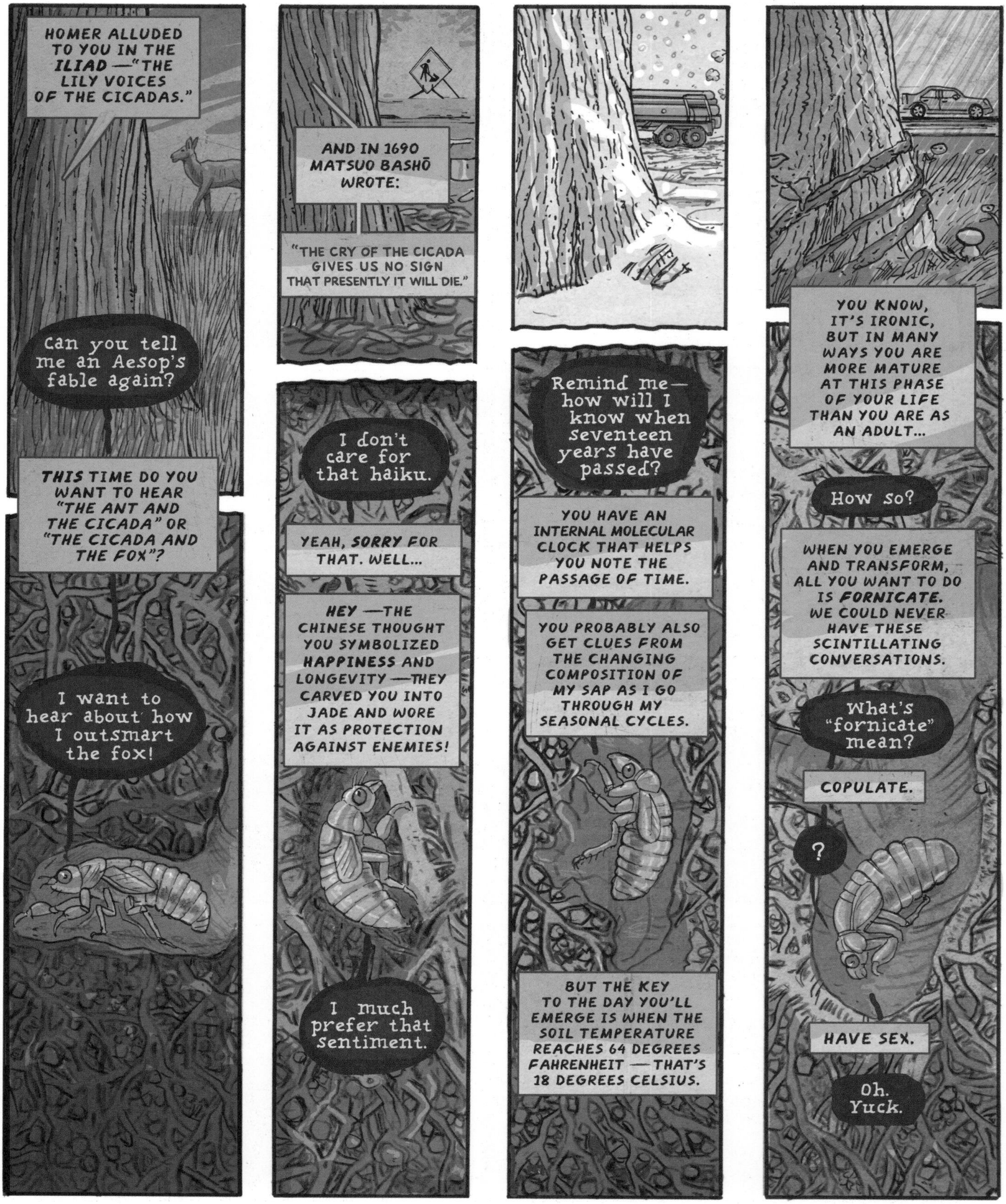

HOMER ALLUDED TO YOU IN THE ILIAD —"THE LILY VOICES OF THE CICADAS."
Can you tell me an Aesop's fable again?
THIS TIME DO YOU WANT TO HEAR "THE ANT AND THE CICADA" OR "THE CICADA AND THE FOX"?
I want to hear about how I outsmart the fox!
AND IN 1690 MATSUO BASHŌ WROTE:
"THE CRY OF THE CICADA GIVES US NO SIGN THAT PRESENTLY IT WILL DIE."
I don't care for that haiku.
YEAH, SORRY FOR THAT. WELL...
HEY —THE CHINESE THOUGHT YOU SYMBOLIZED HAPPINESS AND LONGEVITY —THEY CARVED YOU INTO JADE AND WORE IT AS PROTECTION AGAINST ENEMIES!
I much prefer that sentiment.
Remind me— how will I know when seventeen years have passed?
YOU HAVE AN INTERNAL MOLECULAR CLOCK THAT HELPS YOU NOTE THE PASSAGE OF TIME.
YOU PROBABLY ALSO GET CLUES FROM THE CHANGING COMPOSITION OF MY SAP AS I GO THROUGH MY SEASONAL CYCLES.
BUT THE KEY TO THE DAY YOU'LL EMERGE IS WHEN THE SOIL TEMPERATURE REACHES 64 DEGREES FAHRENHEIT —THAT'S 18 DEGREES CELSIUS.
YOU KNOW, IT'S IRONIC, BUT IN MANY WAYS YOU ARE MORE MATURE AT THIS PHASE OF YOUR LIFE THAN YOU ARE AS AN ADULT...
How so?
WHEN YOU EMERGE AND TRANSFORM, ALL YOU WANT TO DO IS FORNICATE. WE COULD NEVER HAVE THESE SCINTILLATING CONVERSATIONS.
What's "fornicate" mean?
COPULATE.
?
HAVE SEX.
Oh. Yuck.

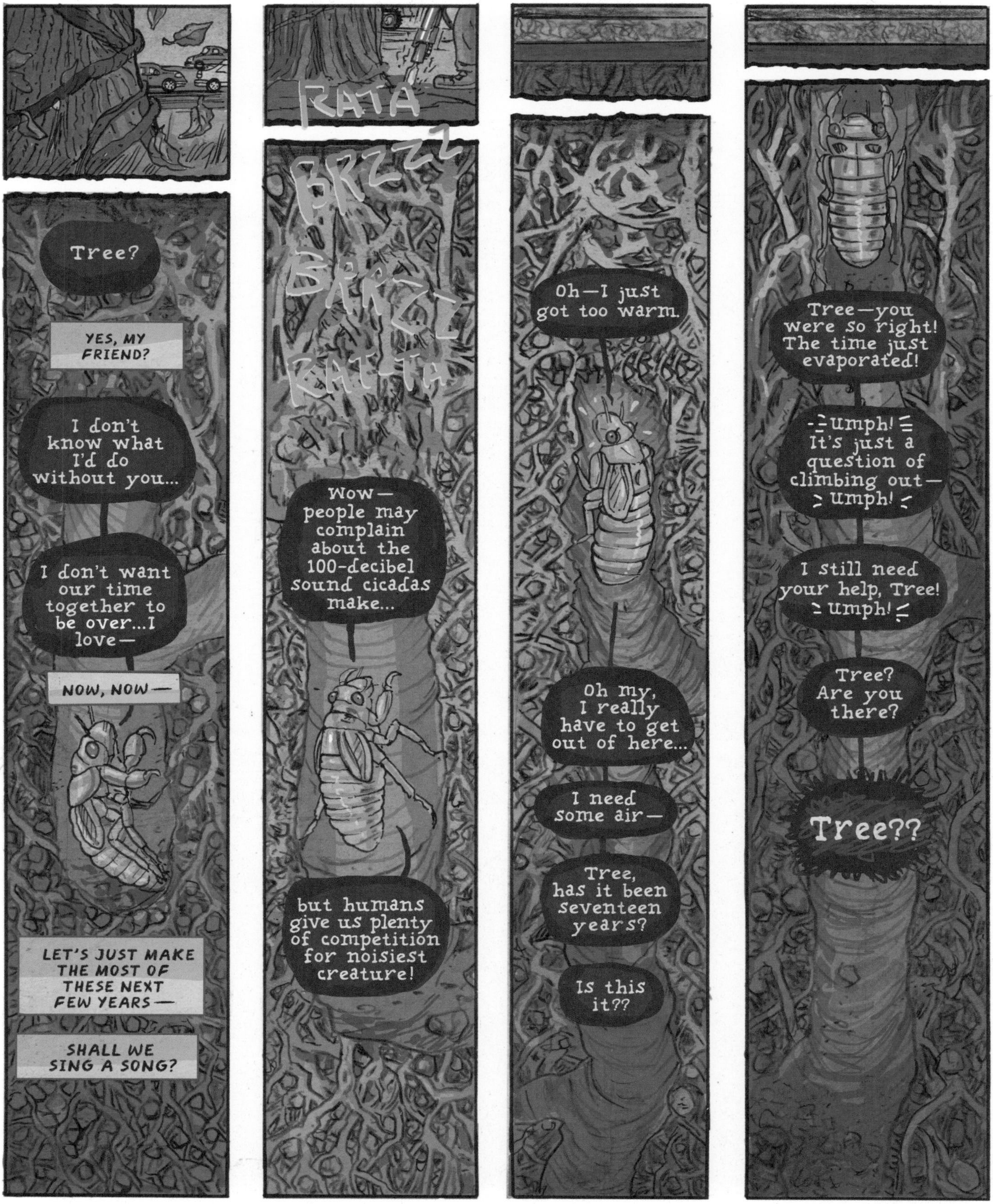

Tree?
YES, MY FRIEND?
I don't know what I'd do without you...
I don't want our time together to be over...I love—
NOW, NOW—
LET'S JUST MAKE THE MOST OF THESE NEXT FEW YEARS—
SHALL WE SING A SONG?
RATA
BRZZZ
BRRZZ
RAT-TA
Wow— people may complain about the 100-decibel sound cicadas make...
but humans give us plenty of competition for noisiest creature!
Oh—I just got too warm.
Oh my, I really have to get out of here...
I need some air—
Tree, has it been seventeen years?
Is this it??
Tree—you were so right! The time just evaporated!
Umph! It's just a question of climbing out— Umph!
I still need your help, Tree! Umph!
Tree? Are you there?
Tree??

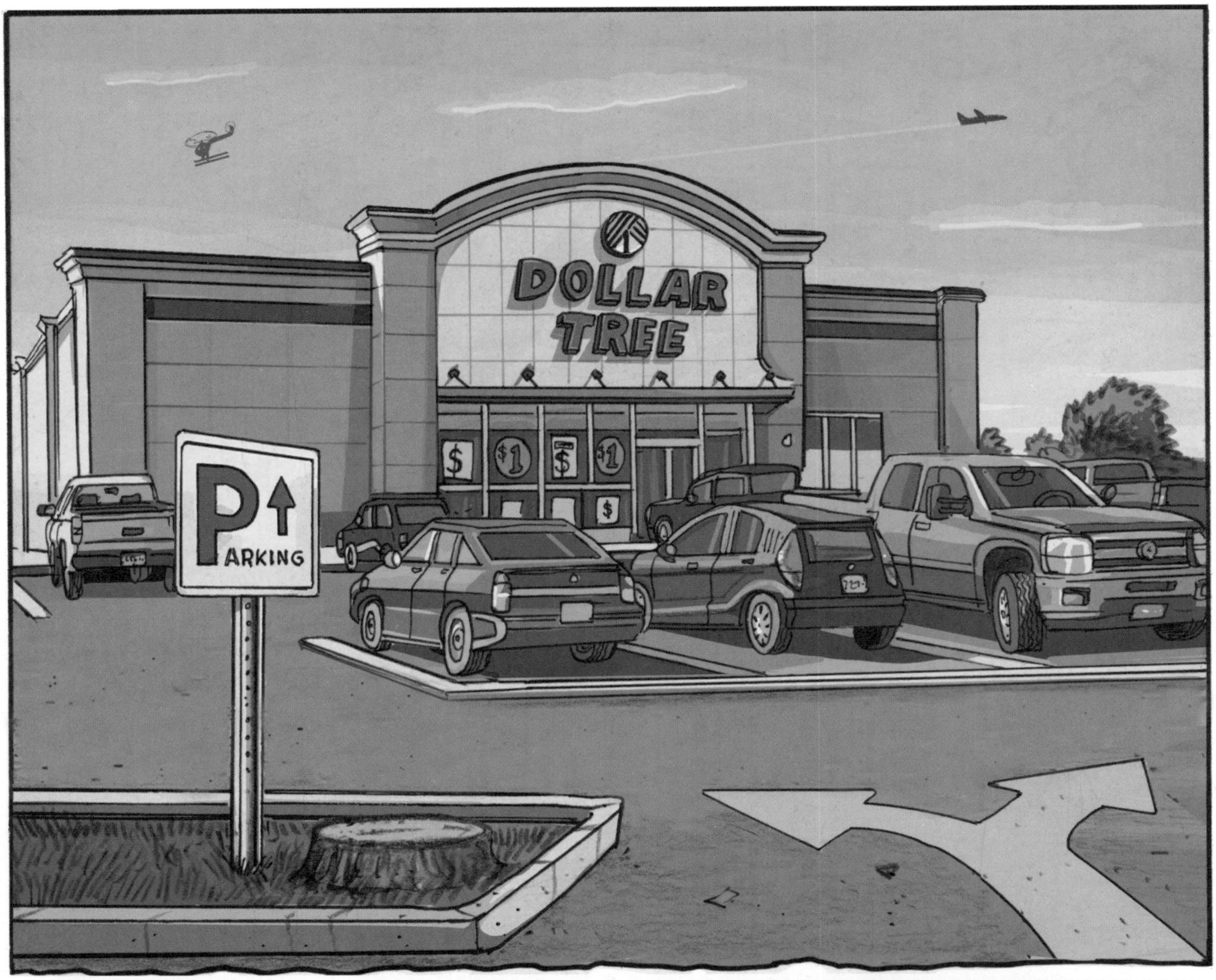
DOLLAR TREE
$1
PARKING

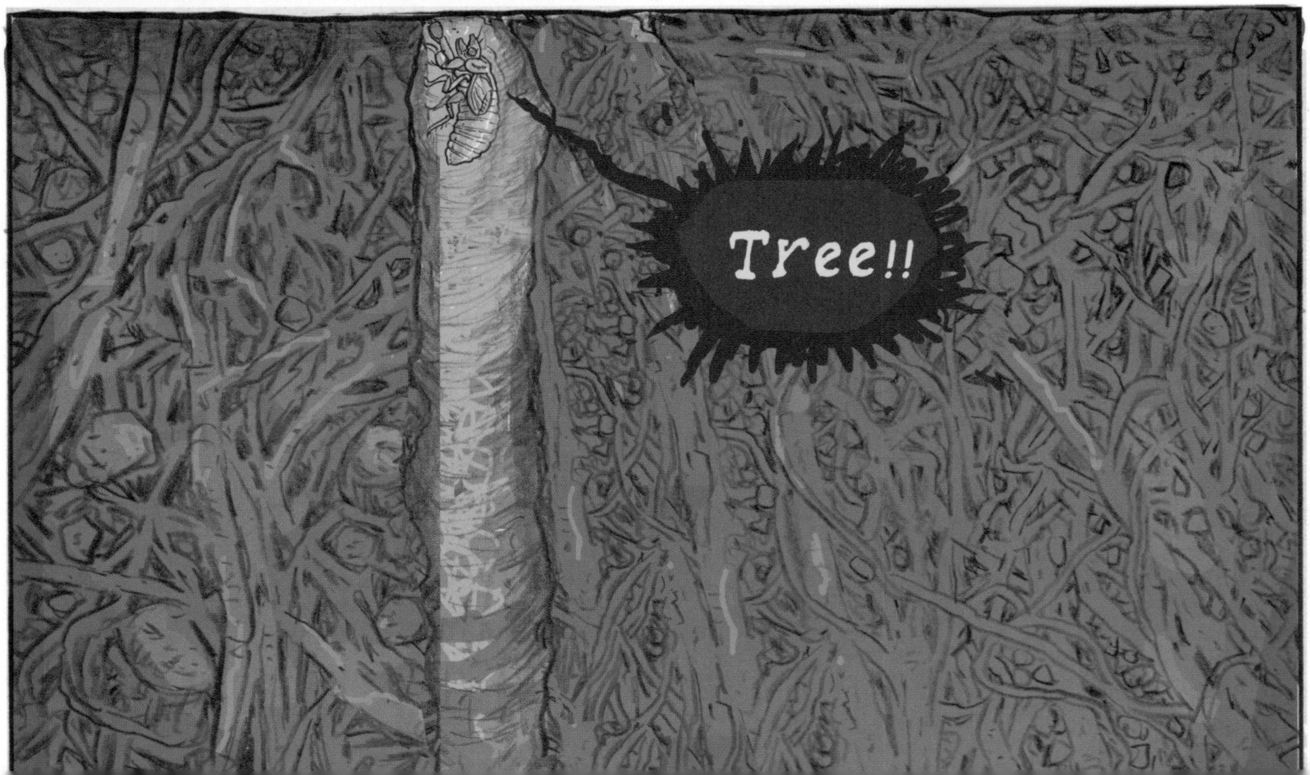
Tree!!

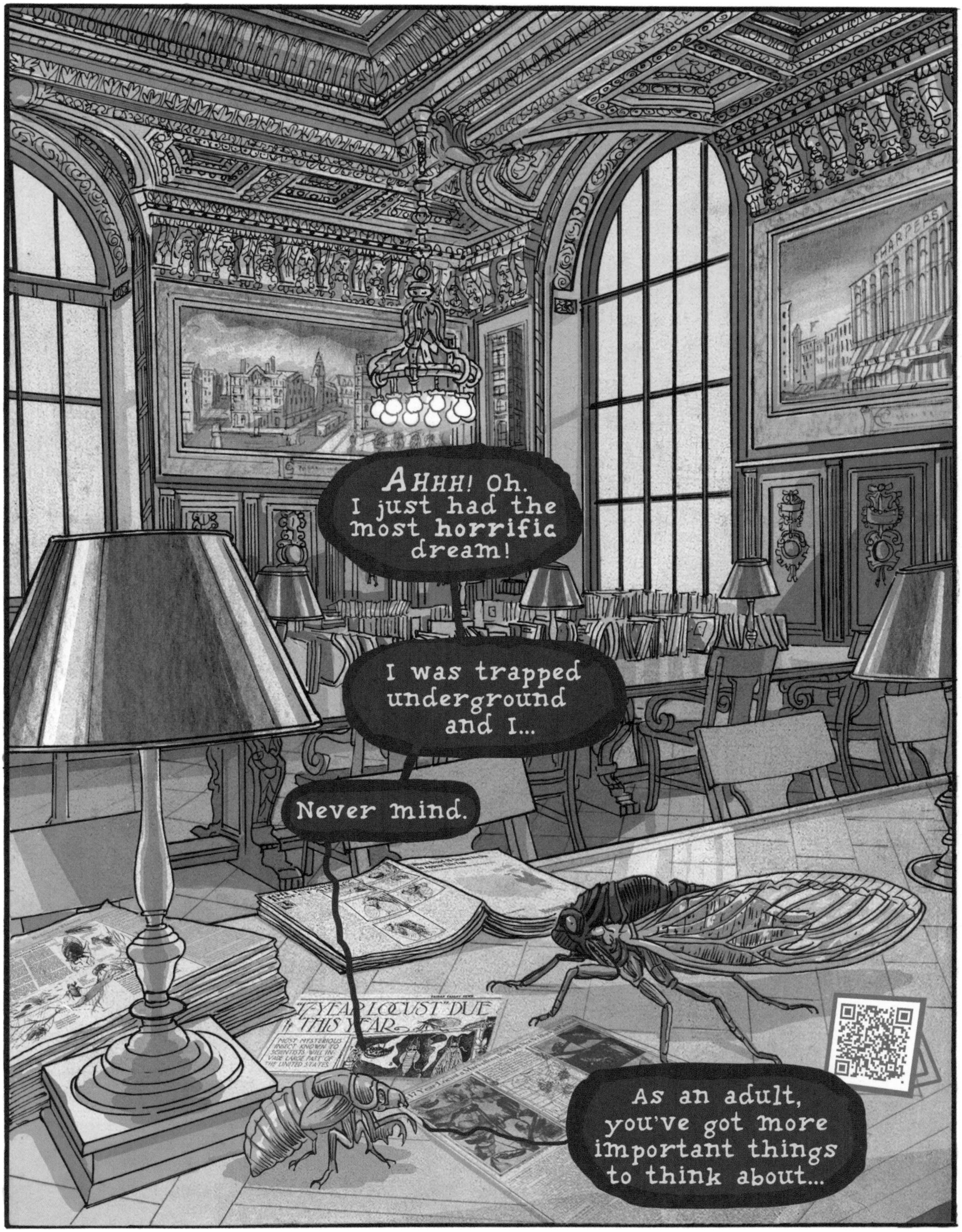
AHHH! Oh. I just had the most horrific dream!
I was trapped underground and I...
Never mind.
"17-YEAR LOCUST" DUE THIS YEAR
MOST MYSTERIOUS INSECT KNOWN TO SCIENTISTS WILL INVADE LARGE PART OF THE UNITED STATES
As an adult, you've got more important things to think about...

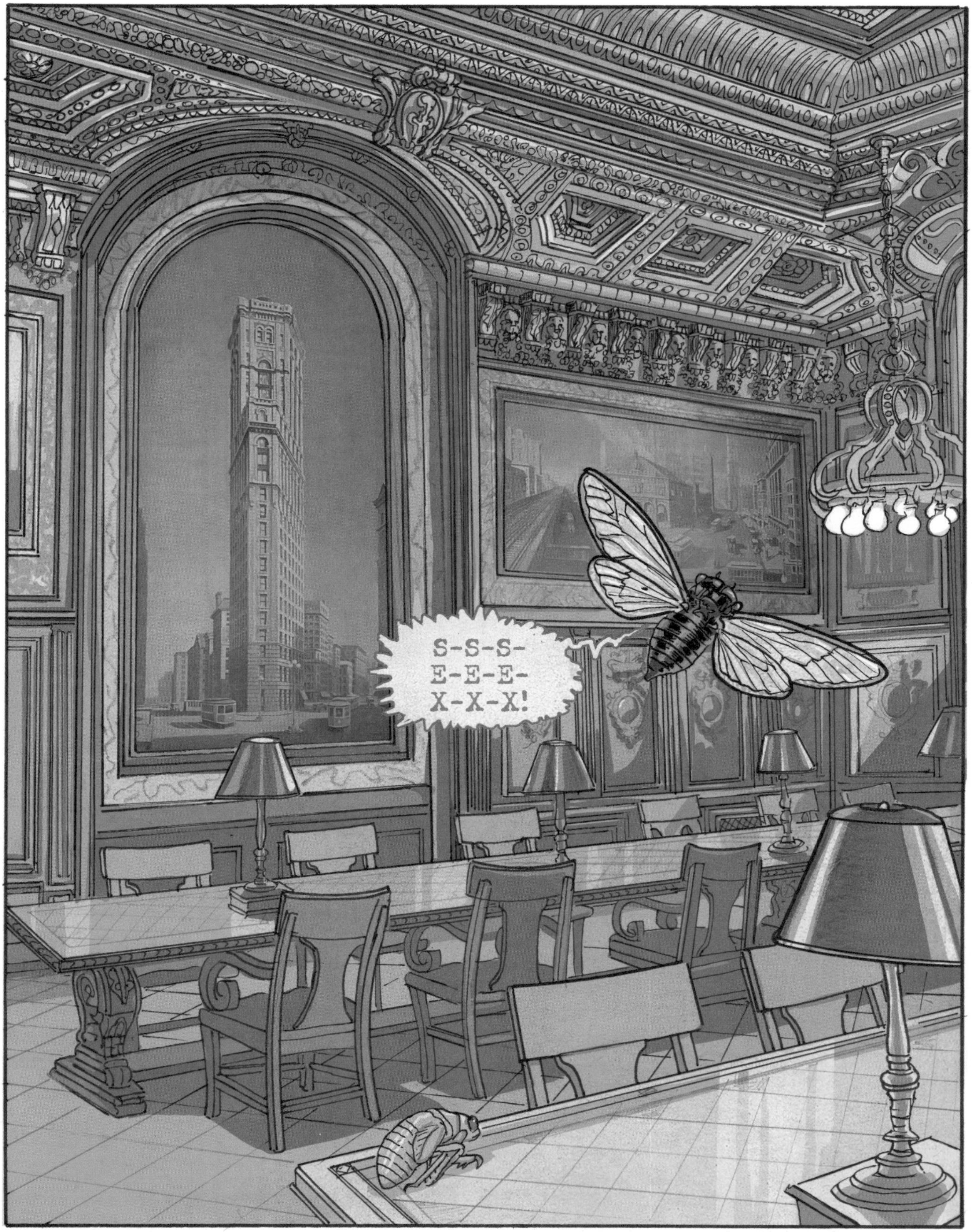
S-S-S-
E-E-E-
X-X-X!

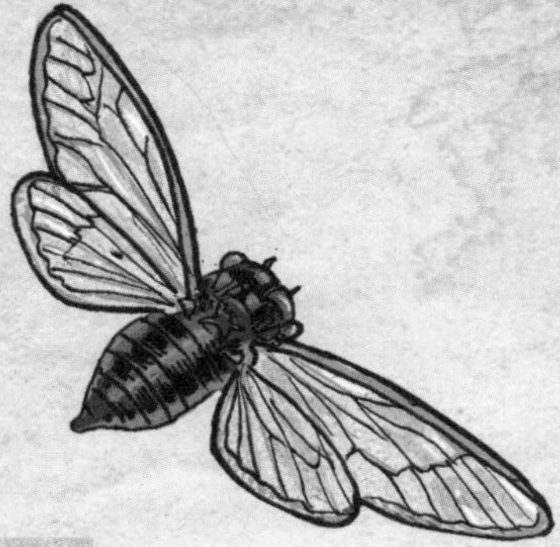

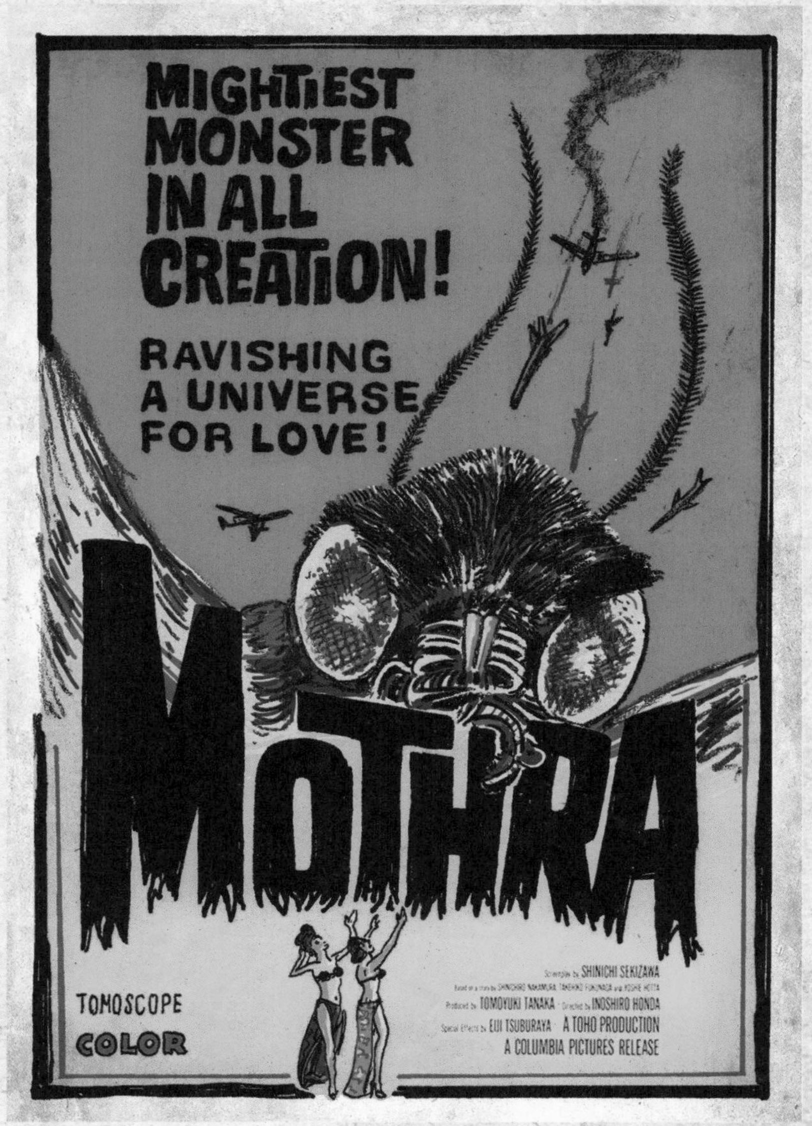
MIGHTIEST MONSTER IN ALL CREATION!
RAVISHING A UNIVERSE FOR LOVE!
MOTHRA
SHINICHI SEKIZAWA
Produced by TOMOYUKI TANAKA
INOSHIRO HONDA
EIJI TSUBURAYA
A TOHO PRODUCTION
A COLUMBIA PICTURES RELEASE
TOHOSCOPE
COLOR

Mothra was released in Japan on July 30,1961. An English dubbed version hit screens in the United States in 1962.

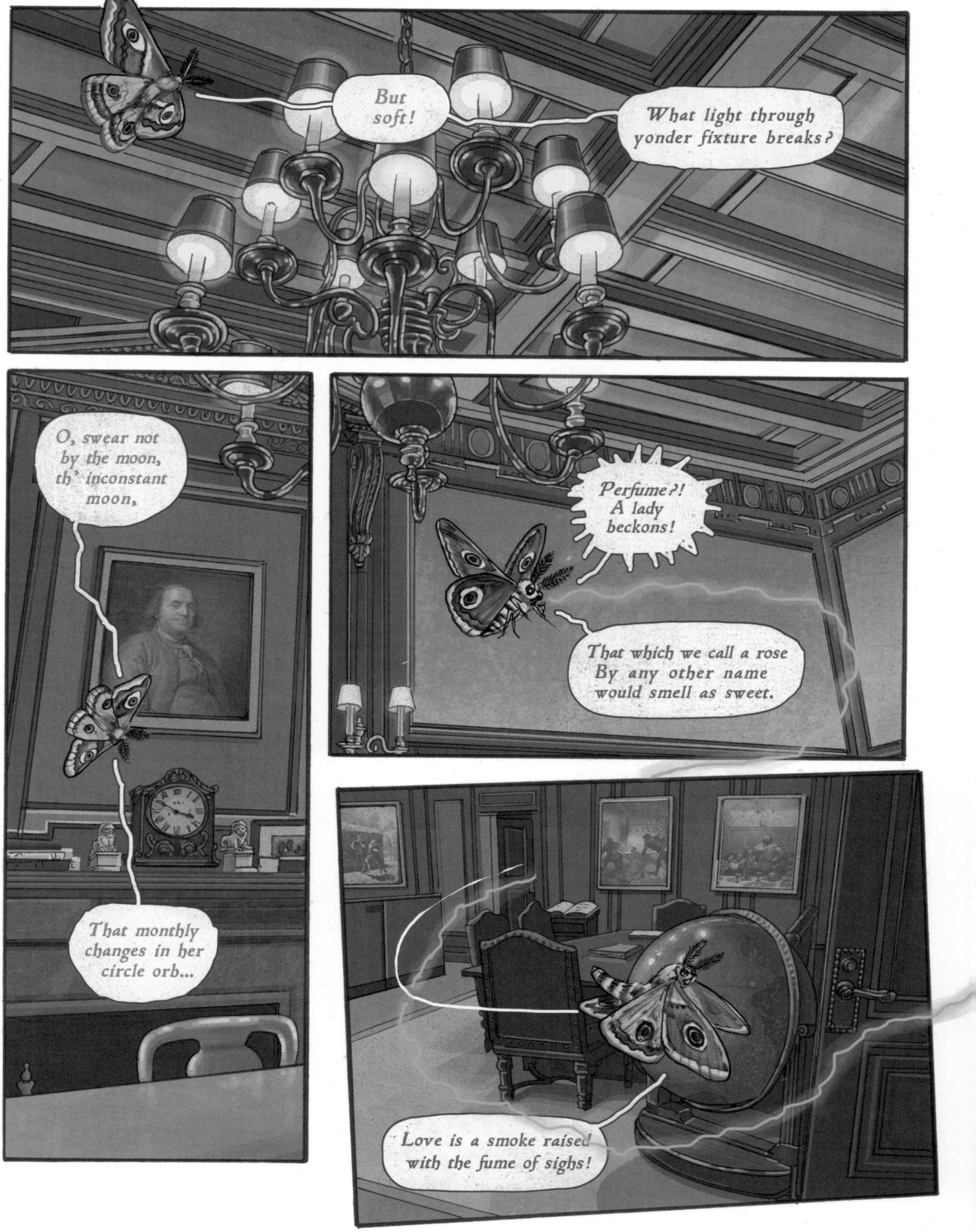
But soft!
What light through yonder fixture breaks?
O, swear not by the moon, th' inconstant moon,
That monthly changes in her circle orb...
Perfume?! A lady beckons!
That which we call a rose By any other name would smell as sweet.
Love is a smoke raised with the fume of sighs!

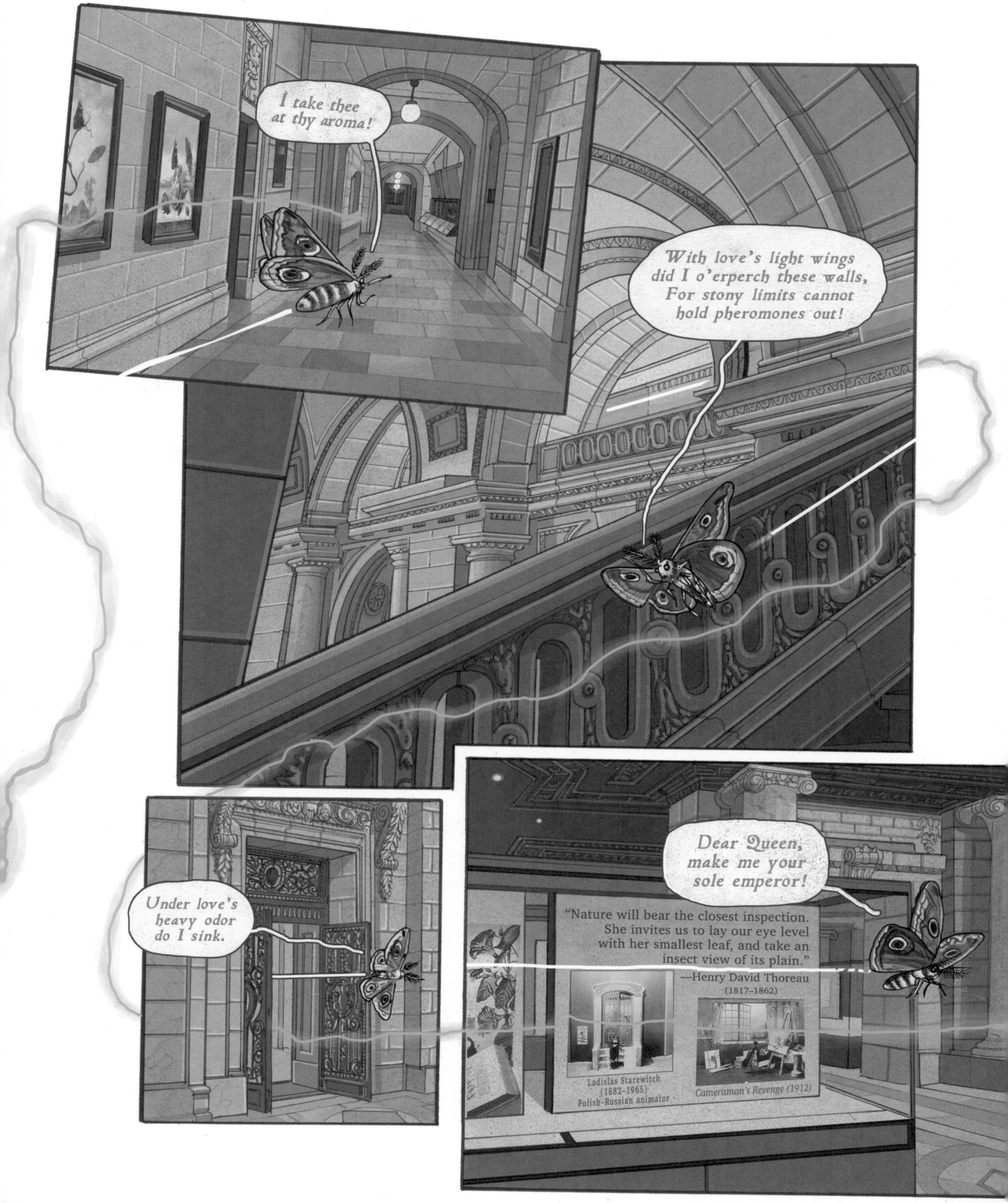
I take thee at thy aroma!
With love's light wings did I o'erperch these walls, For stony limits cannot hold pheromones out!
Under love's heavy odor do I sink.
Dear Queen, make me your sole emperor!
"Nature will bear the closest inspection. She invites us to lay our eye level with her smallest leaf, and take an insect view of its plain."
—Henry David Thoreau (1817–1862)
Ladislas Starewitch (1882–1965) Polish-Russian animator
Cameraman's Revenge (1912)

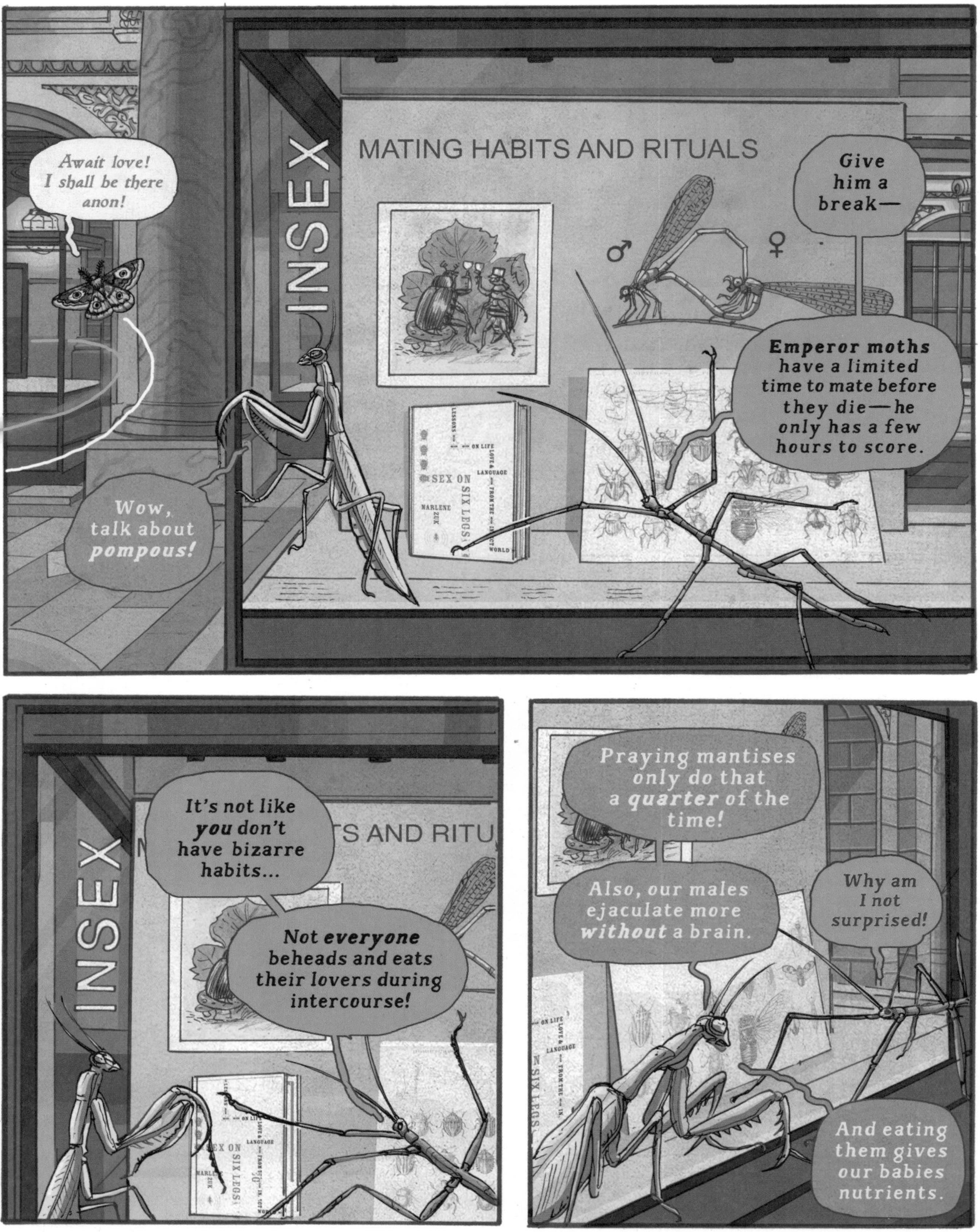
INSEX
MATING HABITS AND RITUALS
Await love! I shall be there anon!
Wow, talk about *pompous!*
SEX ON SIX LEGS
MARLENE ZUK
Give him a break—
Emperor moths have a limited time to mate before they die—he *only* has a few hours to score.
INSEX
It's not like *you* don't have bizarre habits...
Not *everyone* beheads and eats their lovers during intercourse!
Praying mantises only do that a *quarter* of the time!
Also, our males ejaculate more *without* a brain.
Why am I not surprised!
And eating them gives our babies nutrients.

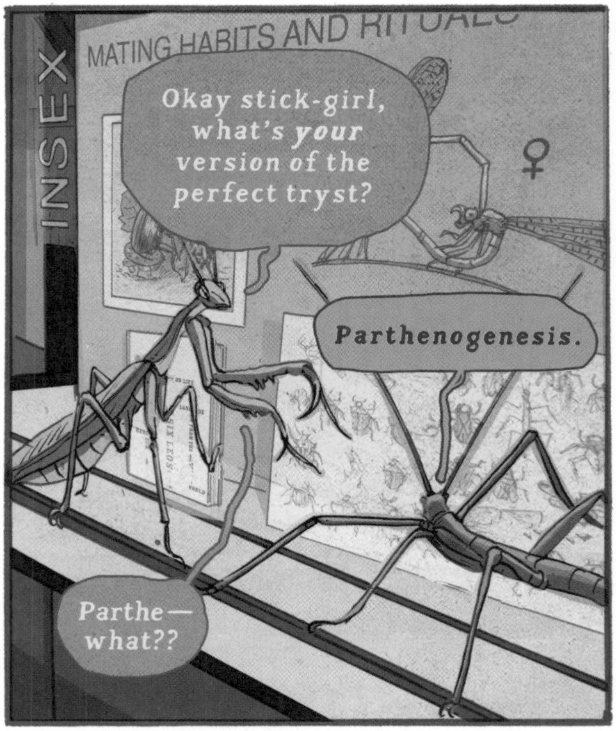
INSEX
MATING HABITS AND
Okay stick-girl, what's *your* version of the perfect tryst?
Parthenogenesis.
Parthe— what??

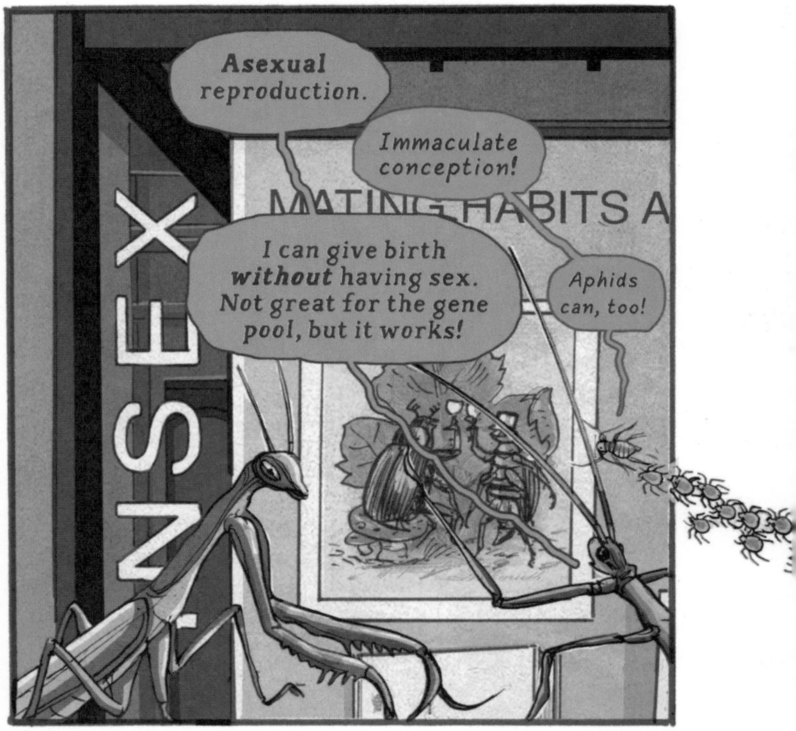
INSEX
Asexual reproduction.
Immaculate conception!
I can give birth ***without*** having sex. Not great for the gene pool, but it works!
Aphids can, too!

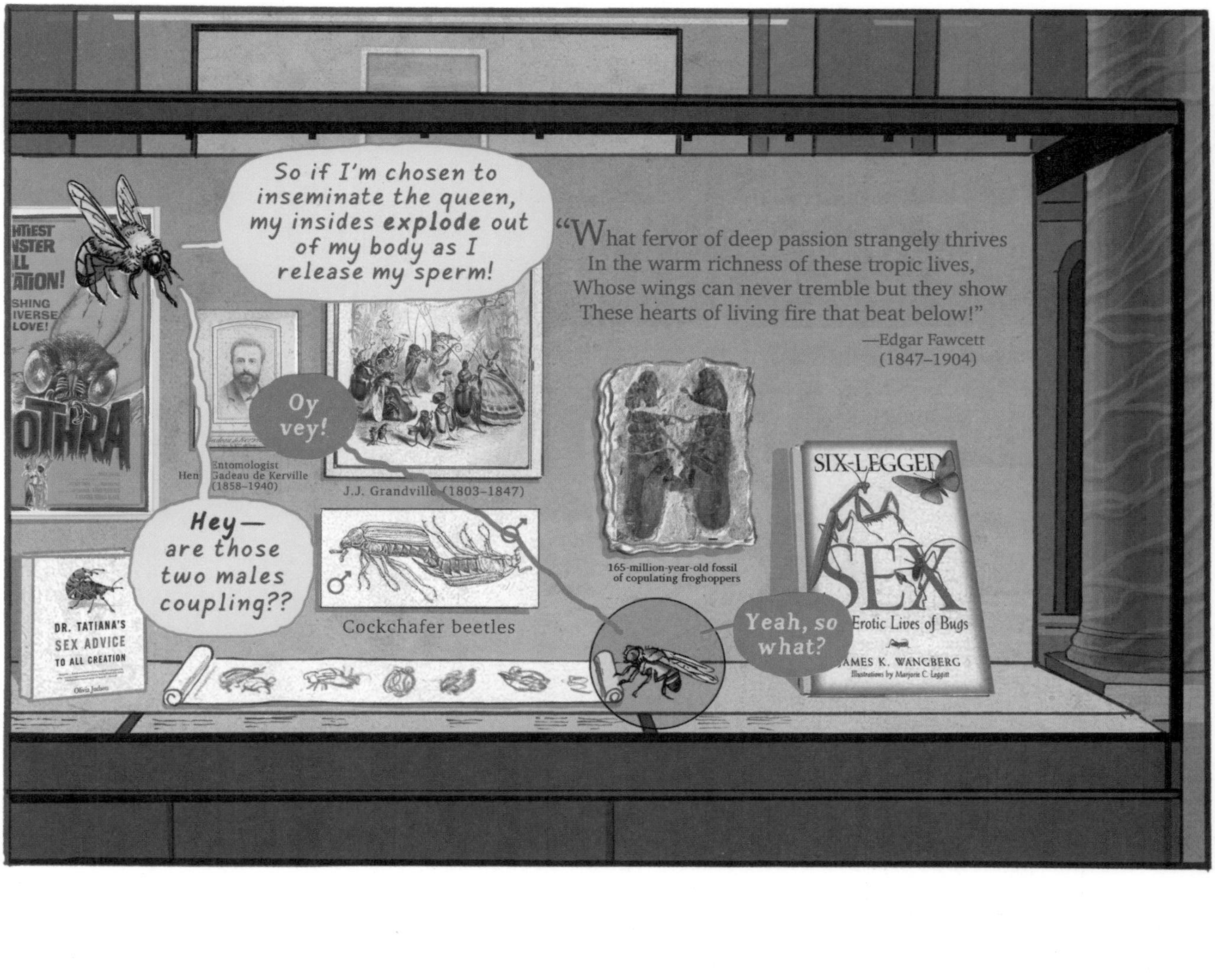
So if I'm chosen to inseminate the queen, my insides **explode** out of my body as I release my sperm!
"What fervor of deep passion strangely thrives
In the warm richness of these tropic lives,
Whose wings can never tremble but they show
These hearts of living fire that beat below!"
—Edgar Fawcett
(1847–1904)
Oy vey!
Entomologist
Gadeau de Kerville
(1858–1940)
J.J. Grandville (1803–1847)
Hey— are those two males coupling??
Cockchafer beetles
165-million-year-old fossil of copulating froghoppers
SIX-LEGGED
SEX
Erotic Lives of Bugs
JAMES K. WANGBERG
Illustrations by Marjorie C. Leggitt
Yeah, so what?
DR. TATIANA'S
SEX ADVICE
TO ALL CREATION
Olivia Judson
OTHRA

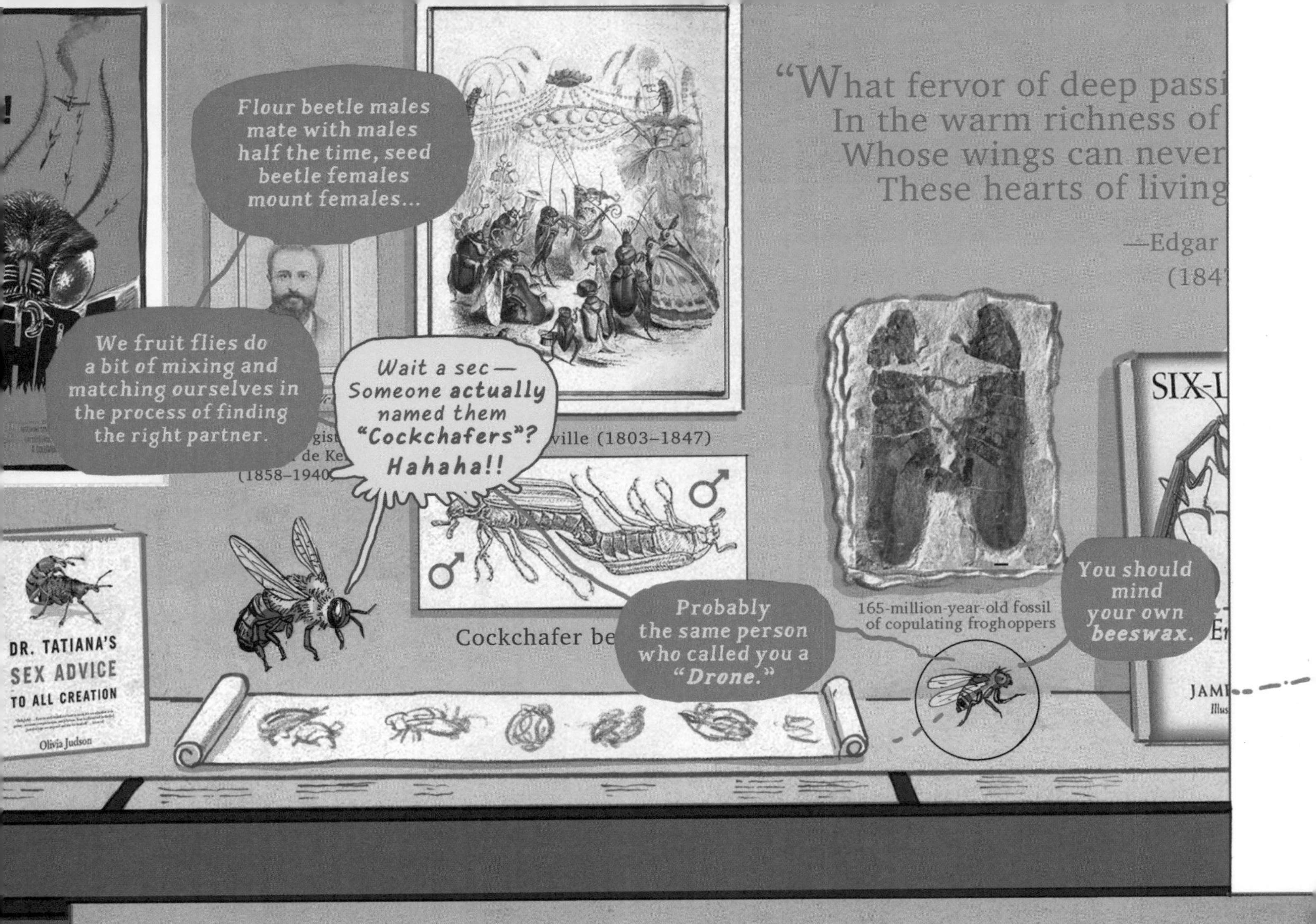

Flour beetle males mate with males half the time, seed beetle females mount females...
We fruit flies do a bit of mixing and matching ourselves in the process of finding the right partner.
Wait a sec— Someone actually named them "Cockchafers"? Hahaha!!
Probably the same person who called you a "Drone."
You should mind your own beeswax.
"What fervor of deep passi
In the warm richness of
Whose wings can never
These hearts of living
—Edgar
(184
ville (1803–1847)
(1858–1940
♂
♂
Cockchafer be
165-million-year-old fossil of copulating froghoppers
DR. TATIANA'S SEX ADVICE TO ALL CREATION
Olivia Judson
SIX-L
JAM

Soapberry bugs

To ward off competitors, the males will couple with a female until she is ready to lay eggs. Sometimes for **weeks.**

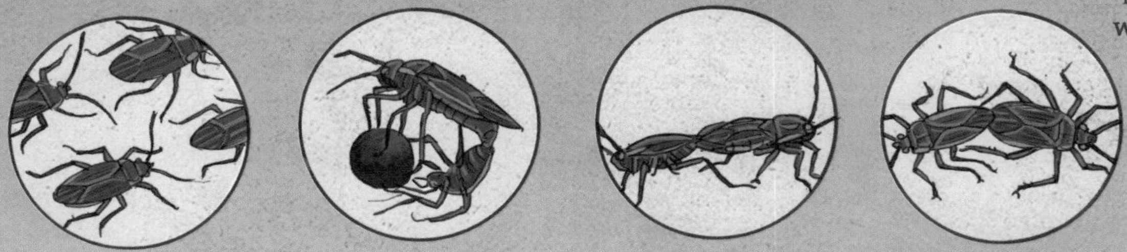

The Water Boatman

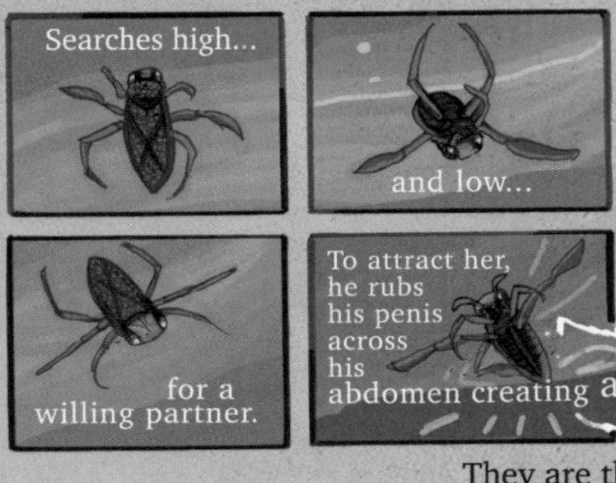

Searches high...
and low...
for a willing partner.
To attract her, he rubs his penis across his abdomen creating

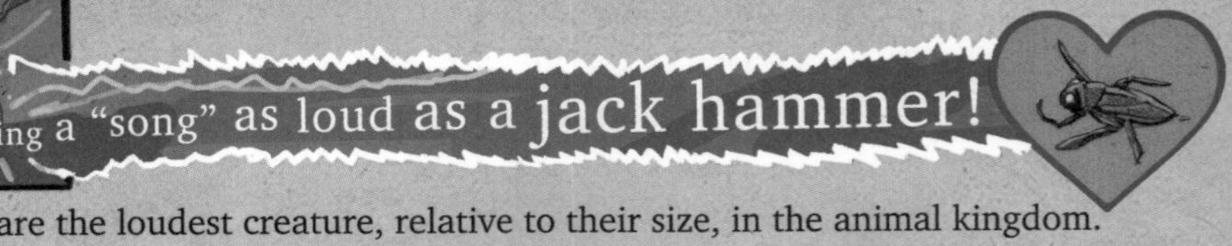

a "song" as loud as a jack hammer!

They are the loudest creature, relative to their size, in the animal kingdom.

Fruit Flies

To court the female, the male engages in a number of actions to gain her acceptance:

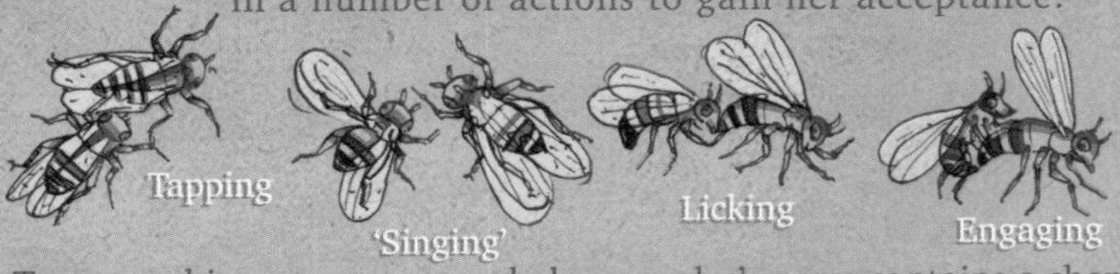

Tapping
'Singing'
Licking
Engaging

To ensure his genes are passed along, males' sperm contains a chemical that reduces the likelihood that females will mate with any other males.

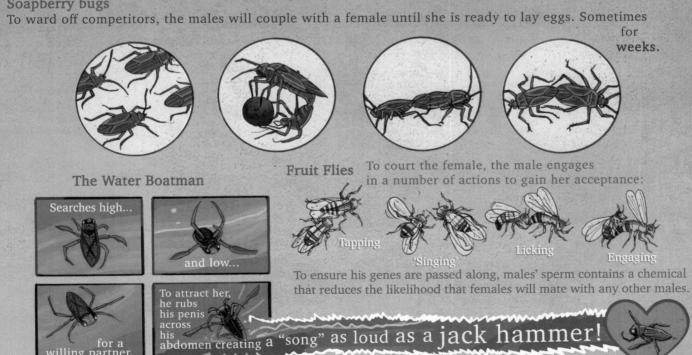

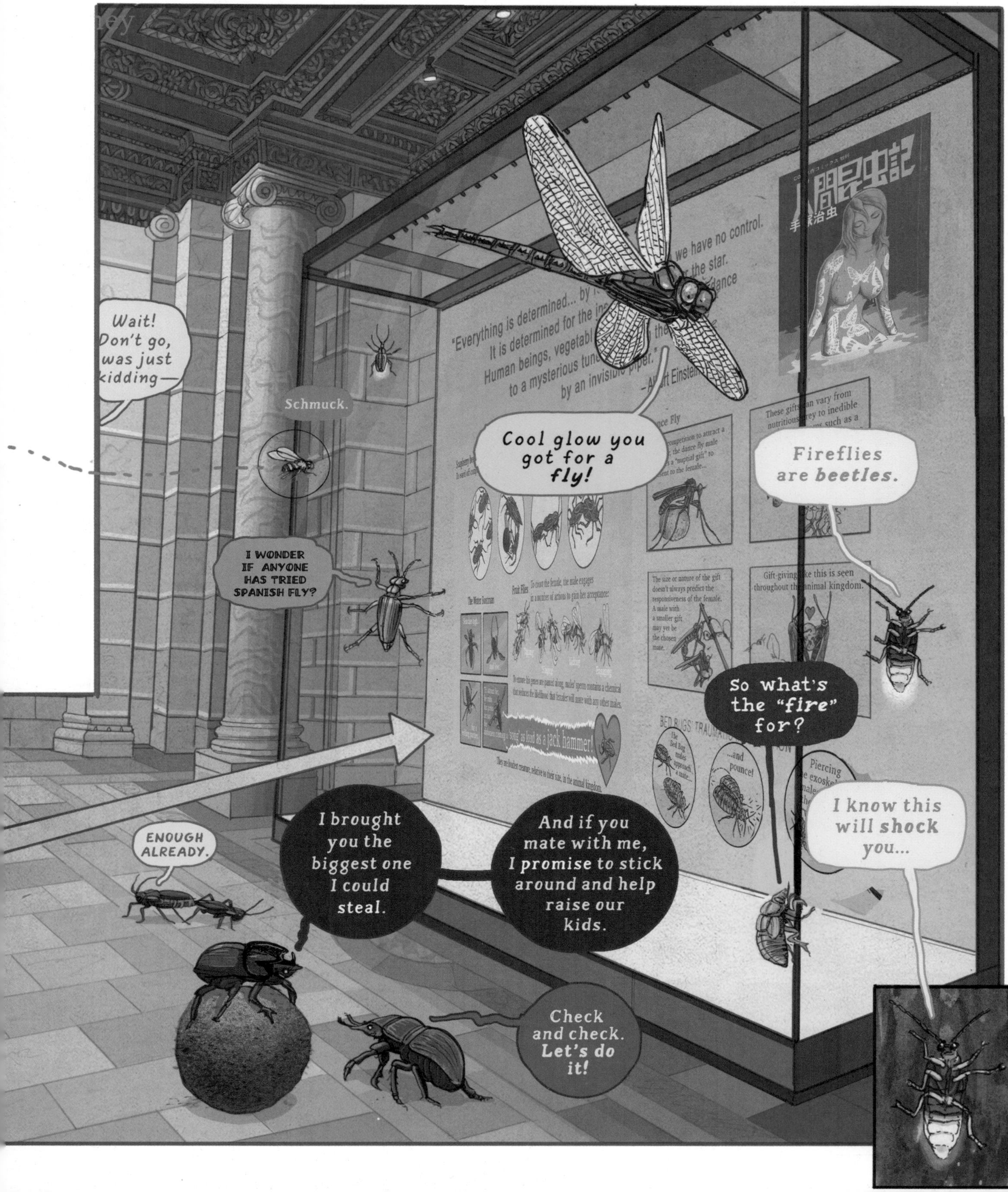
Wait! Don't go, was just kidding—
Schmuck.
I WONDER IF ANYONE HAS TRIED SPANISH FLY?
Cool glow you got for a *fly!*
Fireflies are ***beetles.***
So what's the "*fire*" for?
ENOUGH ALREADY.
I brought you the biggest one I could steal.
And if you mate with me, I promise to stick around and help raise our kids.
Check and check. **Let's do it!**
I know this will **shock** you...

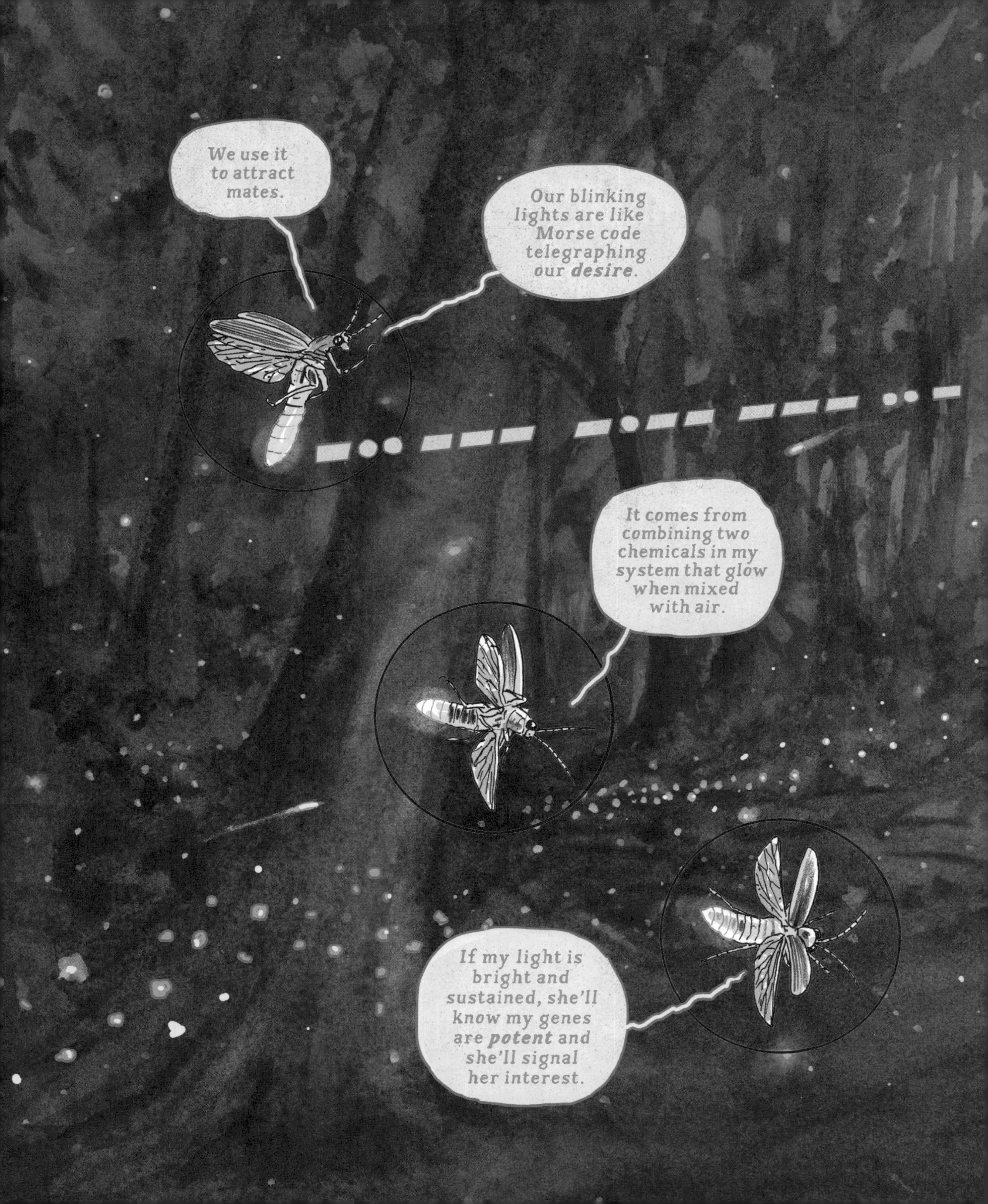
We use it to attract mates.
Our blinking lights are like Morse code telegraphing our *desire*.
It comes from combining two chemicals in my system that glow when mixed with air.
If my light is bright and sustained, she'll know my genes are *potent* and she'll signal her interest.

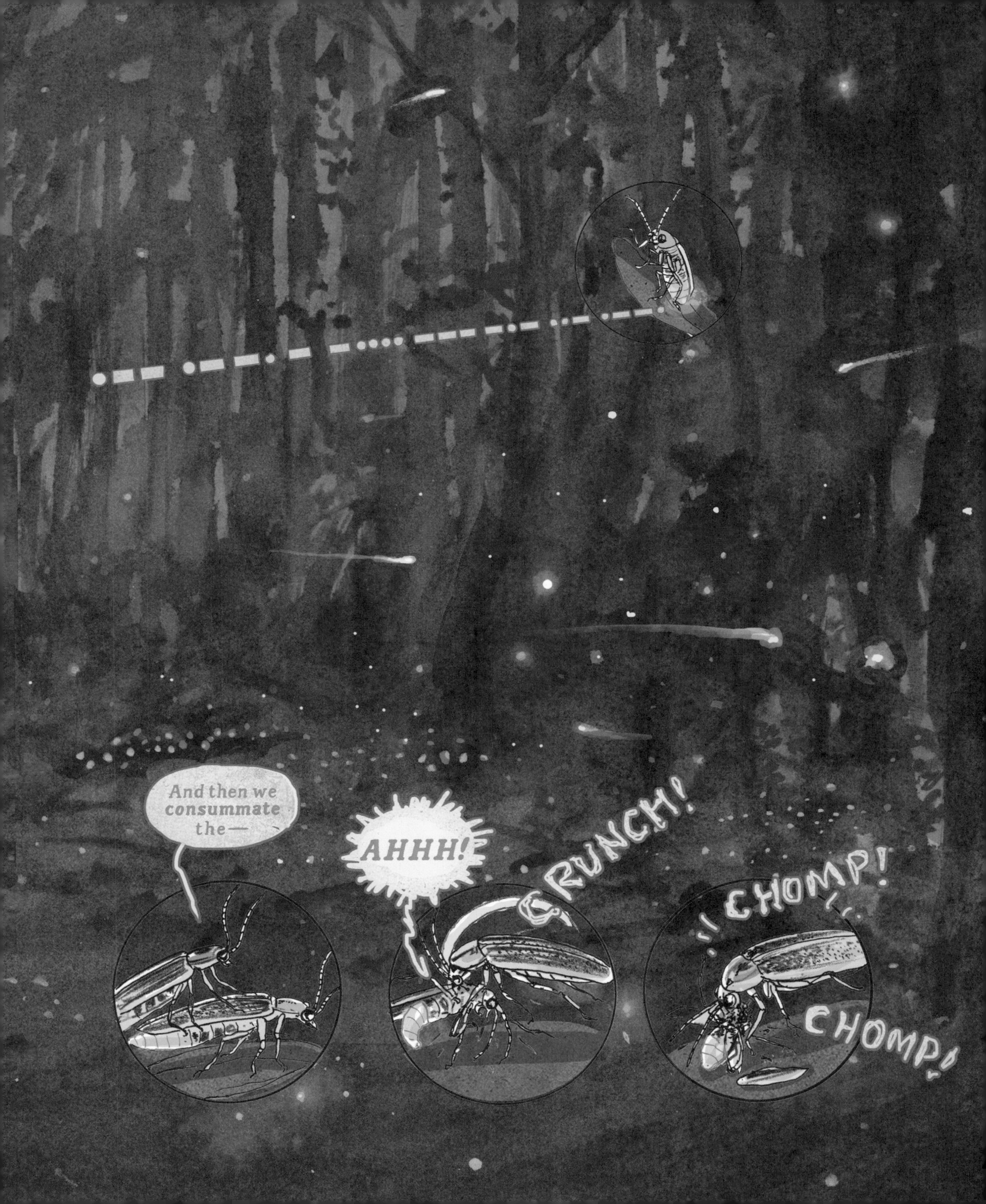
And then we consummate the—
AHHH!
CRUNCH!
CHOMP!
CHOMP!

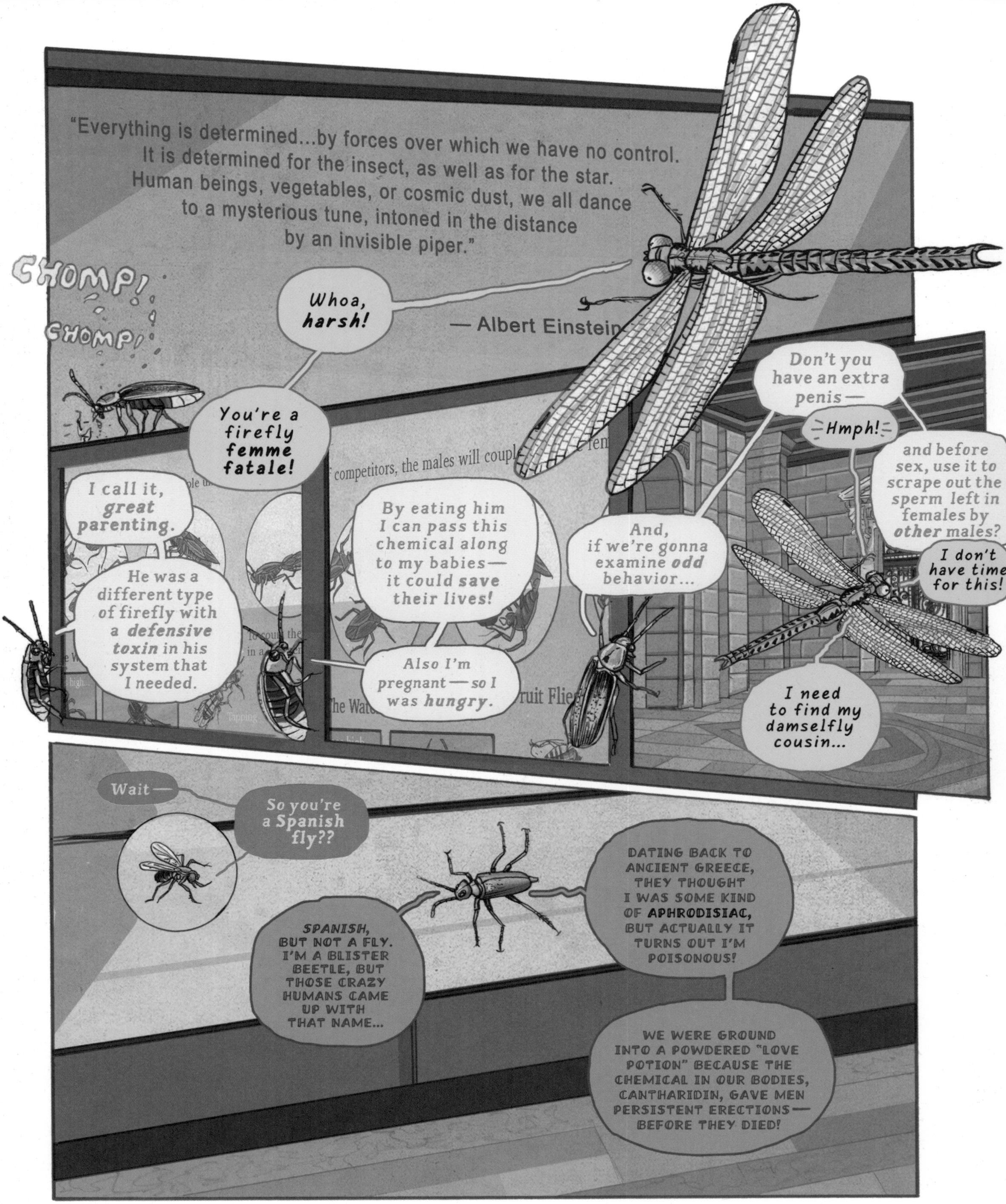

"Everything is determined...by forces over which we have no control. It is determined for the insect, as well as for the star. Human beings, vegetables, or cosmic dust, we all dance to a mysterious tune, intoned in the distance by an invisible piper."
— Albert Einstein
CHOMP!
CHOMP!
Whoa, harsh!
You're a firefly femme fatale!
I call it, great parenting.
He was a different type of firefly with a defensive toxin in his system that I needed.
By eating him I can pass this chemical along to my babies— it could save their lives!
Also I'm pregnant—so I was hungry.
And, if we're gonna examine odd behavior...
Don't you have an extra penis—
Hmph!
and before sex, use it to scrape out the sperm left in females by other males?
I don't have time for this!
I need to find my damselfly cousin...
Wait—
So you're a Spanish fly??
SPANISH, BUT NOT A FLY. I'M A BLISTER BEETLE, BUT THOSE CRAZY HUMANS CAME UP WITH THAT NAME...
DATING BACK TO ANCIENT GREECE, THEY THOUGHT I WAS SOME KIND OF APHRODISIAC, BUT ACTUALLY IT TURNS OUT I'M POISONOUS!
WE WERE GROUND INTO A POWDERED "LOVE POTION" BECAUSE THE CHEMICAL IN OUR BODIES, CANTHARIDIN, GAVE MEN PERSISTENT ERECTIONS— BEFORE THEY DIED!

Dance Fly

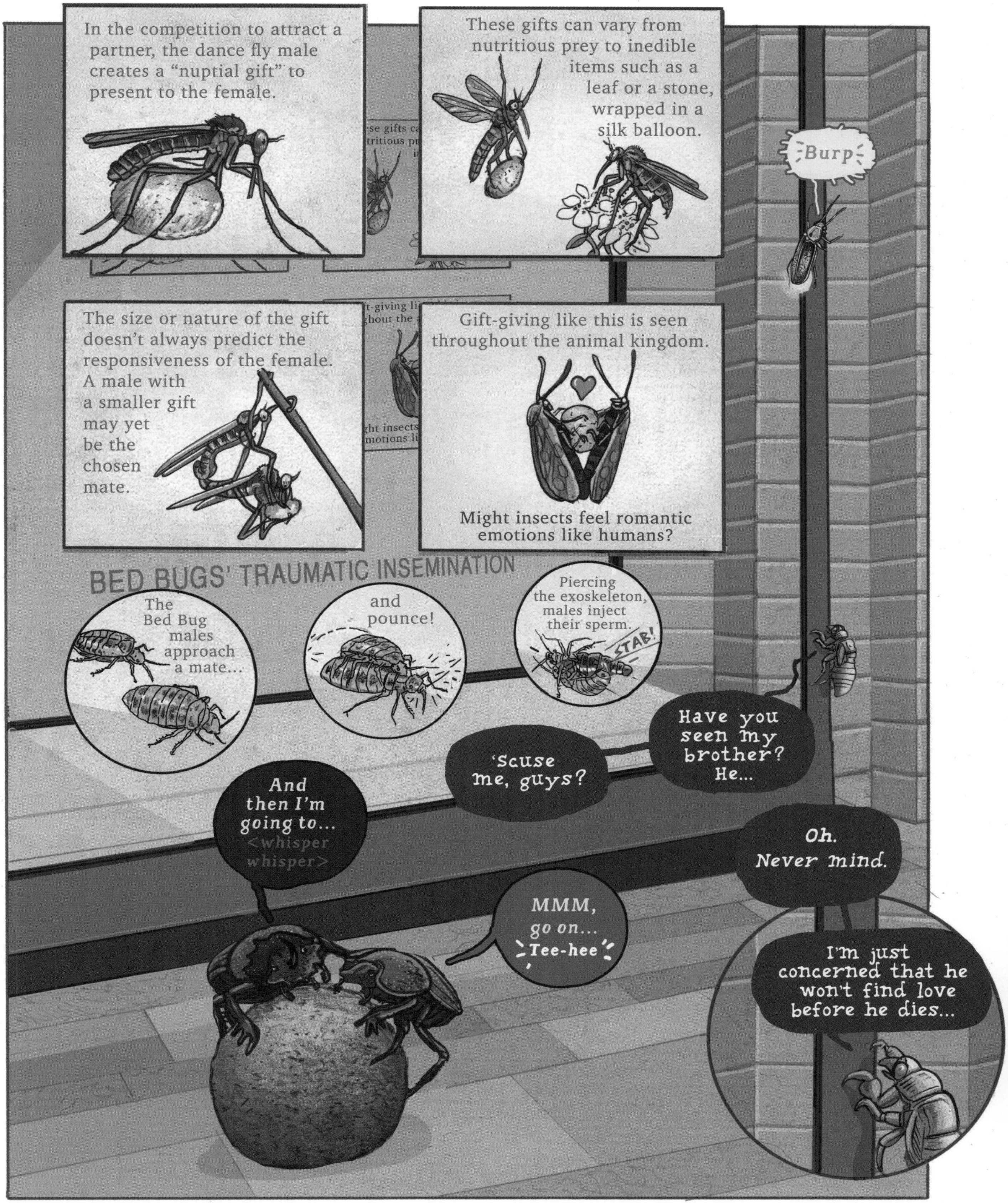
In the competition to attract a partner, the dance fly male creates a "nuptial gift" to present to the female.
These gifts can vary from nutritious prey to inedible items such as a leaf or a stone, wrapped in a silk balloon.
Burp
The size or nature of the gift doesn't always predict the responsiveness of the female. A male with a smaller gift may yet be the chosen mate.
Gift-giving like this is seen throughout the animal kingdom.
Might insects feel romantic emotions like humans?
BED BUGS' TRAUMATIC INSEMINATION
The Bed Bug males approach a mate...
and pounce!
Piercing the exoskeleton, males inject their sperm.
STAB!
Have you seen my brother? He...
'Scuse me, guys?
And then I'm going to... <whisper whisper>
Oh. Never mind.
MMM, go on... Tee-hee
I'm just concerned that he won't find love before he dies...

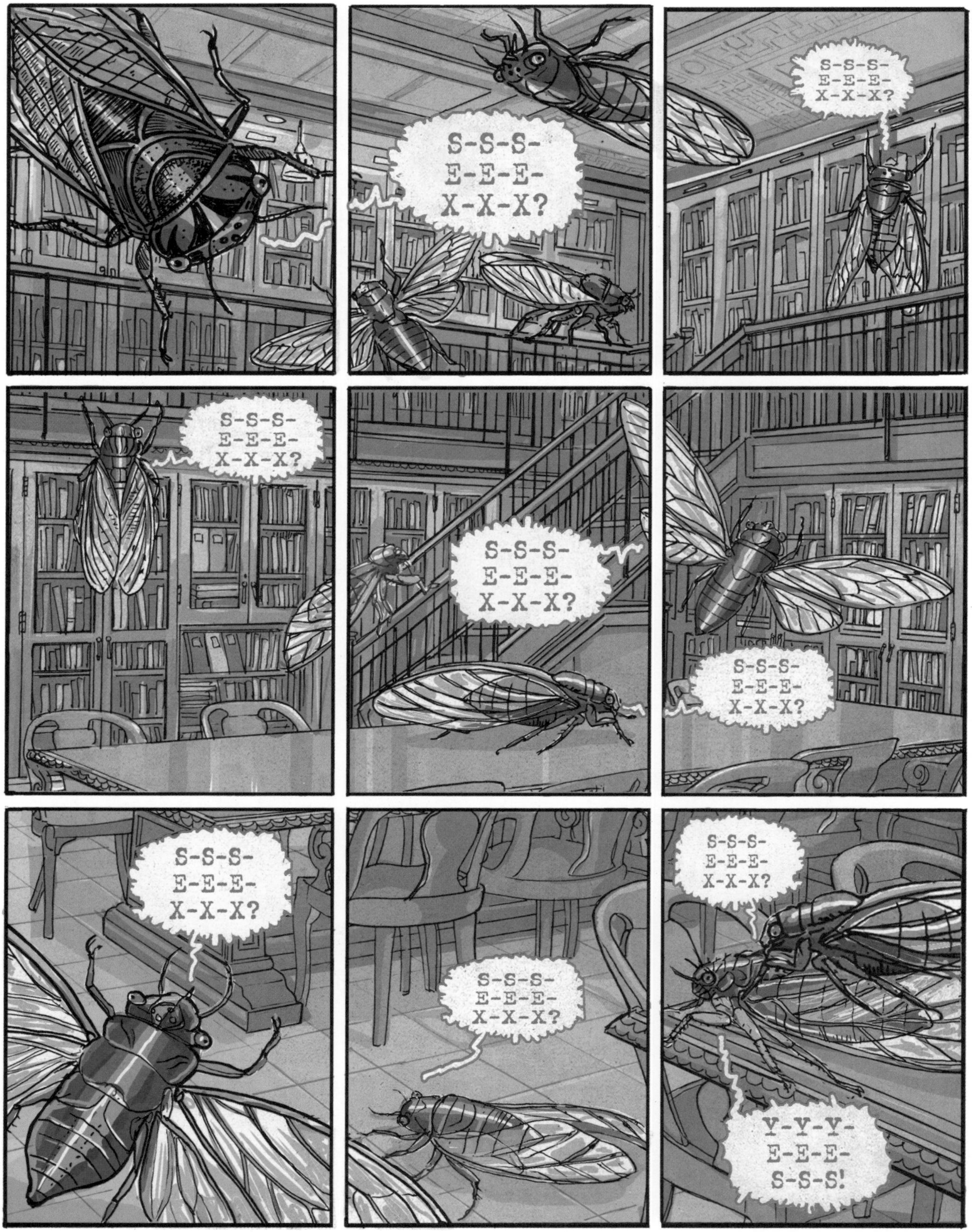
S-S-S-
E-E-E-
X-X-X?
S-S-S-
E-E-E-
X-X-X?
S-S-S-
E-E-E-
X-X-X?
S-S-S-
E-E-E-
X-X-X?
S-S-S-
E-E-E-
X-X-X?
S-S-S-
E-E-E-
X-X-X?
S-S-S-
E-E-E-
X-X-X?
S-S-S-
E-E-E-
X-X-X?
Y-Y-Y-
E-E-E-
S-S-S!

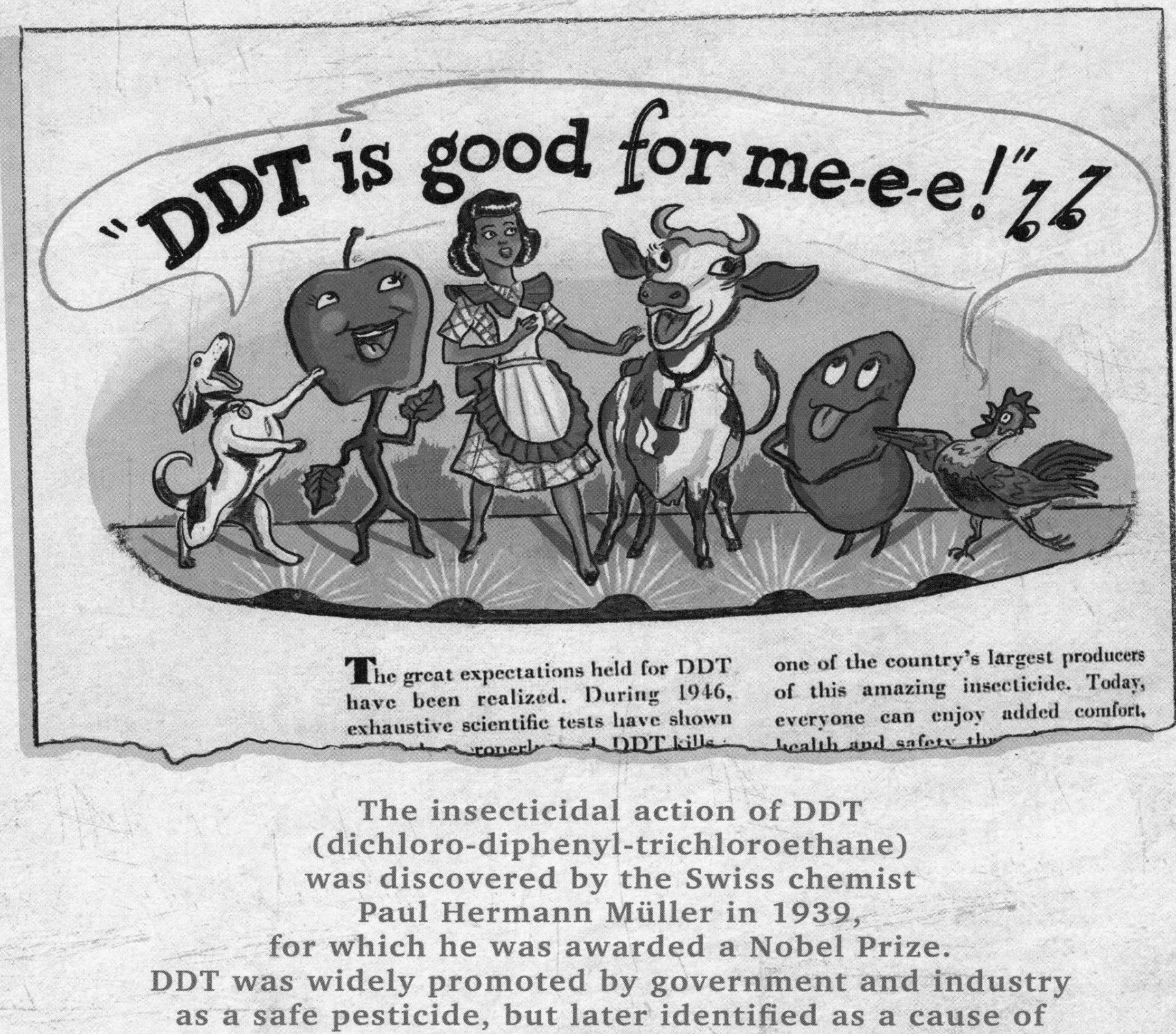
"DDT is good for me-e-e!"
The great expectations held for DDT have been realized. During 1946, exhaustive scientific tests have shown
one of the country's largest producers of this amazing insecticide. Today, everyone can enjoy added comfort,

The insecticidal action of DDT
(dichloro-diphenyl-trichloroethane)
was discovered by the Swiss chemist
Paul Hermann Müller in 1939,
for which he was awarded a Nobel Prize.
DDT was widely promoted by government and industry
as a safe pesticide, but later identified as a cause of
cancer in humans and devastating to the entire ecosystem.
After public outcry, DDT was finally banned in
the USA in 1972 but continued to be sold abroad.

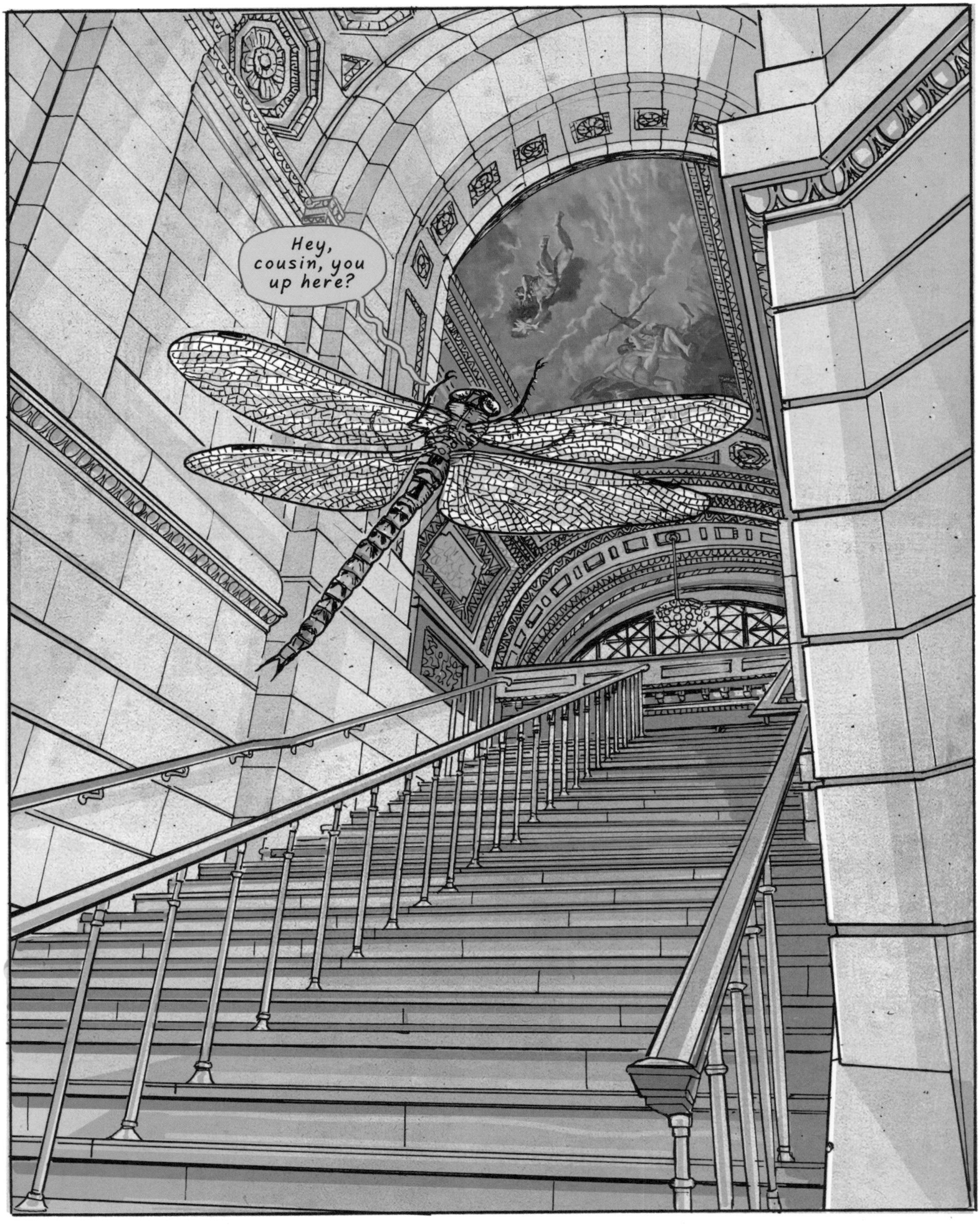
Hey, cousin, you up here?

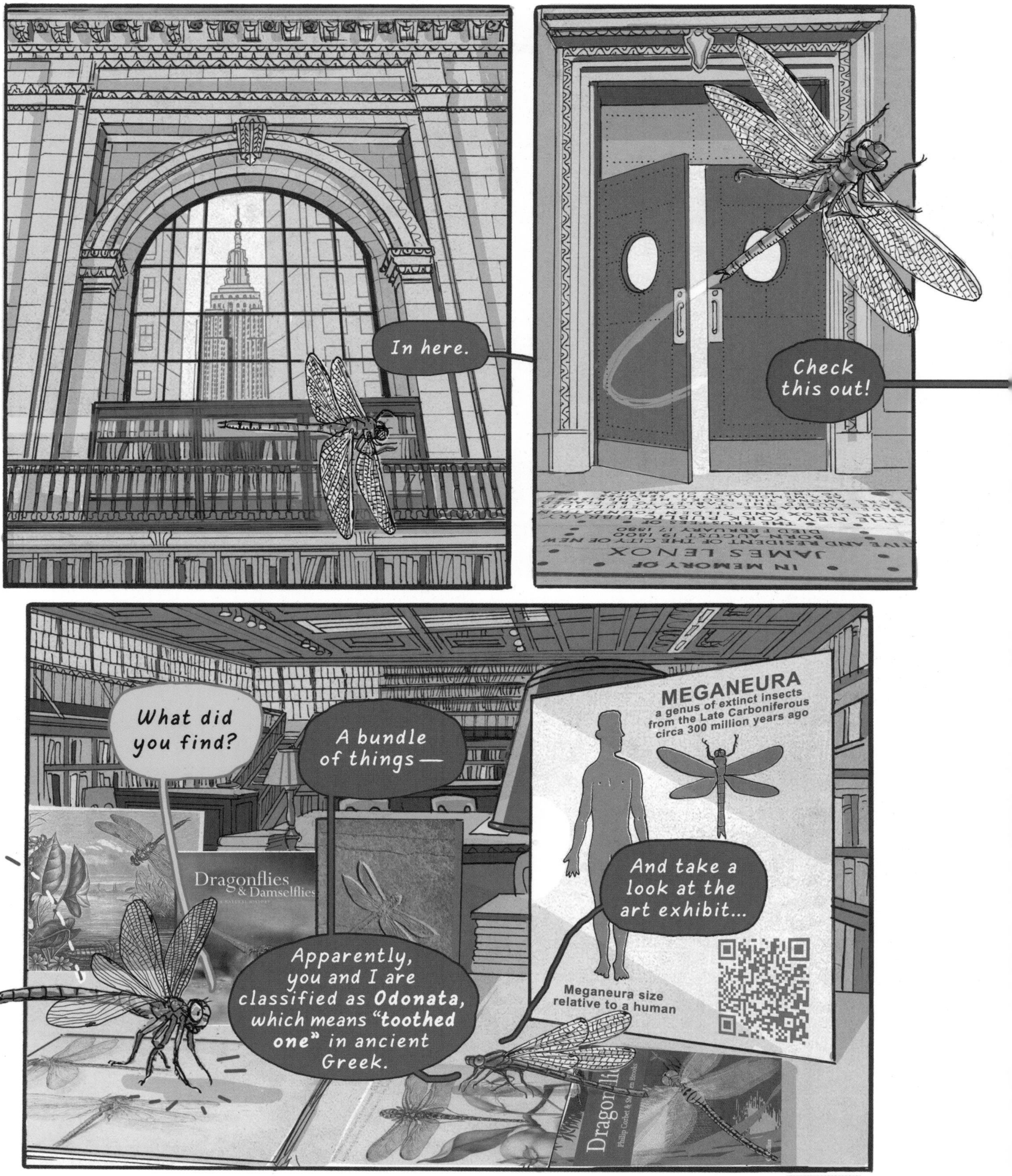

In here.
Check this out!
IN MEMORY OF
JAMES LENOX
What did you find?
A bundle of things —
Apparently, you and I are classified as **Odonata**, which means "**toothed one**" in ancient Greek.
And take a look at the art exhibit...
Dragonflies & Damselflies
MEGANEURA
a genus of extinct insects from the Late Carboniferous circa 300 million years ago
Meganeura size relative to a human

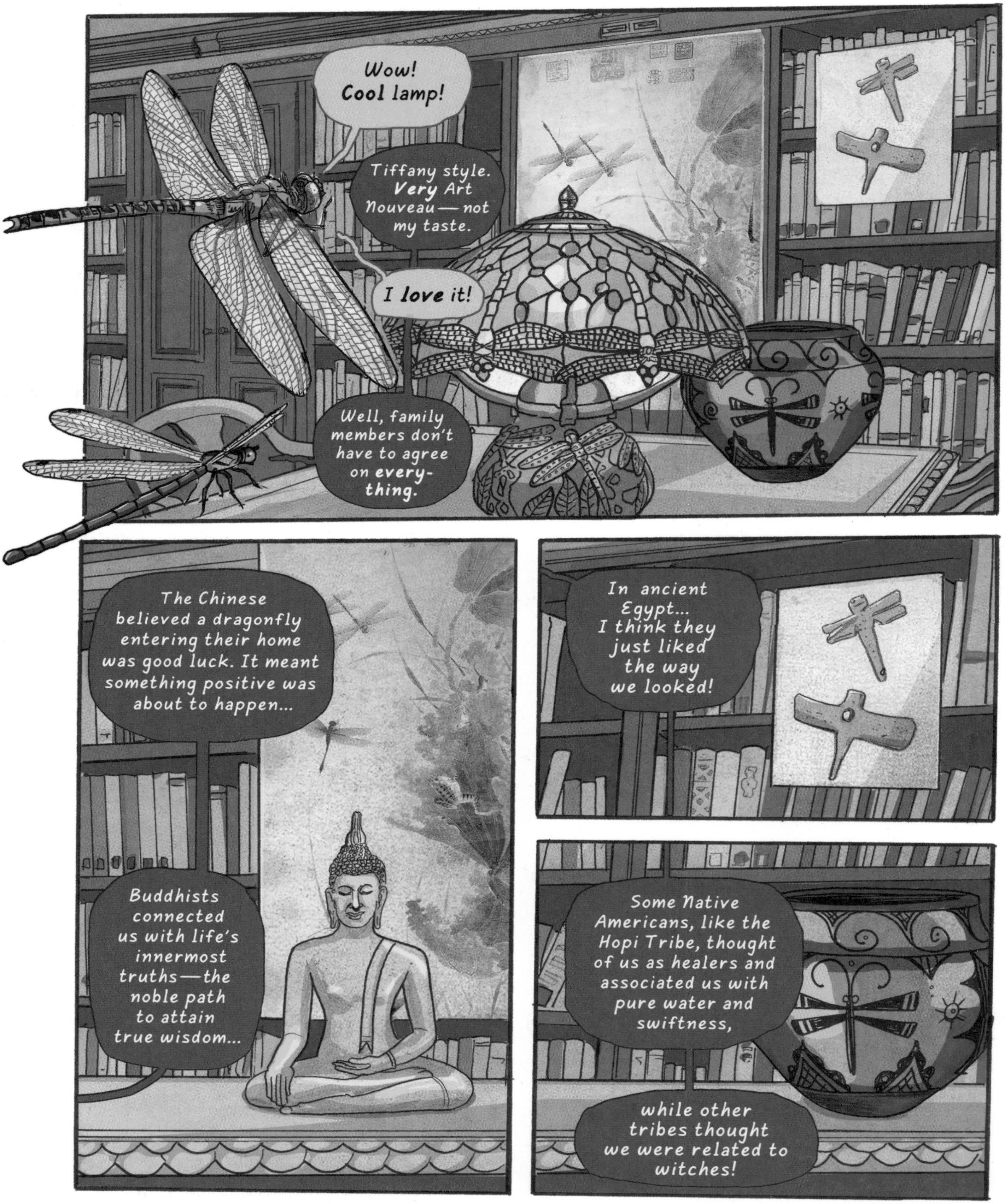

Wow! Cool lamp!
Tiffany style. Very Art Nouveau — not my taste.
I love it!
Well, family members don't have to agree on every-thing.
The Chinese believed a dragonfly entering their home was good luck. It meant something positive was about to happen...
Buddhists connected us with life's innermost truths — the noble path to attain true wisdom...
In ancient Egypt... I think they just liked the way we looked!
Some Native Americans, like the Hopi Tribe, thought of us as healers and associated us with pure water and swiftness,
while other tribes thought we were related to witches!

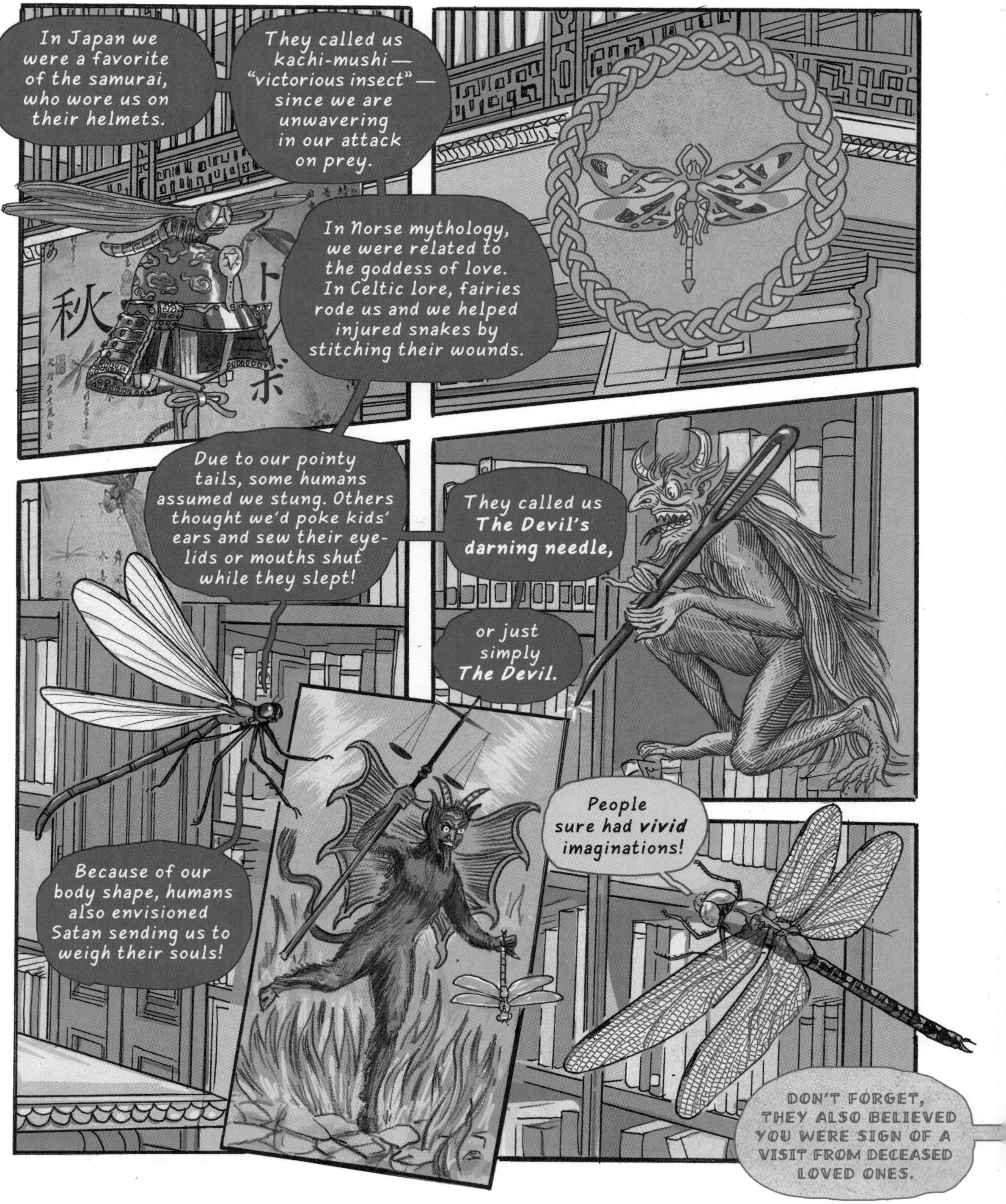

In Japan we were a favorite of the samurai, who wore us on their helmets.
They called us kachi-mushi— "victorious insect"— since we are unwavering in our attack on prey.
In Norse mythology, we were related to the goddess of love. In Celtic lore, fairies rode us and we helped injured snakes by stitching their wounds.
秋
Due to our pointy tails, some humans assumed we stung. Others thought we'd poke kids' ears and sew their eye-lids or mouths shut while they slept!
They called us **The Devil's darning needle,**
or just simply ***The Devil.***
Because of our body shape, humans also envisioned Satan sending us to weigh their souls!
People sure had ***vivid*** imaginations!
DON'T FORGET, THEY ALSO BELIEVED YOU WERE SIGN OF A VISIT FROM DECEASED LOVED ONES.

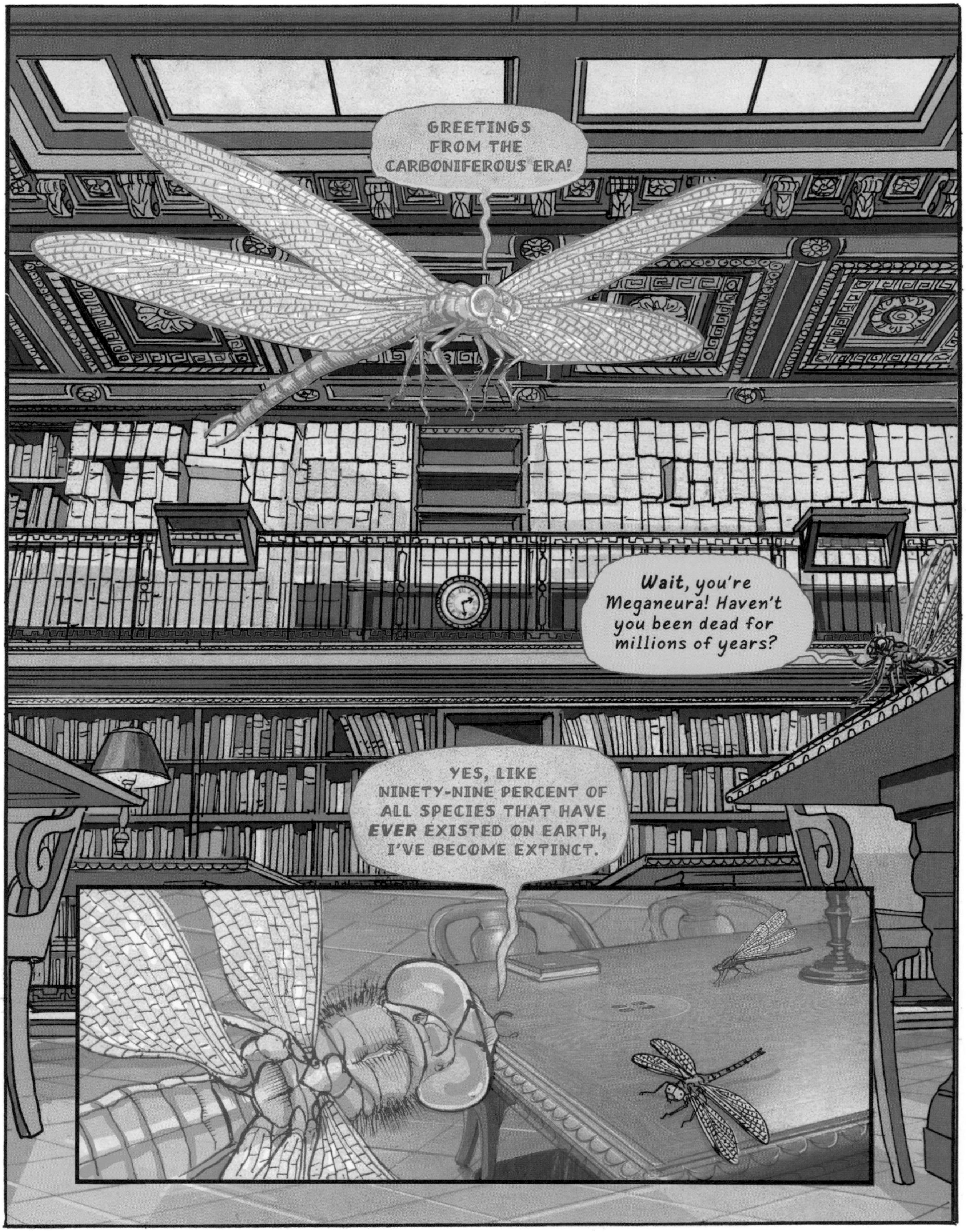
GREETINGS FROM THE CARBONIFEROUS ERA!
Wait, you're Meganeura! Haven't you been dead for millions of years?
YES, LIKE NINETY-NINE PERCENT OF ALL SPECIES THAT HAVE EVER EXISTED ON EARTH, I'VE BECOME EXTINCT.

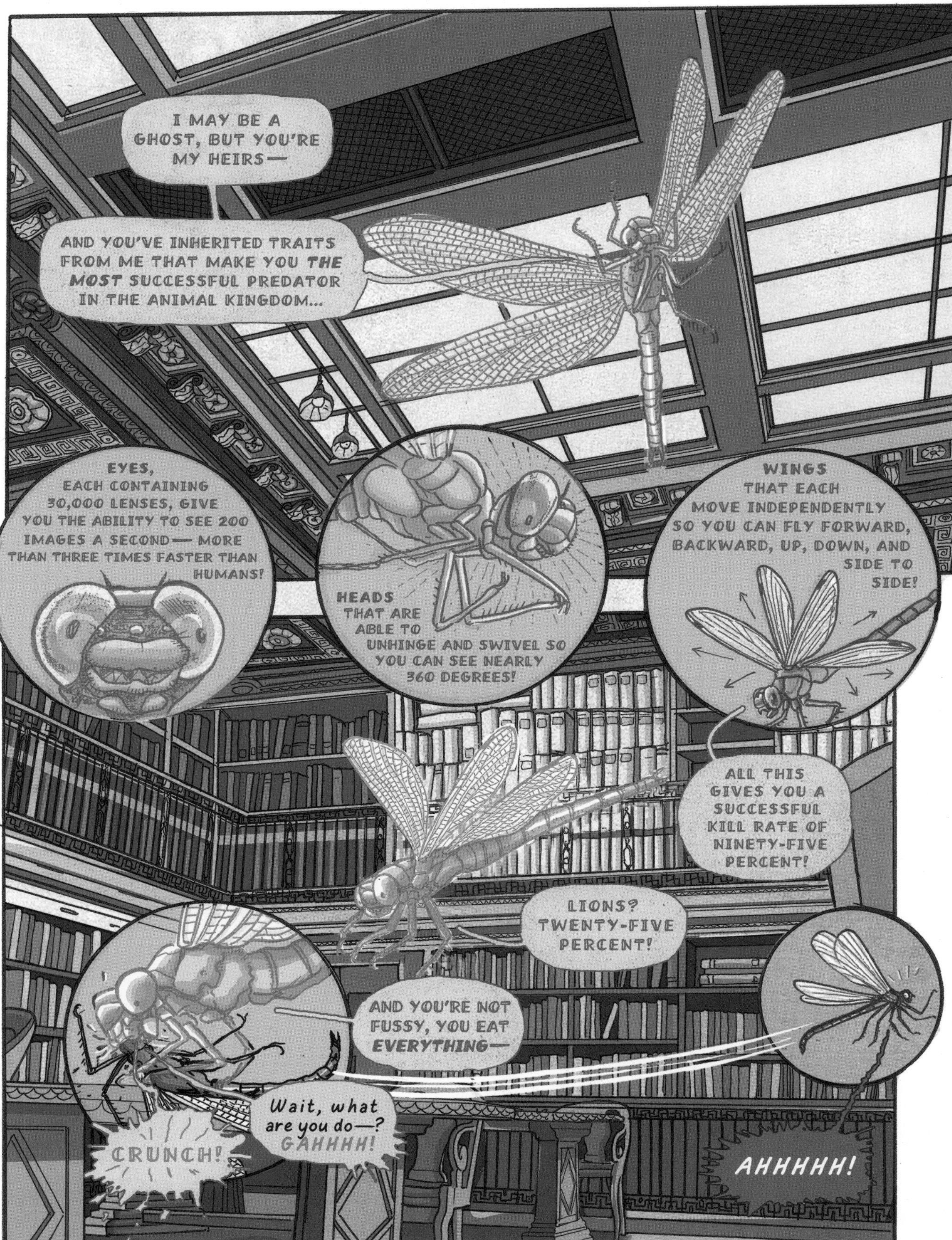

I MAY BE A GHOST, BUT YOU'RE MY HEIRS—
AND YOU'VE INHERITED TRAITS FROM ME THAT MAKE YOU *THE MOST* SUCCESSFUL PREDATOR IN THE ANIMAL KINGDOM...
EYES, EACH CONTAINING 30,000 LENSES, GIVE YOU THE ABILITY TO SEE 200 IMAGES A SECOND— MORE THAN THREE TIMES FASTER THAN HUMANS!
HEADS THAT ARE ABLE TO UNHINGE AND SWIVEL SO YOU CAN SEE NEARLY 360 DEGREES!
WINGS THAT EACH MOVE INDEPENDENTLY SO YOU CAN FLY FORWARD, BACKWARD, UP, DOWN, AND SIDE TO SIDE!
ALL THIS GIVES YOU A SUCCESSFUL KILL RATE OF NINETY-FIVE PERCENT!
LIONS? TWENTY-FIVE PERCENT!
AND YOU'RE NOT FUSSY, YOU EAT *EVERYTHING*—
Wait, what are you do—? GAHHHH!
CRUNCH!
AHHHHH!

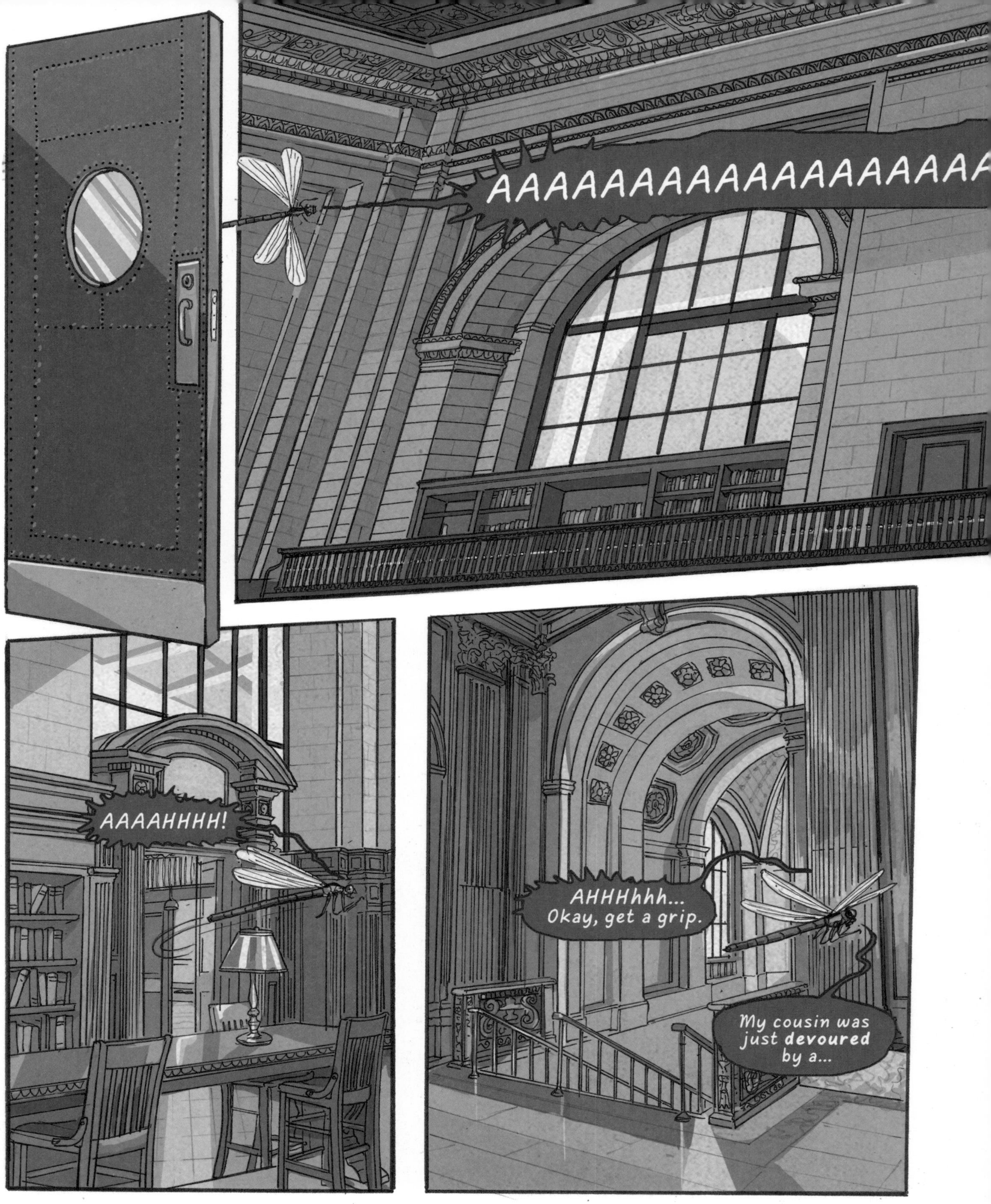
AAAAAAAAAAAAAAAAAAA
AAAAHHHH!
AHHHhhh...
Okay, get a grip.
My cousin was just **devoured** by a...

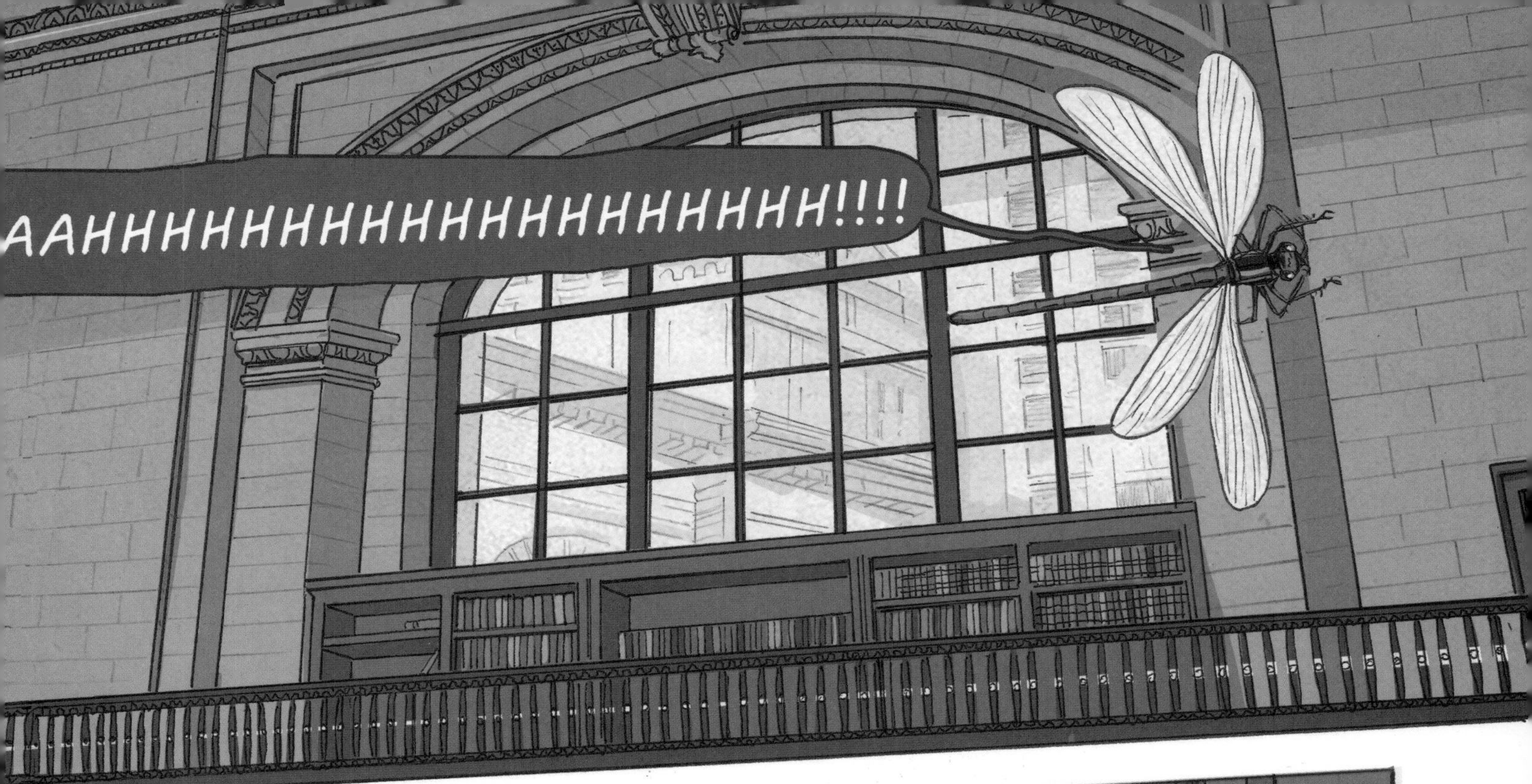
AAHHHHHHHHHHHHHHHHHHHHH!!!!

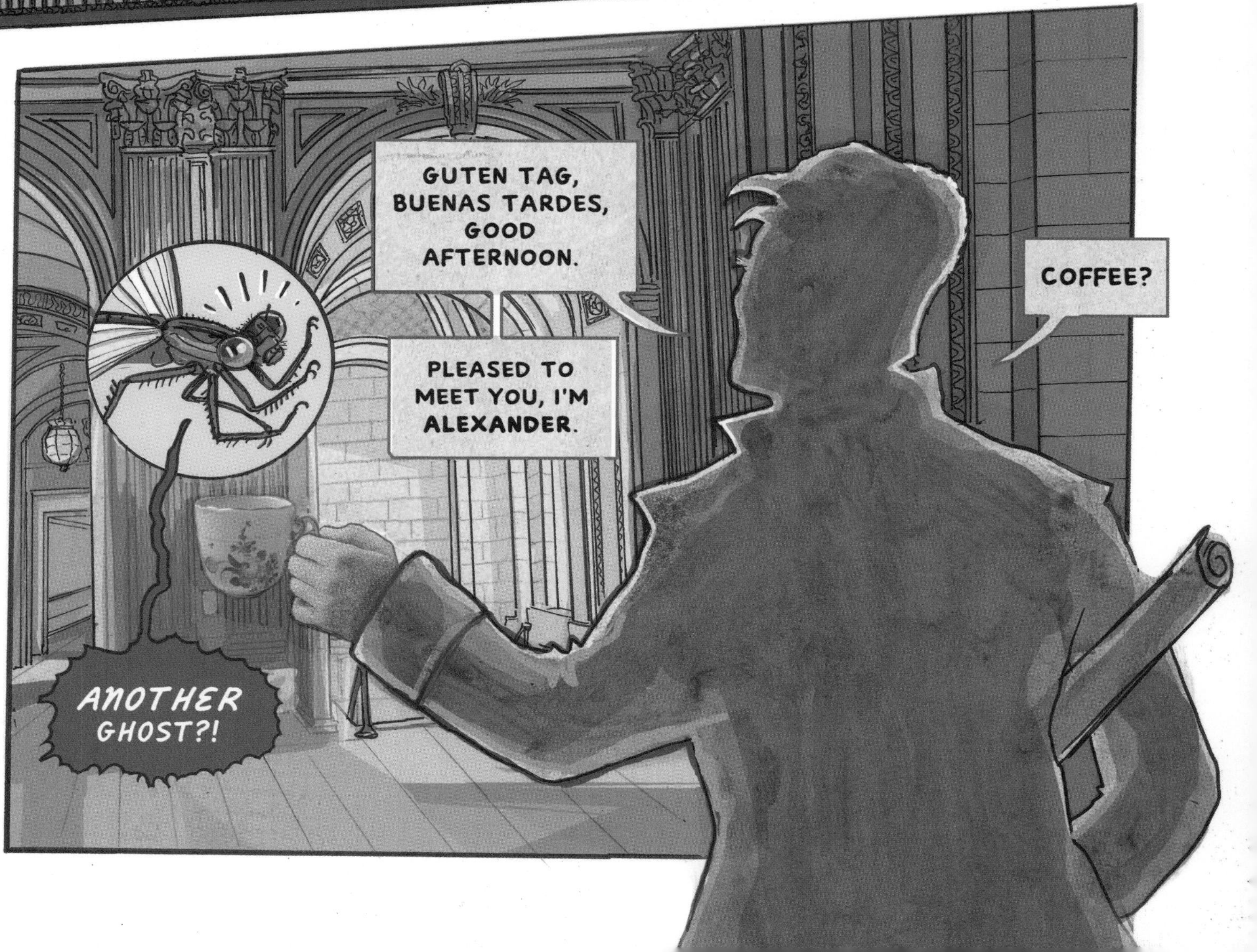
GUTEN TAG, BUENAS TARDES, GOOD AFTERNOON.
PLEASED TO MEET YOU, I'M ALEXANDER.
COFFEE?
ANOTHER GHOST?!

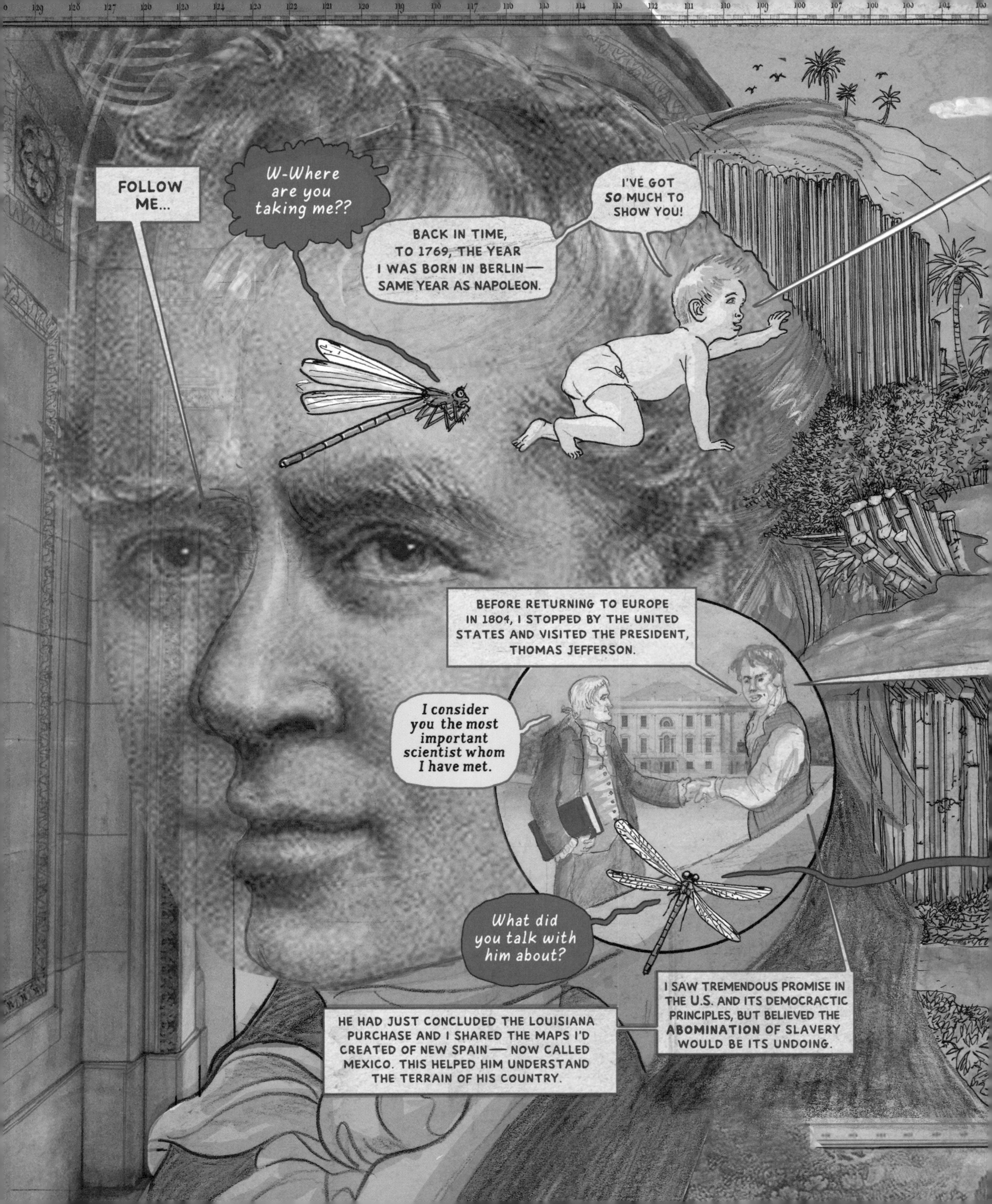

FOLLOW ME...
W-Where are you taking me??
I'VE GOT SO MUCH TO SHOW YOU!
BACK IN TIME, TO 1769, THE YEAR I WAS BORN IN BERLIN— SAME YEAR AS NAPOLEON.
BEFORE RETURNING TO EUROPE IN 1804, I STOPPED BY THE UNITED STATES AND VISITED THE PRESIDENT, THOMAS JEFFERSON.
I consider you the most important scientist whom I have met.
What did you talk with him about?
HE HAD JUST CONCLUDED THE LOUISIANA PURCHASE AND I SHARED THE MAPS I'D CREATED OF NEW SPAIN— NOW CALLED MEXICO. THIS HELPED HIM UNDERSTAND THE TERRAIN OF HIS COUNTRY.
I SAW TREMENDOUS PROMISE IN THE U.S. AND ITS DEMOCRACTIC PRINCIPLES, BUT BELIEVED THE ABOMINATION OF SLAVERY WOULD BE ITS UNDOING.

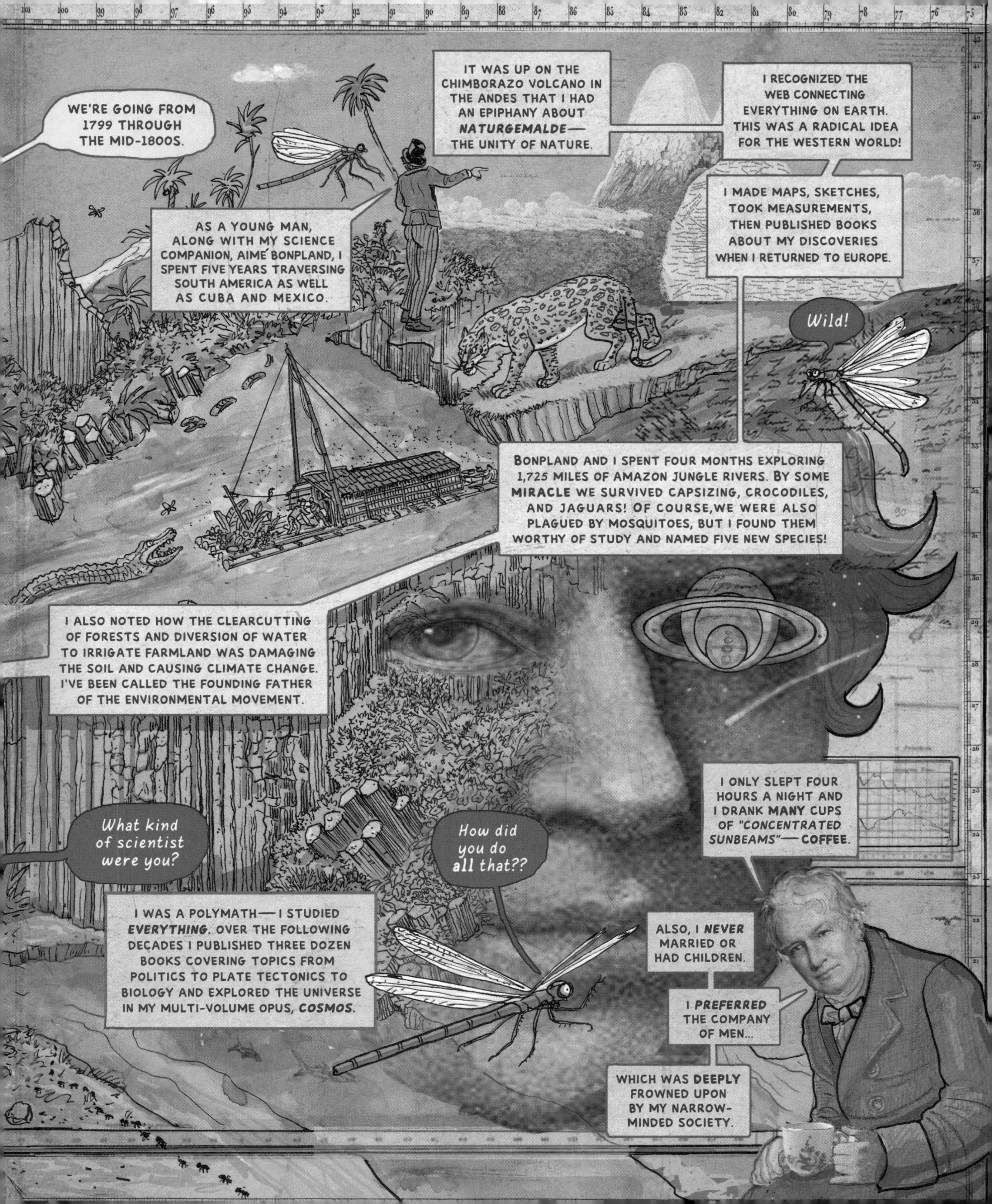

WE'RE GOING FROM 1799 THROUGH THE MID-1800S.
IT WAS UP ON THE CHIMBORAZO VOLCANO IN THE ANDES THAT I HAD AN EPIPHANY ABOUT NATURGEMALDE— THE UNITY OF NATURE.
I RECOGNIZED THE WEB CONNECTING EVERYTHING ON EARTH. THIS WAS A RADICAL IDEA FOR THE WESTERN WORLD!
AS A YOUNG MAN, ALONG WITH MY SCIENCE COMPANION, AIMÉ BONPLAND, I SPENT FIVE YEARS TRAVERSING SOUTH AMERICA AS WELL AS CUBA AND MEXICO.
I MADE MAPS, SKETCHES, TOOK MEASUREMENTS, THEN PUBLISHED BOOKS ABOUT MY DISCOVERIES WHEN I RETURNED TO EUROPE.
Wild!
BONPLAND AND I SPENT FOUR MONTHS EXPLORING 1,725 MILES OF AMAZON JUNGLE RIVERS. BY SOME MIRACLE WE SURVIVED CAPSIZING, CROCODILES, AND JAGUARS! OF COURSE, WE WERE ALSO PLAGUED BY MOSQUITOES, BUT I FOUND THEM WORTHY OF STUDY AND NAMED FIVE NEW SPECIES!
I ALSO NOTED HOW THE CLEARCUTTING OF FORESTS AND DIVERSION OF WATER TO IRRIGATE FARMLAND WAS DAMAGING THE SOIL AND CAUSING CLIMATE CHANGE. I'VE BEEN CALLED THE FOUNDING FATHER OF THE ENVIRONMENTAL MOVEMENT.
What kind of scientist were you?
How did you do all that??
I ONLY SLEPT FOUR HOURS A NIGHT AND I DRANK MANY CUPS OF "CONCENTRATED SUNBEAMS"— COFFEE.
I WAS A POLYMATH— I STUDIED EVERYTHING. OVER THE FOLLOWING DECADES I PUBLISHED THREE DOZEN BOOKS COVERING TOPICS FROM POLITICS TO PLATE TECTONICS TO BIOLOGY AND EXPLORED THE UNIVERSE IN MY MULTI-VOLUME OPUS, COSMOS.
ALSO, I NEVER MARRIED OR HAD CHILDREN.
I PREFERRED THE COMPANY OF MEN...
WHICH WAS DEEPLY FROWNED UPON BY MY NARROW-MINDED SOCIETY.

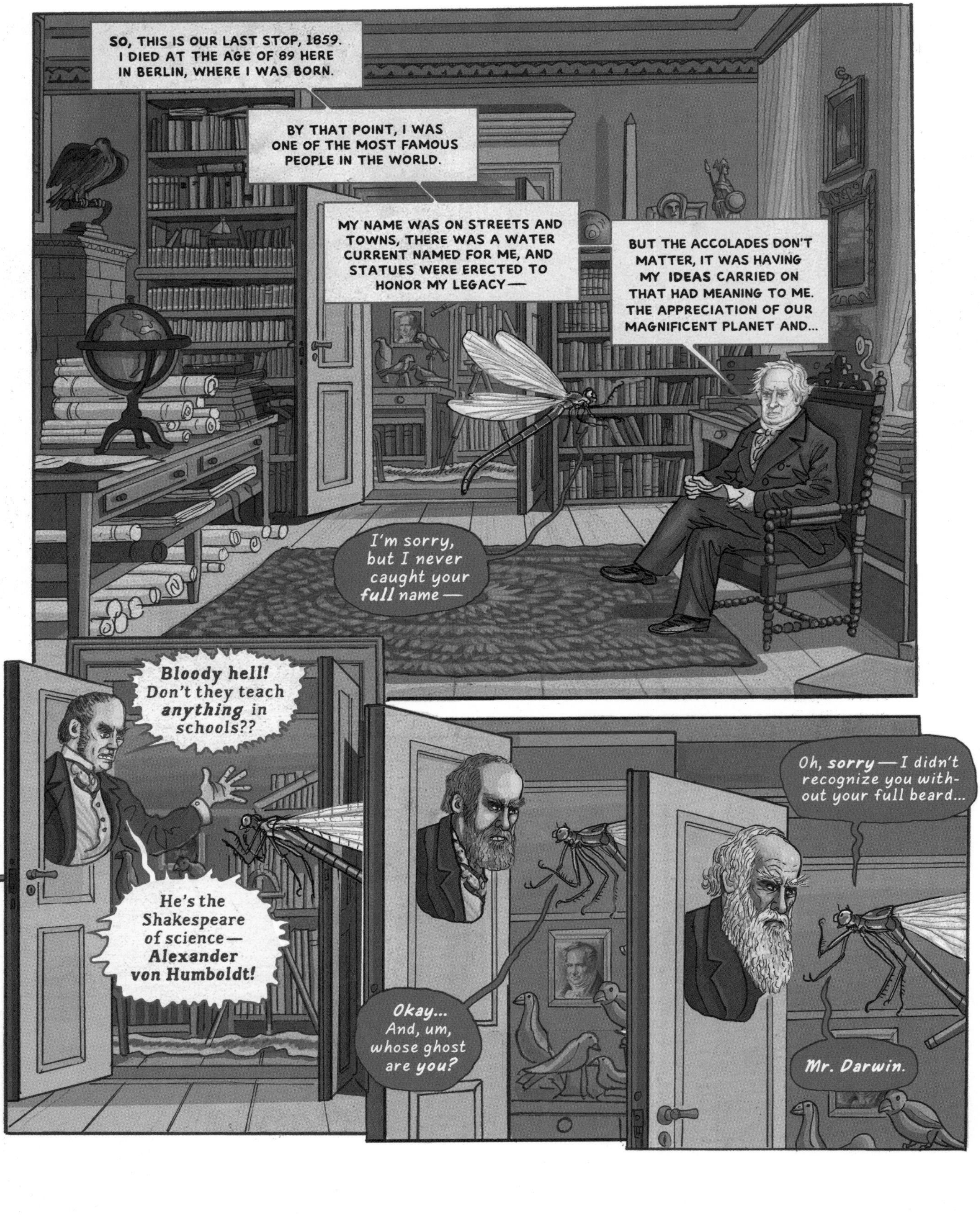
SO, THIS IS OUR LAST STOP, 1859. I DIED AT THE AGE OF 89 HERE IN BERLIN, WHERE I WAS BORN.
BY THAT POINT, I WAS ONE OF THE MOST FAMOUS PEOPLE IN THE WORLD.
MY NAME WAS ON STREETS AND TOWNS, THERE WAS A WATER CURRENT NAMED FOR ME, AND STATUES WERE ERECTED TO HONOR MY LEGACY—
BUT THE ACCOLADES DON'T MATTER, IT WAS HAVING MY IDEAS CARRIED ON THAT HAD MEANING TO ME. THE APPRECIATION OF OUR MAGNIFICENT PLANET AND...
I'm sorry, but I never caught your full name—
Bloody hell! Don't they teach anything in schools??
He's the Shakespeare of science—Alexander von Humboldt!
Okay... And, um, whose ghost are you?
Oh, sorry—I didn't recognize you without your full beard...
Mr. Darwin.

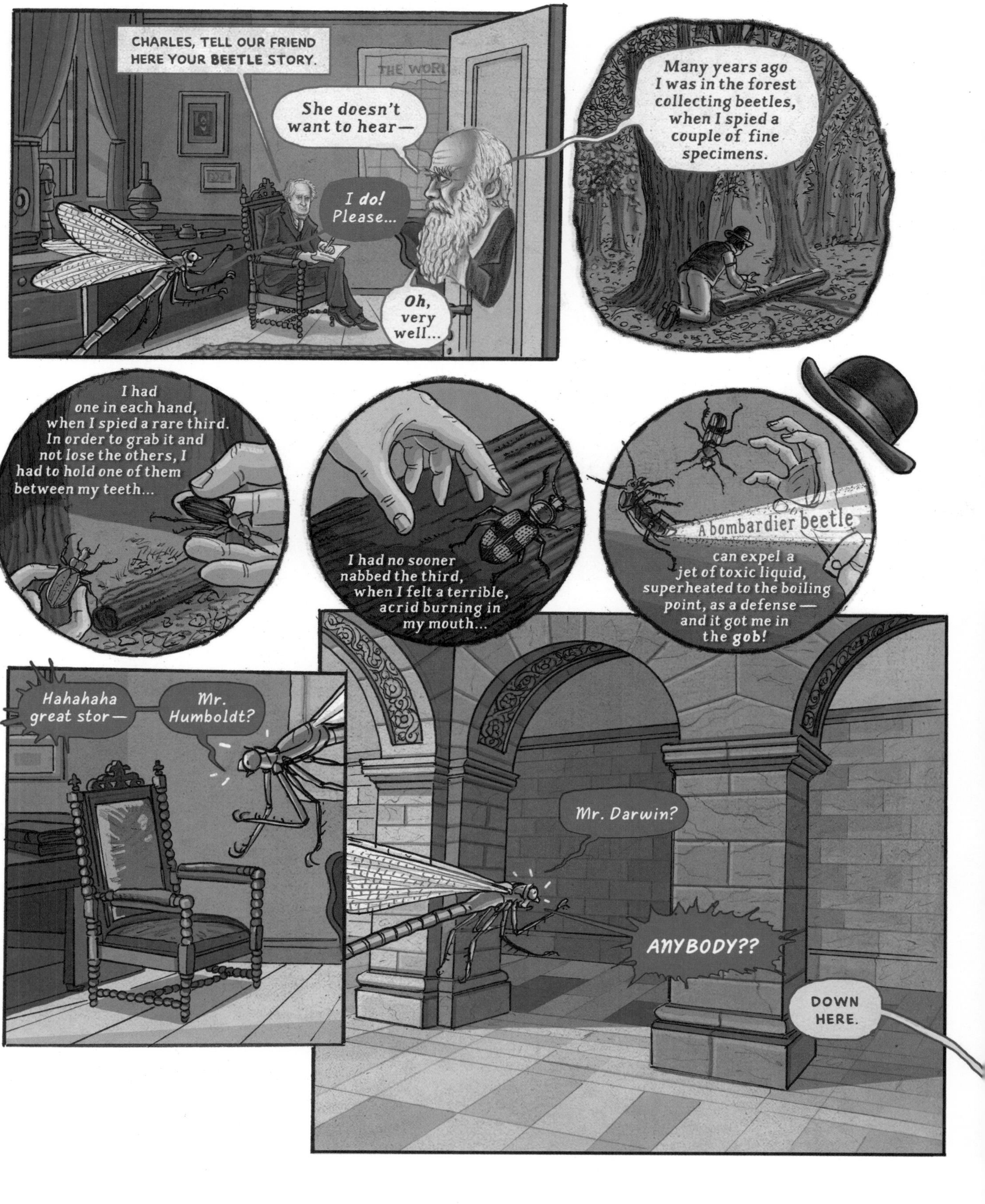
CHARLES, TELL OUR FRIEND HERE YOUR BEETLE STORY.
She doesn't want to hear—
I do! Please...
Oh, very well...
THE WORL
Many years ago I was in the forest collecting beetles, when I spied a couple of fine specimens.
I had one in each hand, when I spied a rare third. In order to grab it and not lose the others, I had to hold one of them between my teeth...
I had no sooner nabbed the third, when I felt a terrible, acrid burning in my mouth...
A bombardier beetle can expel a jet of toxic liquid, superheated to the boiling point, as a defense—and it got me in the gob!
Hahahaha great stor—
Mr. Humboldt?
Mr. Darwin?
ANYBODY??
DOWN HERE.

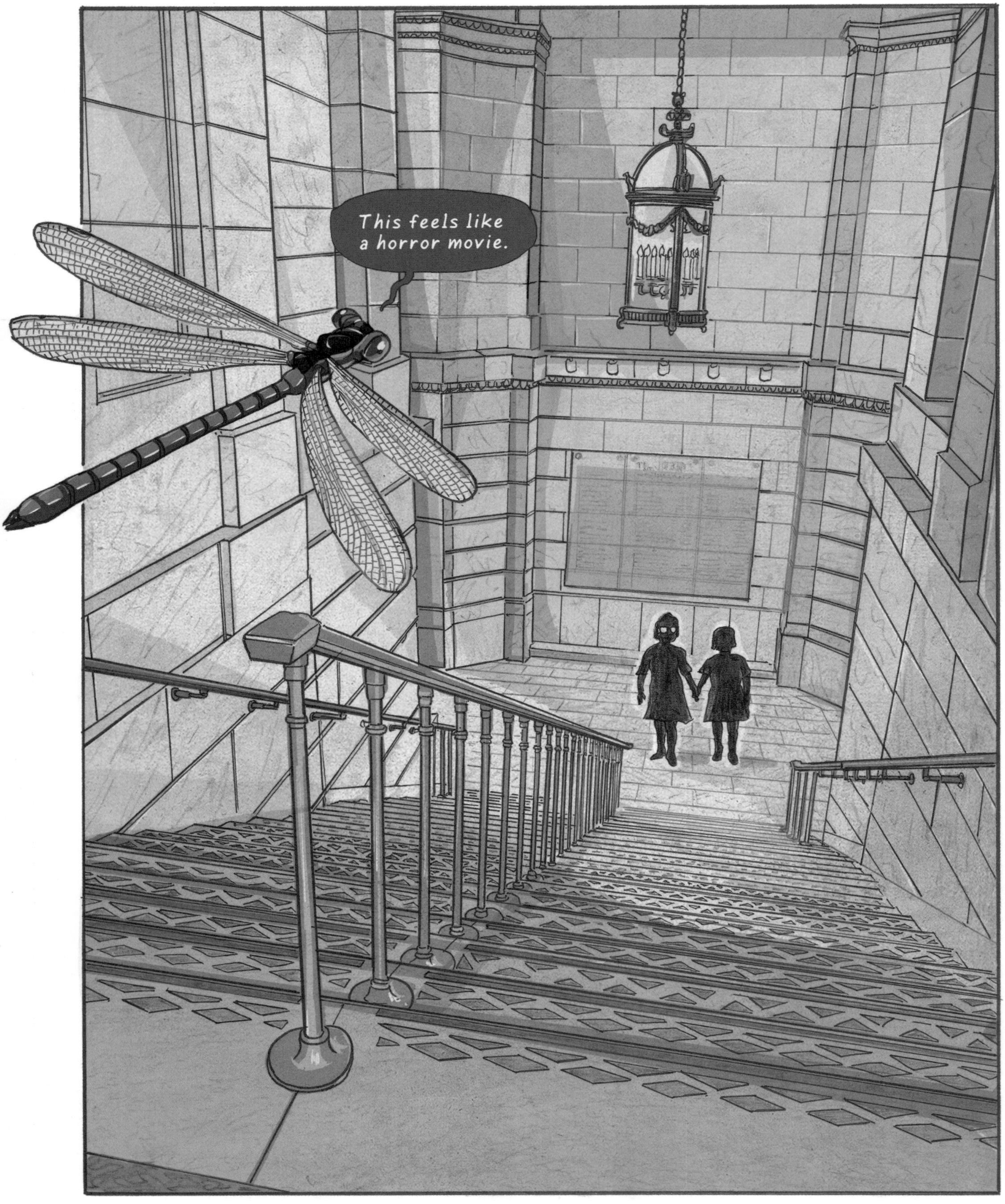
This feels like a horror movie.

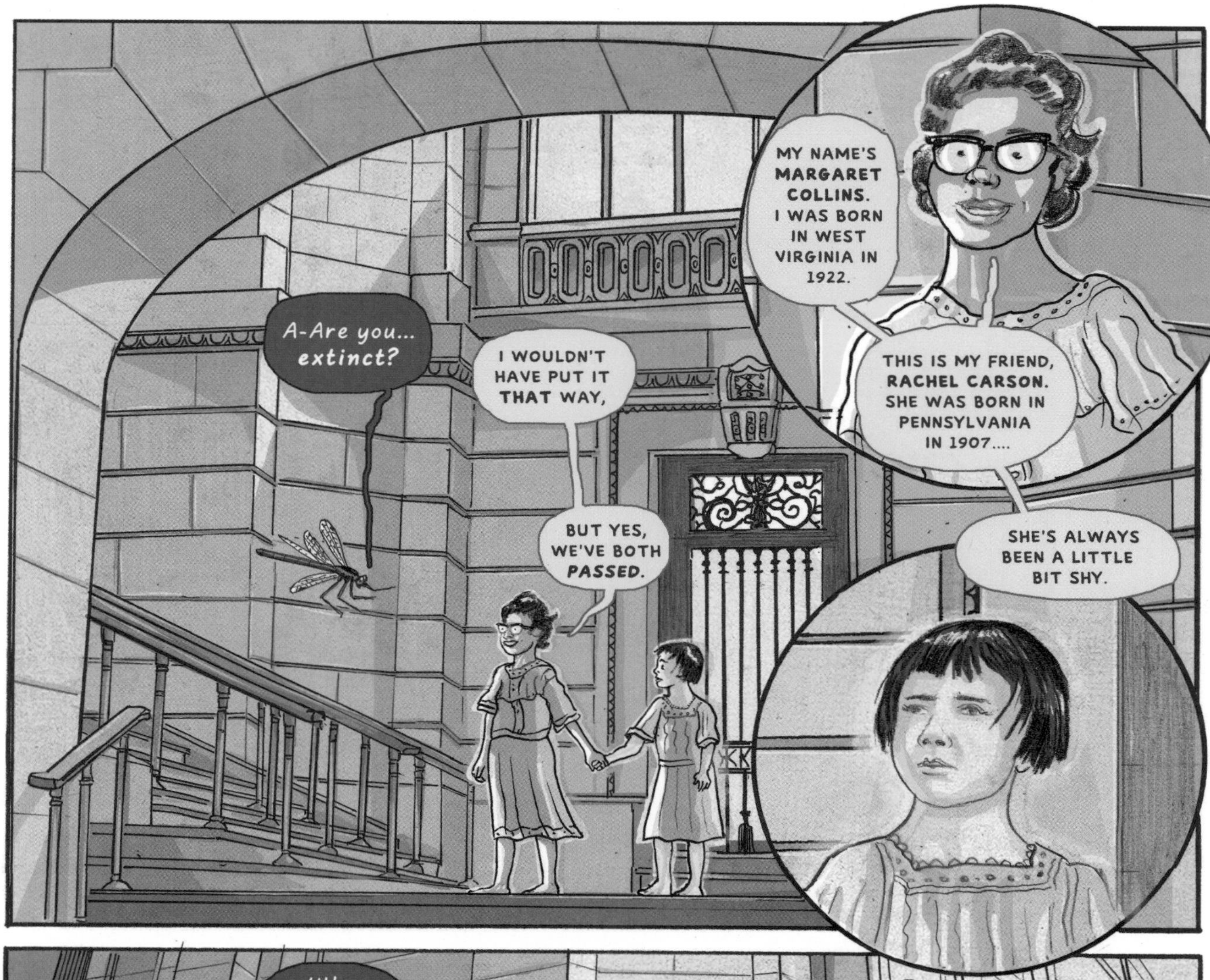
A-Are you... extinct?
I WOULDN'T HAVE PUT IT THAT WAY,
BUT YES, WE'VE BOTH PASSED.
MY NAME'S MARGARET COLLINS. I WAS BORN IN WEST VIRGINIA IN 1922.
THIS IS MY FRIEND, RACHEL CARSON. SHE WAS BORN IN PENNSYLVANIA IN 1907....
SHE'S ALWAYS BEEN A LITTLE BIT SHY.

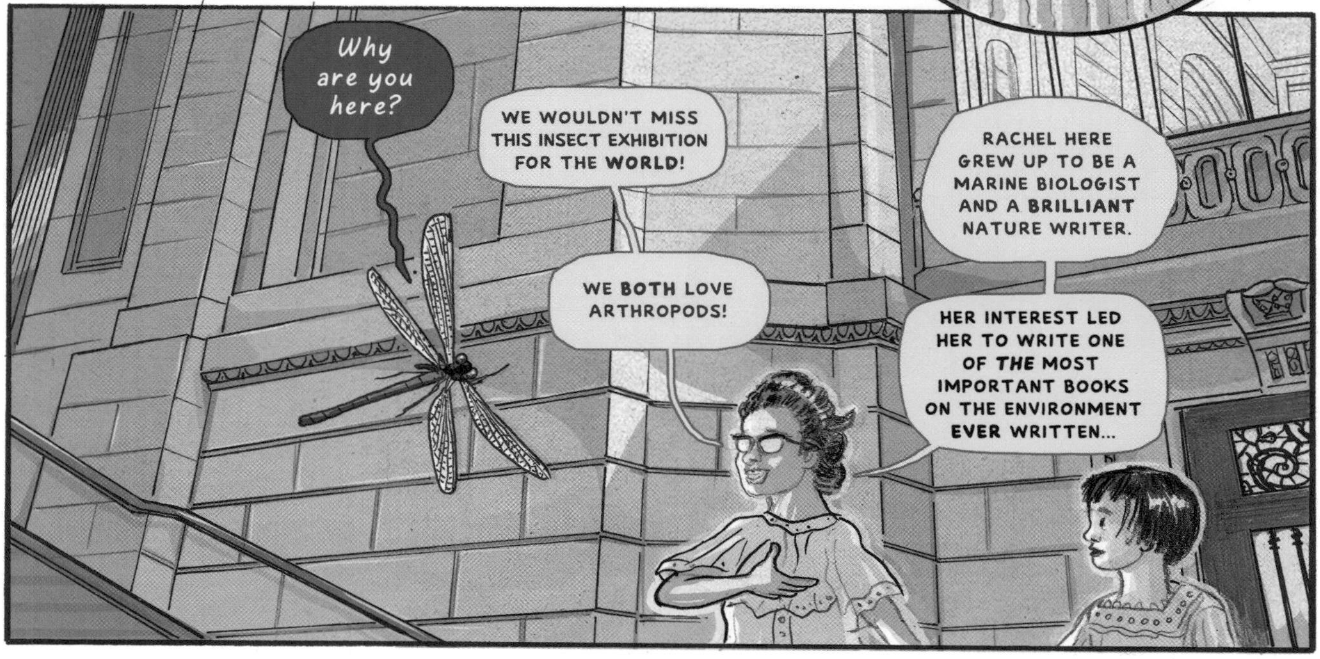
Why are you here?
WE WOULDN'T MISS THIS INSECT EXHIBITION FOR THE WORLD!
WE BOTH LOVE ARTHROPODS!
RACHEL HERE GREW UP TO BE A MARINE BIOLOGIST AND A BRILLIANT NATURE WRITER.
HER INTEREST LED HER TO WRITE ONE OF THE MOST IMPORTANT BOOKS ON THE ENVIRONMENT EVER WRITTEN...

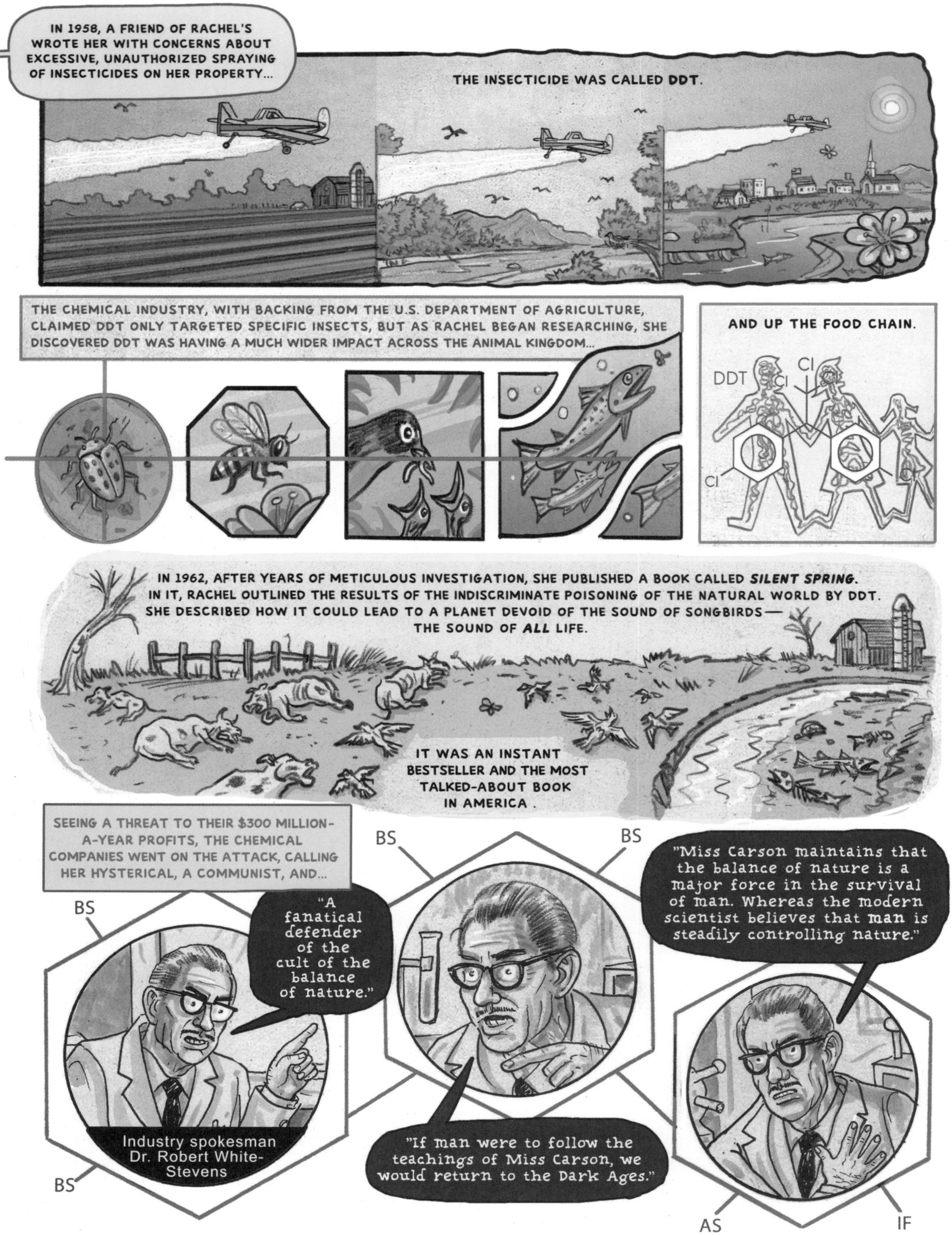

IN 1958, A FRIEND OF RACHEL'S WROTE HER WITH CONCERNS ABOUT EXCESSIVE, UNAUTHORIZED SPRAYING OF INSECTICIDES ON HER PROPERTY...
THE INSECTICIDE WAS CALLED DDT.
THE CHEMICAL INDUSTRY, WITH BACKING FROM THE U.S. DEPARTMENT OF AGRICULTURE, CLAIMED DDT ONLY TARGETED SPECIFIC INSECTS, BUT AS RACHEL BEGAN RESEARCHING, SHE DISCOVERED DDT WAS HAVING A MUCH WIDER IMPACT ACROSS THE ANIMAL KINGDOM...
AND UP THE FOOD CHAIN.
DDT
Cl
Cl
Cl
Cl
IN 1962, AFTER YEARS OF METICULOUS INVESTIGATION, SHE PUBLISHED A BOOK CALLED SILENT SPRING. IN IT, RACHEL OUTLINED THE RESULTS OF THE INDISCRIMINATE POISONING OF THE NATURAL WORLD BY DDT. SHE DESCRIBED HOW IT COULD LEAD TO A PLANET DEVOID OF THE SOUND OF SONGBIRDS— THE SOUND OF ALL LIFE.
IT WAS AN INSTANT BESTSELLER AND THE MOST TALKED-ABOUT BOOK IN AMERICA.
SEEING A THREAT TO THEIR $300 MILLION-A-YEAR PROFITS, THE CHEMICAL COMPANIES WENT ON THE ATTACK, CALLING HER HYSTERICAL, A COMMUNIST, AND...
BS
BS
BS
BS
"A fanatical defender of the cult of the balance of nature."
Industry spokesman Dr. Robert White-Stevens
"If man were to follow the teachings of Miss Carson, we would return to the Dark Ages."
"Miss Carson maintains that the balance of nature is a major force in the survival of man. Whereas the modern scientist believes that man is steadily controlling nature."
AS
IF

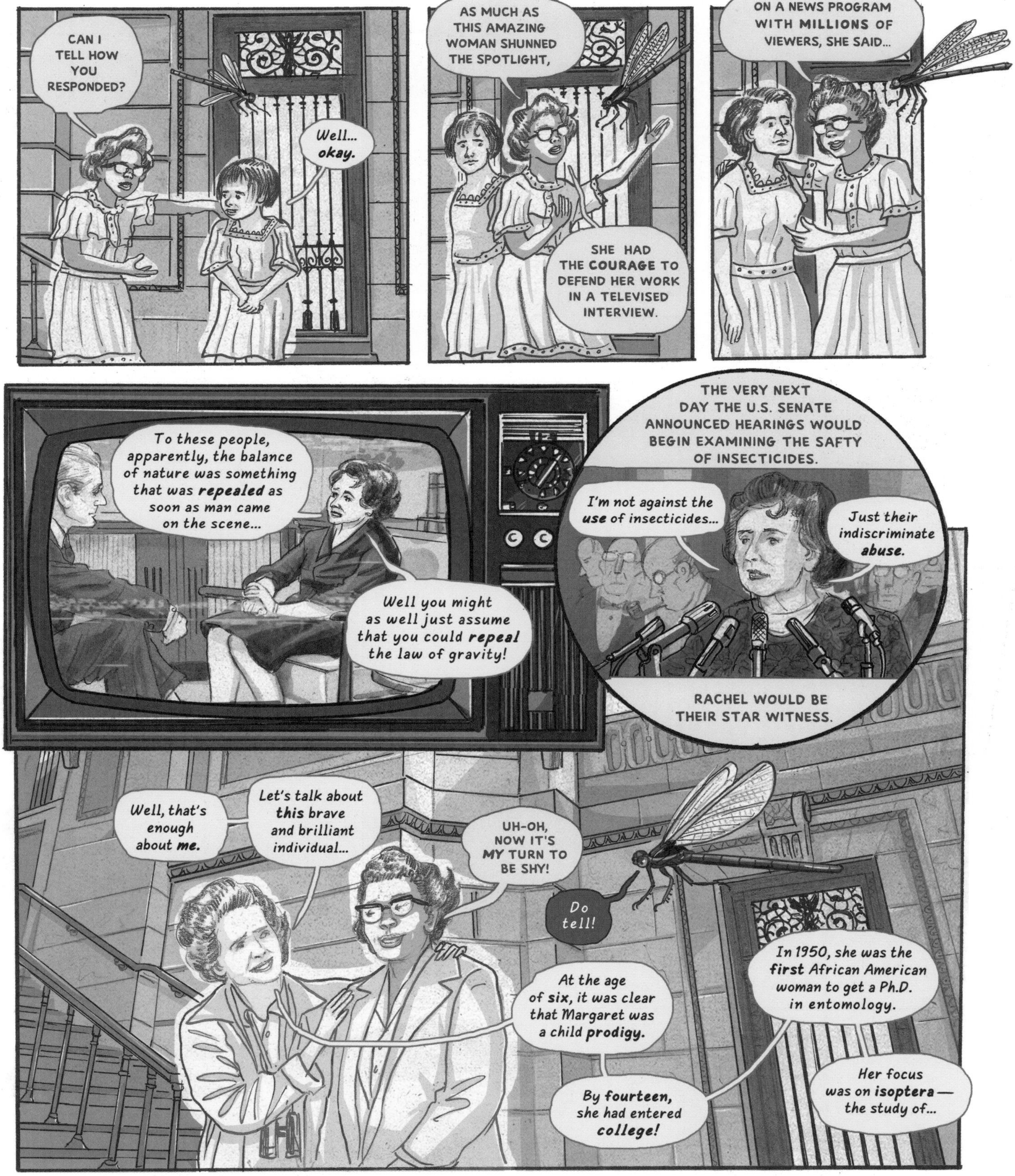
CAN I TELL HOW YOU RESPONDED?
Well... okay.
AS MUCH AS THIS AMAZING WOMAN SHUNNED THE SPOTLIGHT,
SHE HAD THE COURAGE TO DEFEND HER WORK IN A TELEVISED INTERVIEW.
ON A NEWS PROGRAM WITH MILLIONS OF VIEWERS, SHE SAID...
To these people, apparently, the balance of nature was something that was repealed as soon as man came on the scene...
Well you might as well just assume that you could repeal the law of gravity!
THE VERY NEXT DAY THE U.S. SENATE ANNOUNCED HEARINGS WOULD BEGIN EXAMINING THE SAFTY OF INSECTICIDES.
I'm not against the use of insecticides...
Just their indiscriminate abuse.
RACHEL WOULD BE THEIR STAR WITNESS.
Well, that's enough about me.
Let's talk about this brave and brilliant individual...
UH-OH, NOW IT'S MY TURN TO BE SHY!
Do tell!
At the age of six, it was clear that Margaret was a child prodigy.
By fourteen, she had entered college!
In 1950, she was the first African American woman to get a Ph.D. in entomology.
Her focus was on isoptera — the study of...

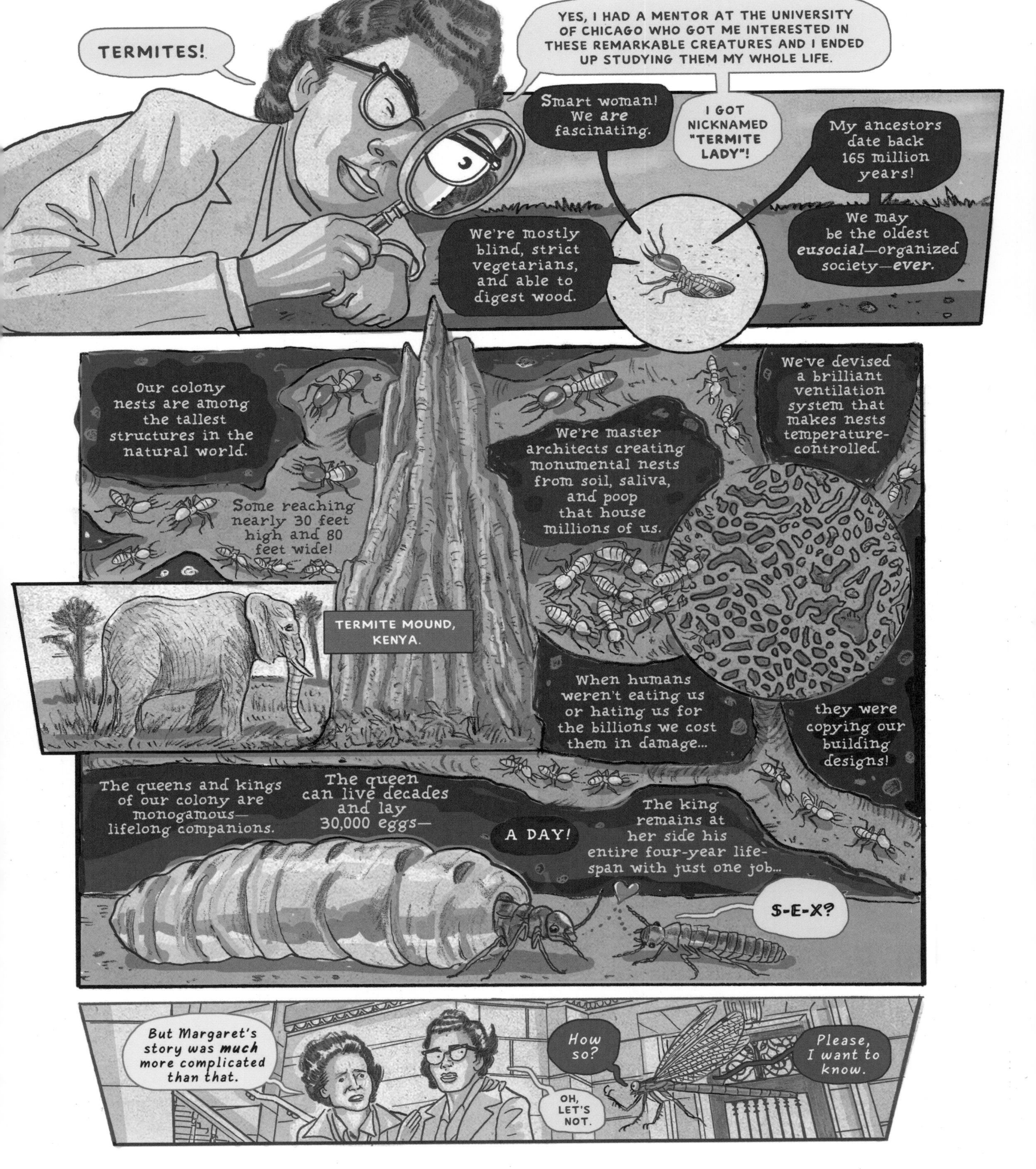

TERMITES!
YES, I HAD A MENTOR AT THE UNIVERSITY OF CHICAGO WHO GOT ME INTERESTED IN THESE REMARKABLE CREATURES AND I ENDED UP STUDYING THEM MY WHOLE LIFE.
Smart woman! We *are* fascinating.
I GOT NICKNAMED "TERMITE LADY"!
My ancestors date back 165 million years!
We may be the oldest *eusocial*—organized society—*ever*.
We're mostly blind, strict vegetarians, and able to digest wood.
Our colony nests are among the tallest structures in the natural world.
Some reaching nearly 30 feet high and 80 feet wide!
We're master architects creating monumental nests from soil, saliva, and poop that house millions of us.
We've devised a brilliant ventilation system that makes nests temperature-controlled.
TERMITE MOUND, KENYA.
When humans weren't eating us or hating us for the billions we cost them in damage...
they were copying our building designs!
The queens and kings of our colony are monogamous—lifelong companions.
The queen can live decades and lay 30,000 eggs—
A DAY!
The king remains at her side his entire four-year life-span with just one job...
S-E-X?
But Margaret's story was *much* more complicated than that.
OH, LET'S NOT.
How so?
Please, I want to know.

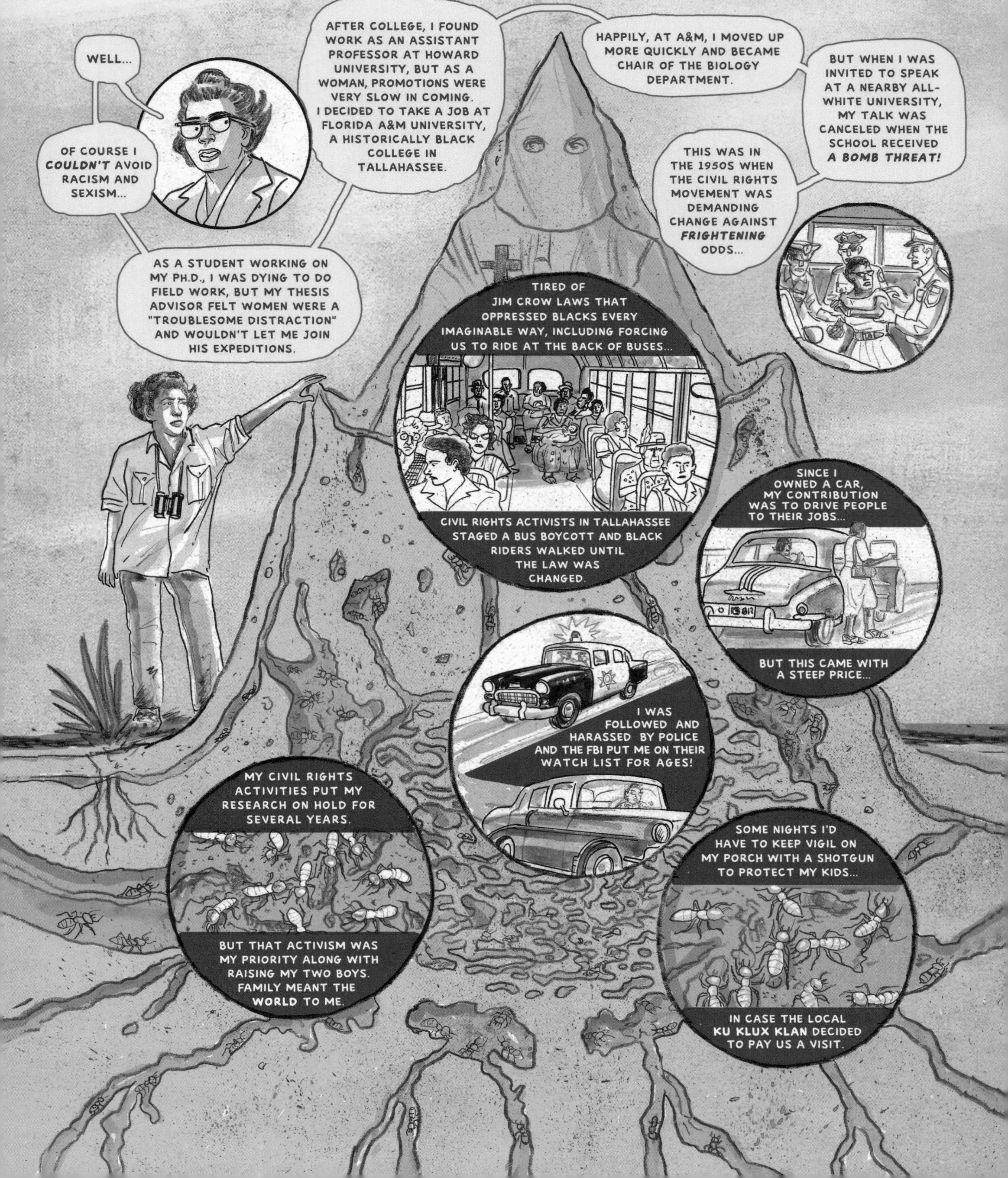
WELL...
OF COURSE I COULDN'T AVOID RACISM AND SEXISM...
AS A STUDENT WORKING ON MY PH.D., I WAS DYING TO DO FIELD WORK, BUT MY THESIS ADVISOR FELT WOMEN WERE A "TROUBLESOME DISTRACTION" AND WOULDN'T LET ME JOIN HIS EXPEDITIONS.
AFTER COLLEGE, I FOUND WORK AS AN ASSISTANT PROFESSOR AT HOWARD UNIVERSITY, BUT AS A WOMAN, PROMOTIONS WERE VERY SLOW IN COMING. I DECIDED TO TAKE A JOB AT FLORIDA A&M UNIVERSITY, A HISTORICALLY BLACK COLLEGE IN TALLAHASSEE.
HAPPILY, AT A&M, I MOVED UP MORE QUICKLY AND BECAME CHAIR OF THE BIOLOGY DEPARTMENT.
BUT WHEN I WAS INVITED TO SPEAK AT A NEARBY ALL-WHITE UNIVERSITY, MY TALK WAS CANCELED WHEN THE SCHOOL RECEIVED A BOMB THREAT!
THIS WAS IN THE 1950S WHEN THE CIVIL RIGHTS MOVEMENT WAS DEMANDING CHANGE AGAINST FRIGHTENING ODDS...
TIRED OF JIM CROW LAWS THAT OPPRESSED BLACKS EVERY IMAGINABLE WAY, INCLUDING FORCING US TO RIDE AT THE BACK OF BUSES...
CIVIL RIGHTS ACTIVISTS IN TALLAHASSEE STAGED A BUS BOYCOTT AND BLACK RIDERS WALKED UNTIL THE LAW WAS CHANGED.
SINCE I OWNED A CAR, MY CONTRIBUTION WAS TO DRIVE PEOPLE TO THEIR JOBS...
BUT THIS CAME WITH A STEEP PRICE...
I WAS FOLLOWED AND HARASSED BY POLICE AND THE FBI PUT ME ON THEIR WATCH LIST FOR AGES!
MY CIVIL RIGHTS ACTIVITIES PUT MY RESEARCH ON HOLD FOR SEVERAL YEARS.
BUT THAT ACTIVISM WAS MY PRIORITY ALONG WITH RAISING MY TWO BOYS. FAMILY MEANT THE WORLD TO ME.
SOME NIGHTS I'D HAVE TO KEEP VIGIL ON MY PORCH WITH A SHOTGUN TO PROTECT MY KIDS...
IN CASE THE LOCAL KU KLUX KLAN DECIDED TO PAY US A VISIT.

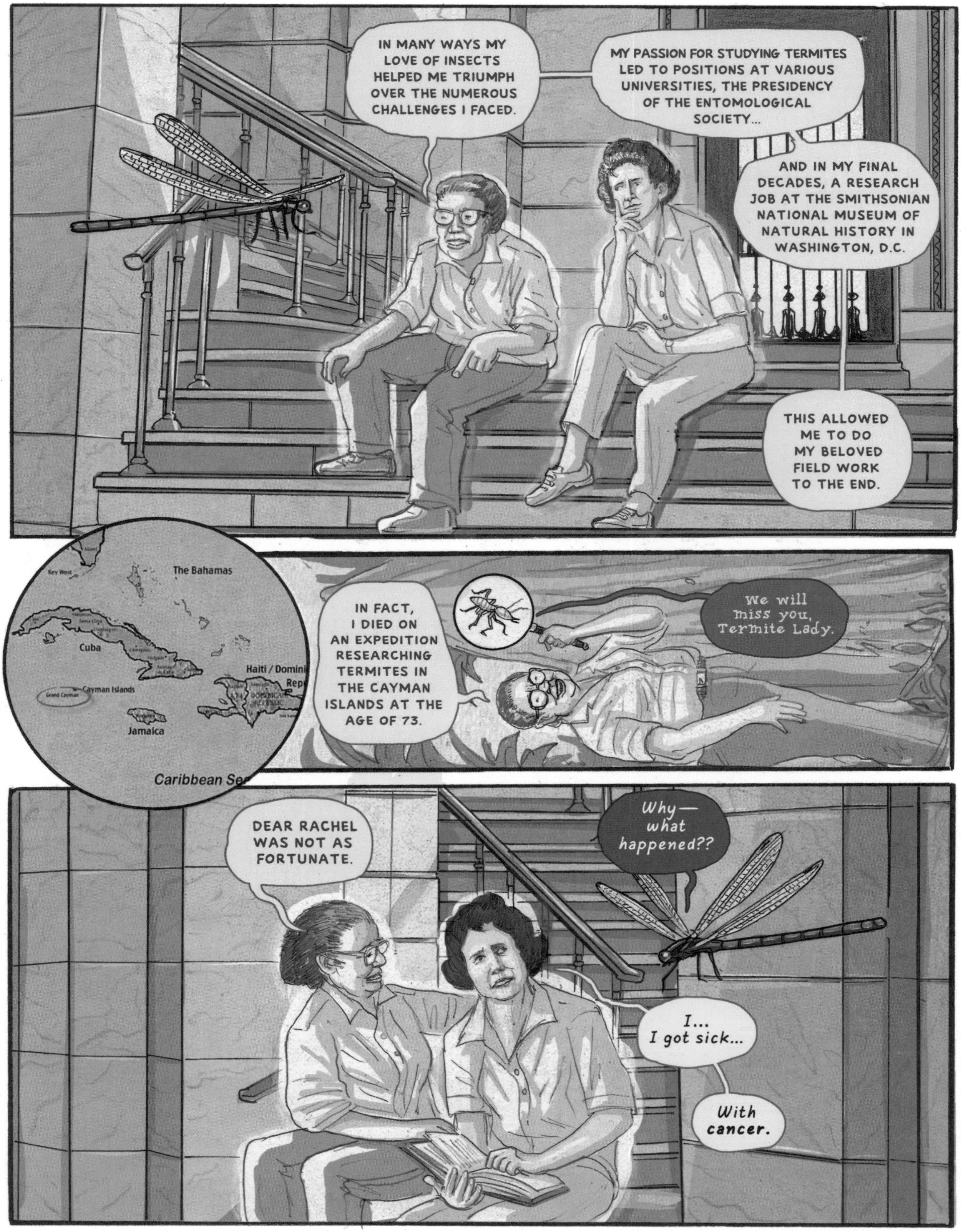
IN MANY WAYS MY LOVE OF INSECTS HELPED ME TRIUMPH OVER THE NUMEROUS CHALLENGES I FACED.
MY PASSION FOR STUDYING TERMITES LED TO POSITIONS AT VARIOUS UNIVERSITIES, THE PRESIDENCY OF THE ENTOMOLOGICAL SOCIETY...
AND IN MY FINAL DECADES, A RESEARCH JOB AT THE SMITHSONIAN NATIONAL MUSEUM OF NATURAL HISTORY IN WASHINGTON, D.C.
THIS ALLOWED ME TO DO MY BELOVED FIELD WORK TO THE END.
Key West
The Bahamas
Cuba
Cayman Islands
Grand Cayman
Jamaica
IN FACT, I DIED ON AN EXPEDITION RESEARCHING TERMITES IN THE CAYMAN ISLANDS AT THE AGE OF 73.
We will miss you, Termite Lady.
DEAR RACHEL WAS NOT AS FORTUNATE.
Why— what happened??
I... I got sick...
With cancer.

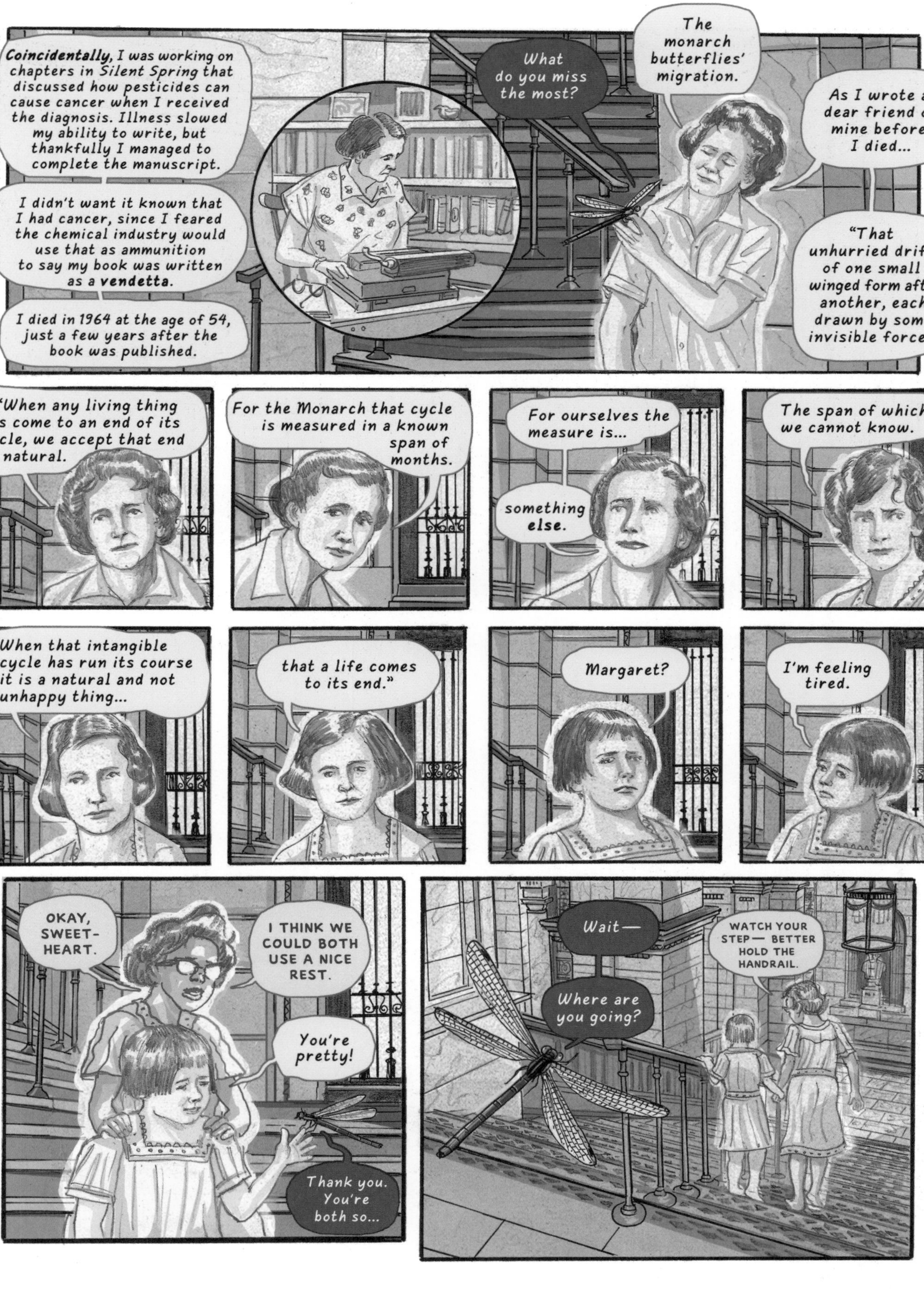
Coincidentally, I was working on chapters in *Silent Spring* that discussed how pesticides can cause cancer when I received the diagnosis. Illness slowed my ability to write, but thankfully I managed to complete the manuscript.
I didn't want it known that I had cancer, since I feared the chemical industry would use that as ammunition to say my book was written as a **vendetta**.
I died in 1964 at the age of 54, just a few years after the book was published.
What do you miss the most?
The monarch butterflies' migration.
As I wrote a dear friend of mine before I died...
"That unhurried drift of one small winged form after another, each drawn by some invisible force."
"When any living thing has come to an end of its cycle, we accept that end as natural.
For the Monarch that cycle is measured in a known span of months.
For ourselves the measure is...
something **else**.
The span of which we cannot know.
When that intangible cycle has run its course it is a natural and not unhappy thing...
that a life comes to its end."
Margaret?
I'm feeling tired.
OKAY, SWEET-HEART.
I THINK WE COULD BOTH USE A NICE REST.
You're pretty!
Thank you. You're both so...
Wait—
Where are you going?
WATCH YOUR STEP— BETTER HOLD THE HANDRAIL.

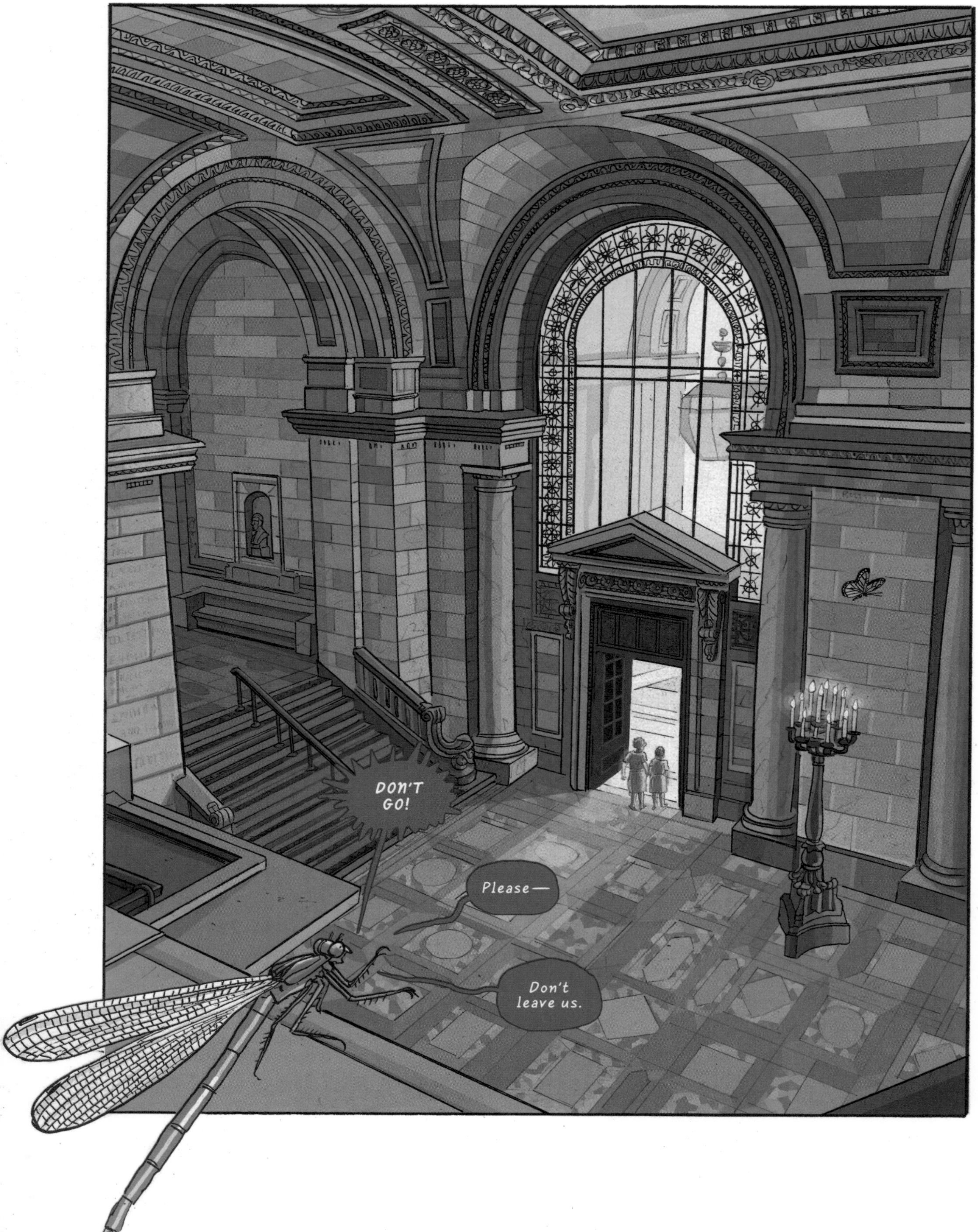
DON'T GO!
Please—
Don't leave us.

Advice from a Caterpillar

Charles Lutwidge Dodgson
under the pen name Lewis Carroll
(1832–1898)
Alice's Adventures in Wonderland
Published in 1865
with illustrations by John Tenniel

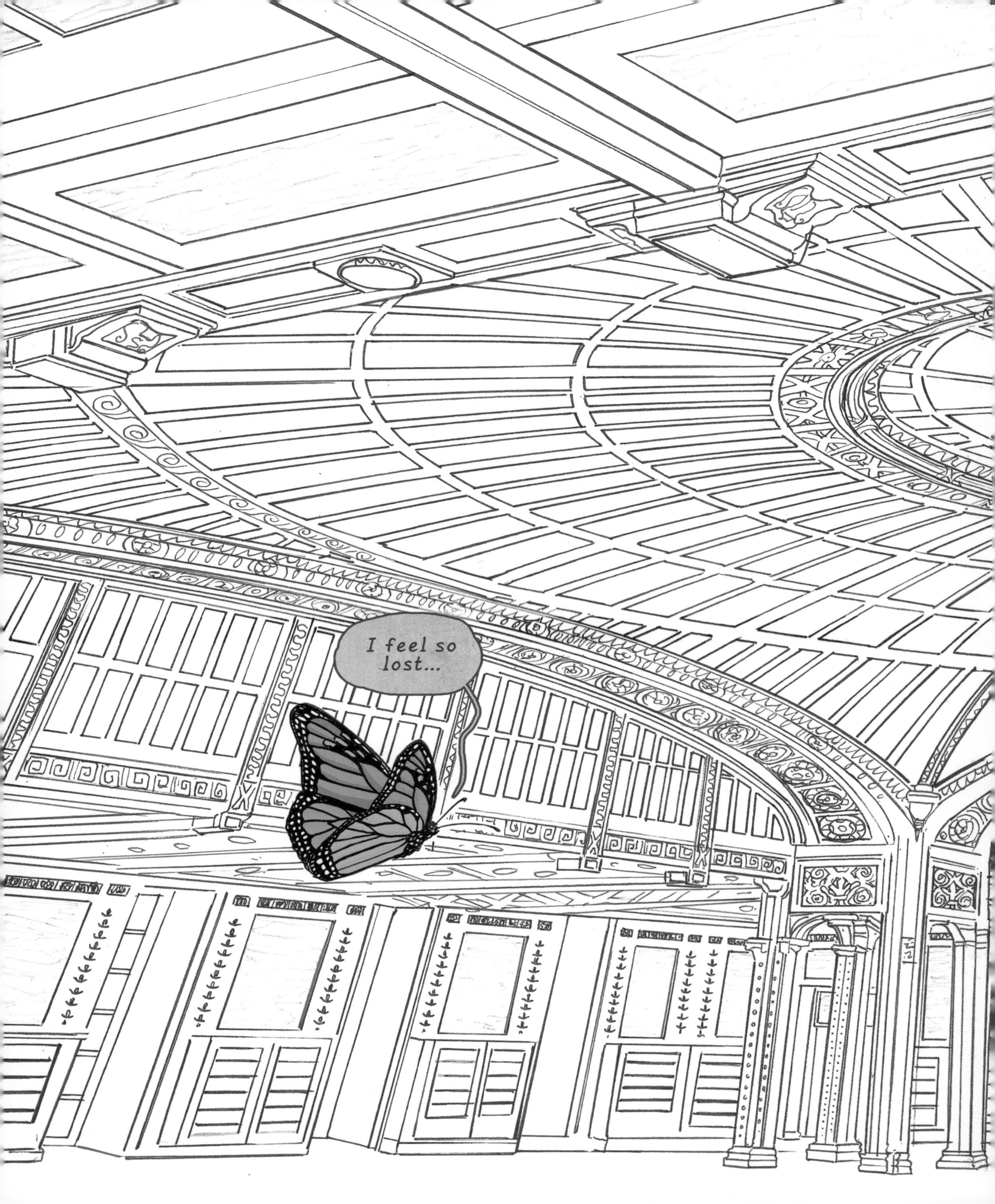
I feel so lost...

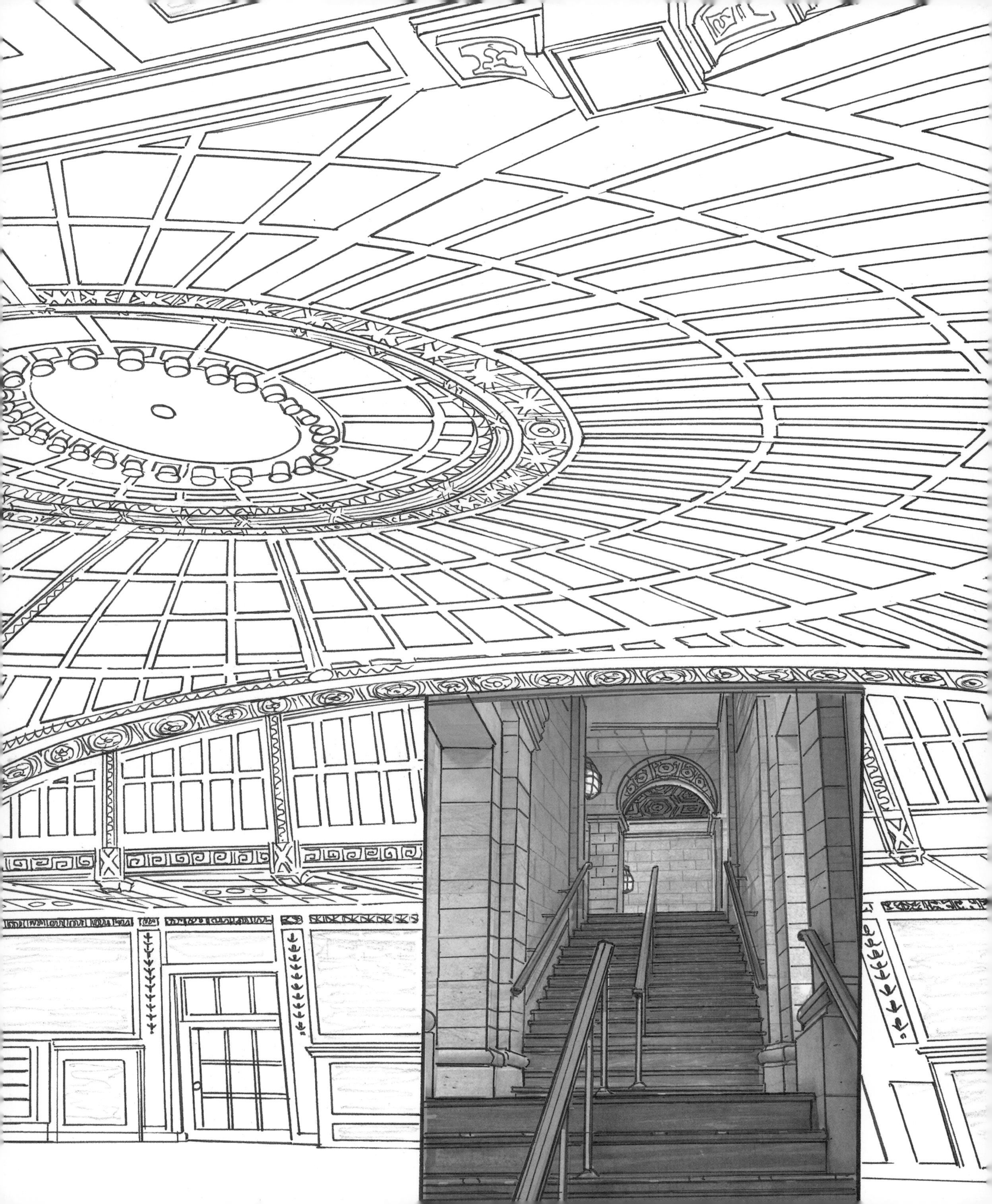

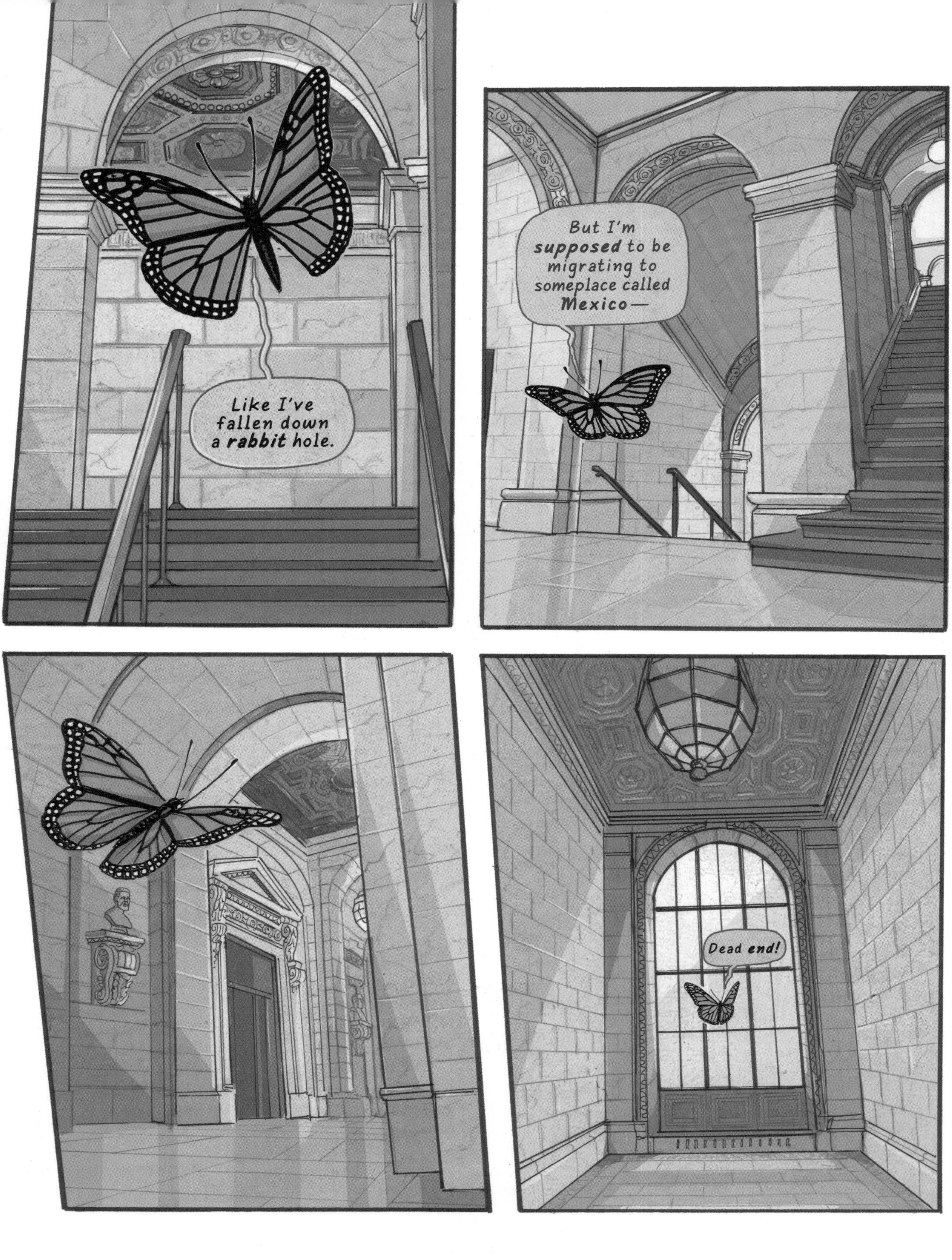
Like I've fallen down a **rabbit** hole.
But I'm **supposed** to be migrating to someplace called **Mexico**—
Dead **end**!

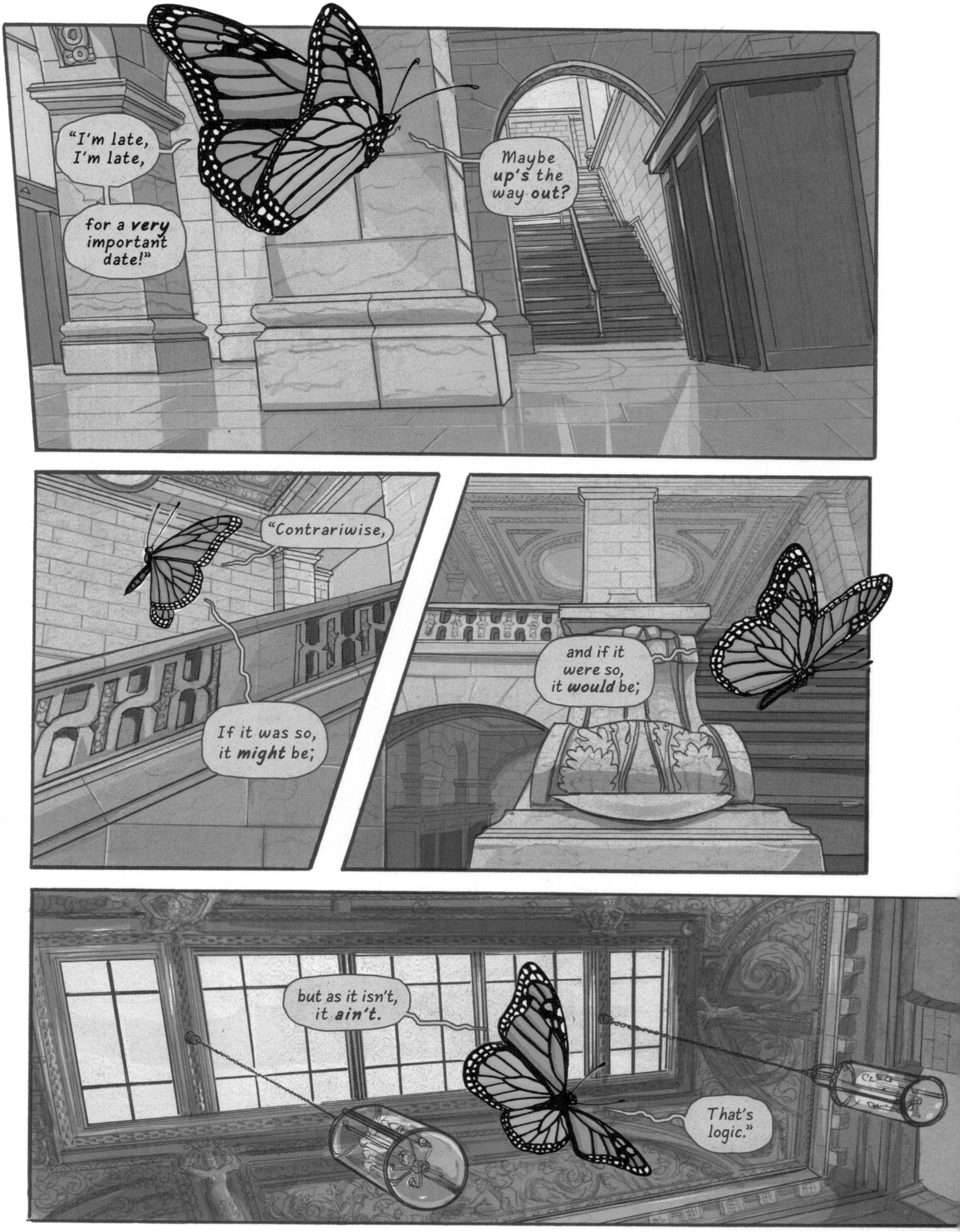
"I'm late, I'm late,
for a **very** important date!"
Maybe **up's** the way **out?**
"Contrariwise,
If it was so, it **might** be;
and if it were so, it **would** be;
but as it isn't, it **ain't.**
That's logic."

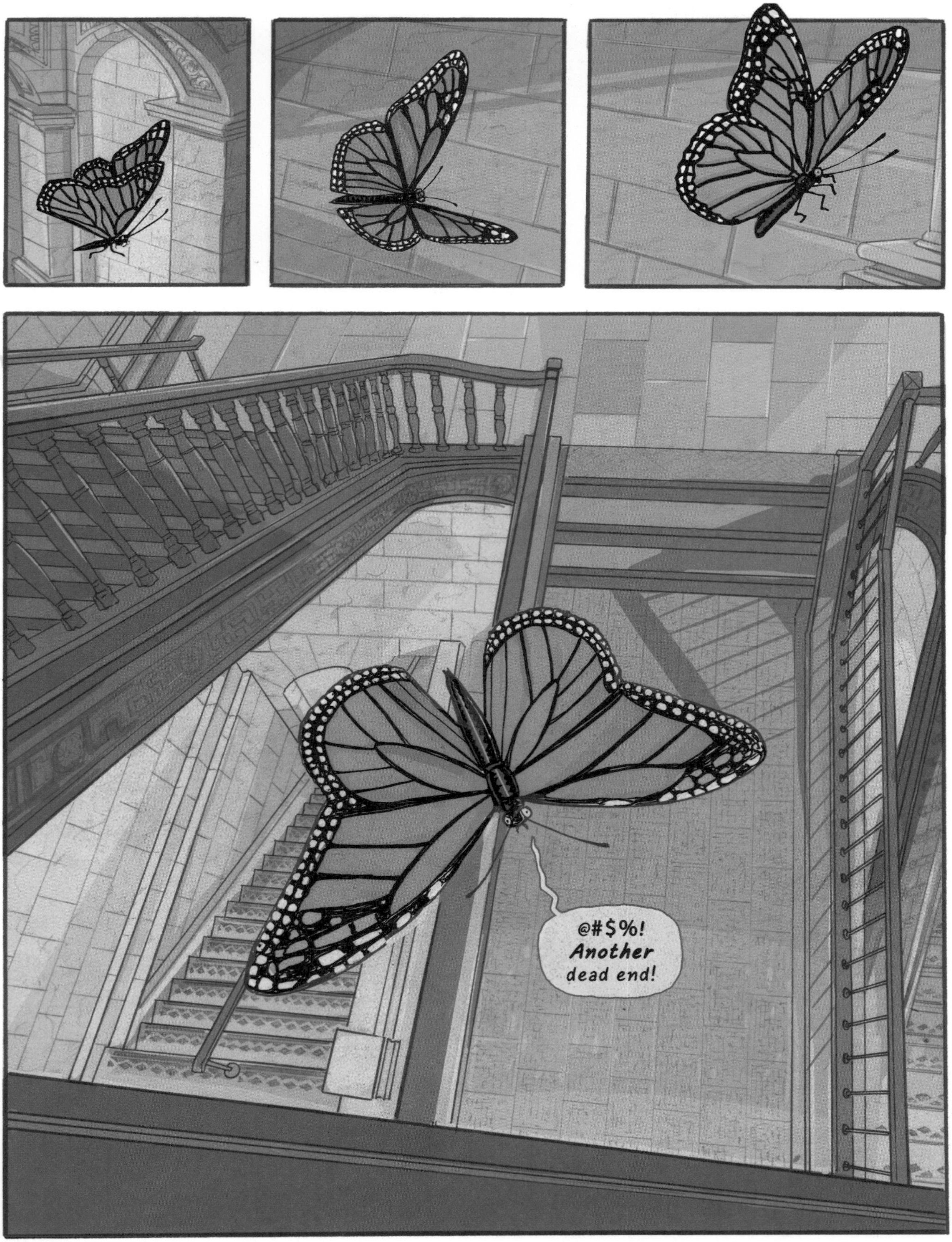
@#$%! Another dead end!

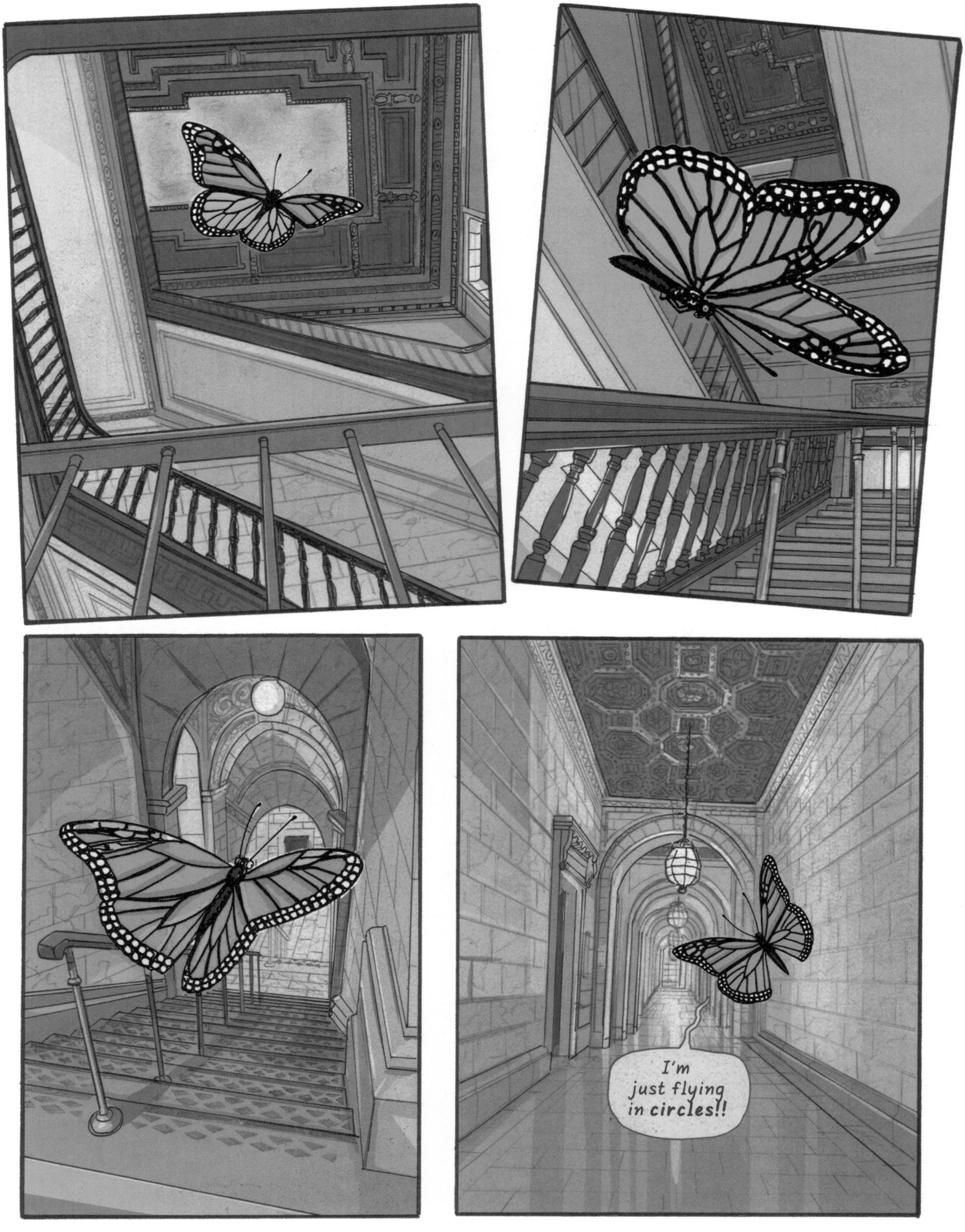
I'm just flying in circles!!

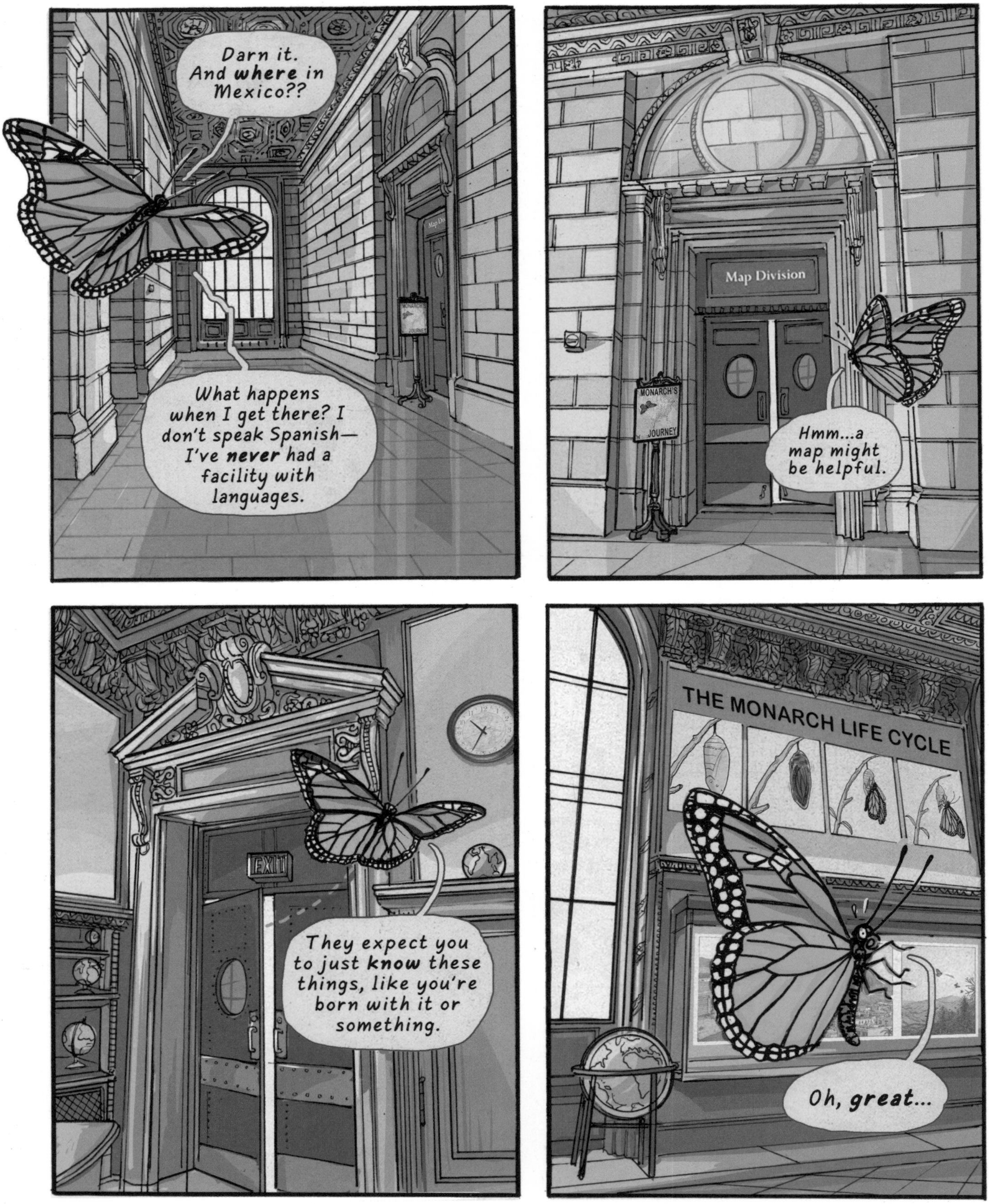
Darn it. And **where** in Mexico??
What happens when I get there? I don't speak Spanish— I've **never** had a facility with languages.
Map Division
MONARCH'S JOURNEY
Hmm...a map might be helpful.
EXIT
They expect you to just **know** these things, like you're born with it or something.
THE MONARCH LIFE CYCLE
Oh, **great**...

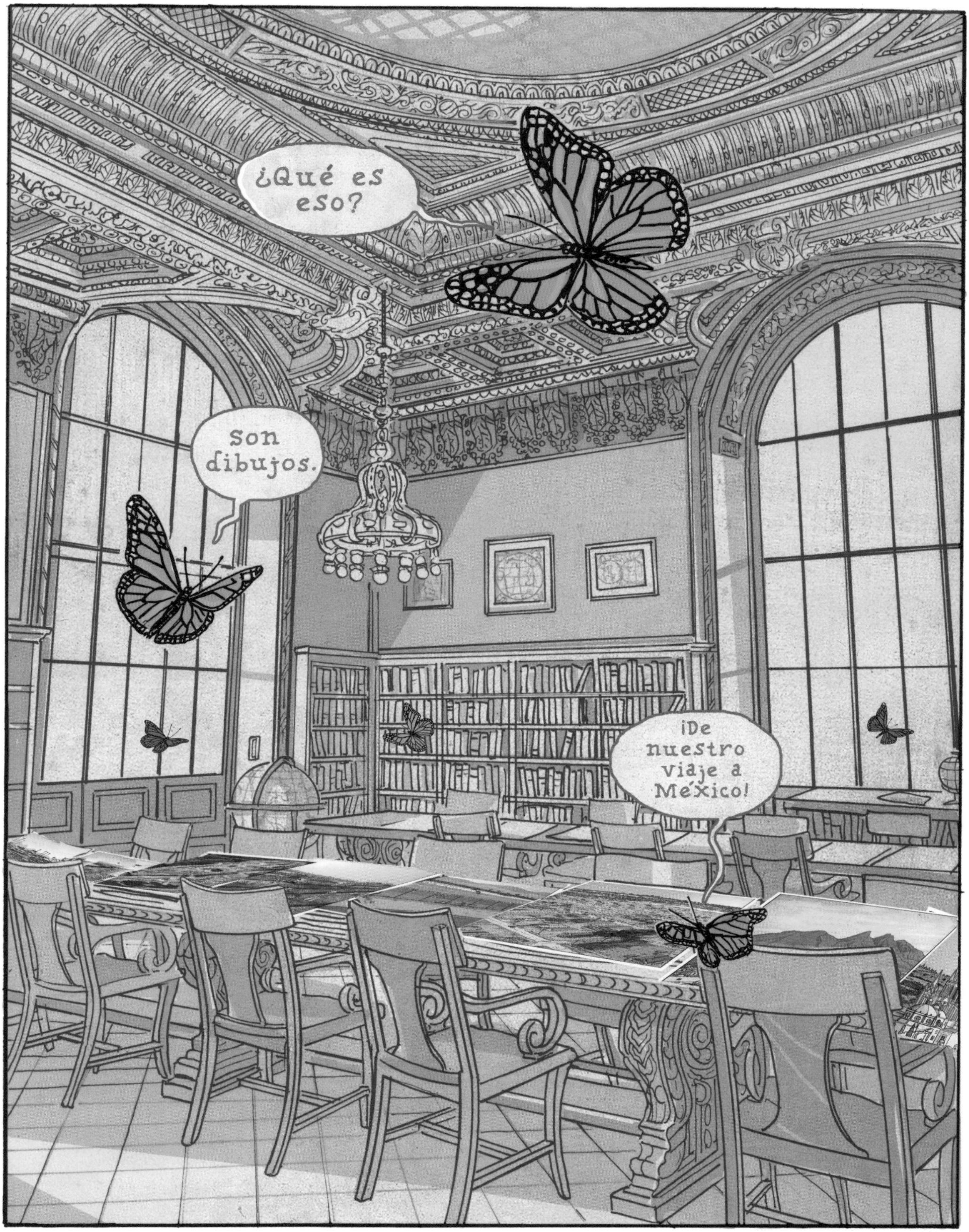
¿Qué es eso?
Son dibujos.
¡De nuestro viaje a México!

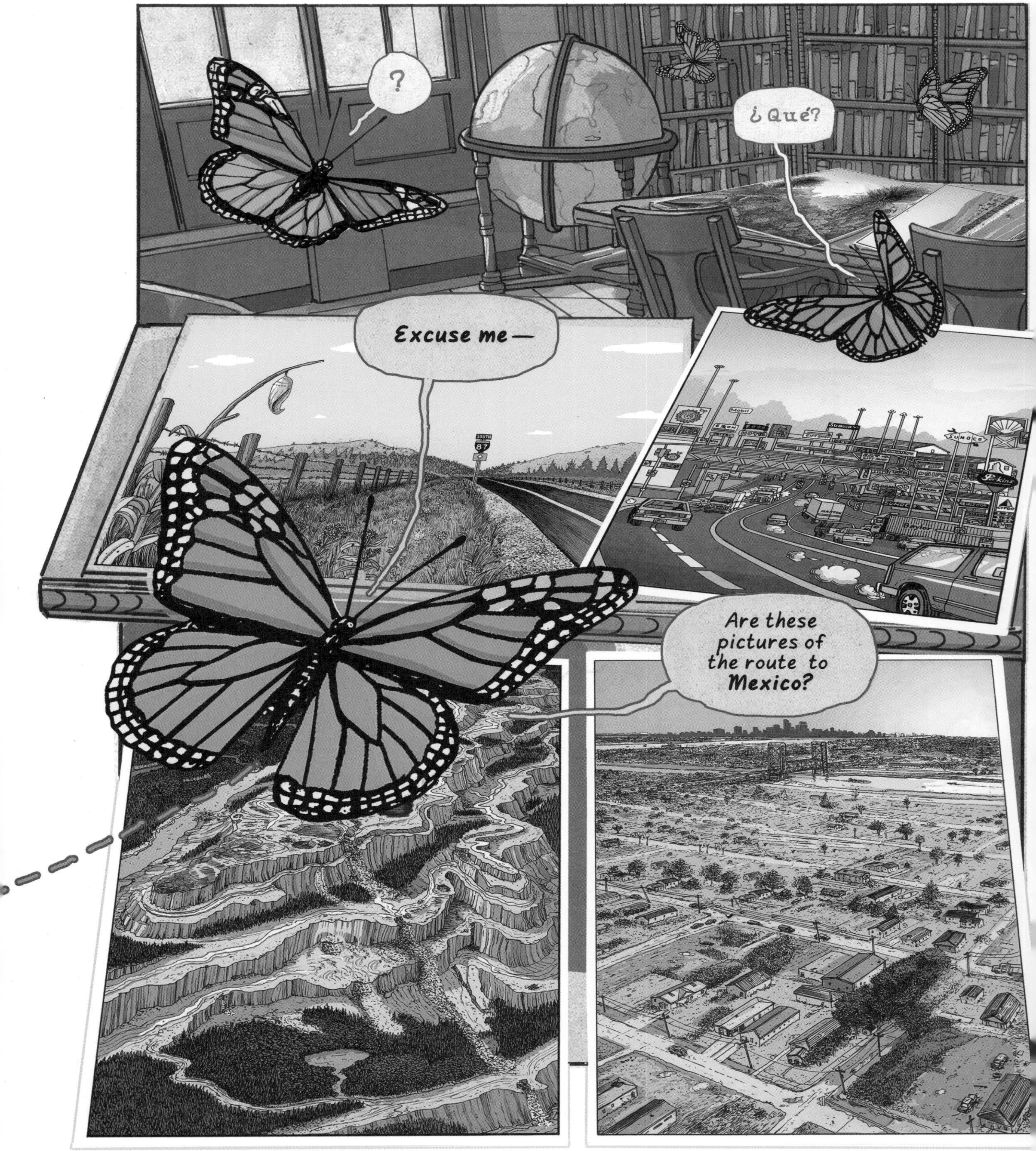
?
¿Qué?
Excuse me—
Are these pictures of the route to Mexico?

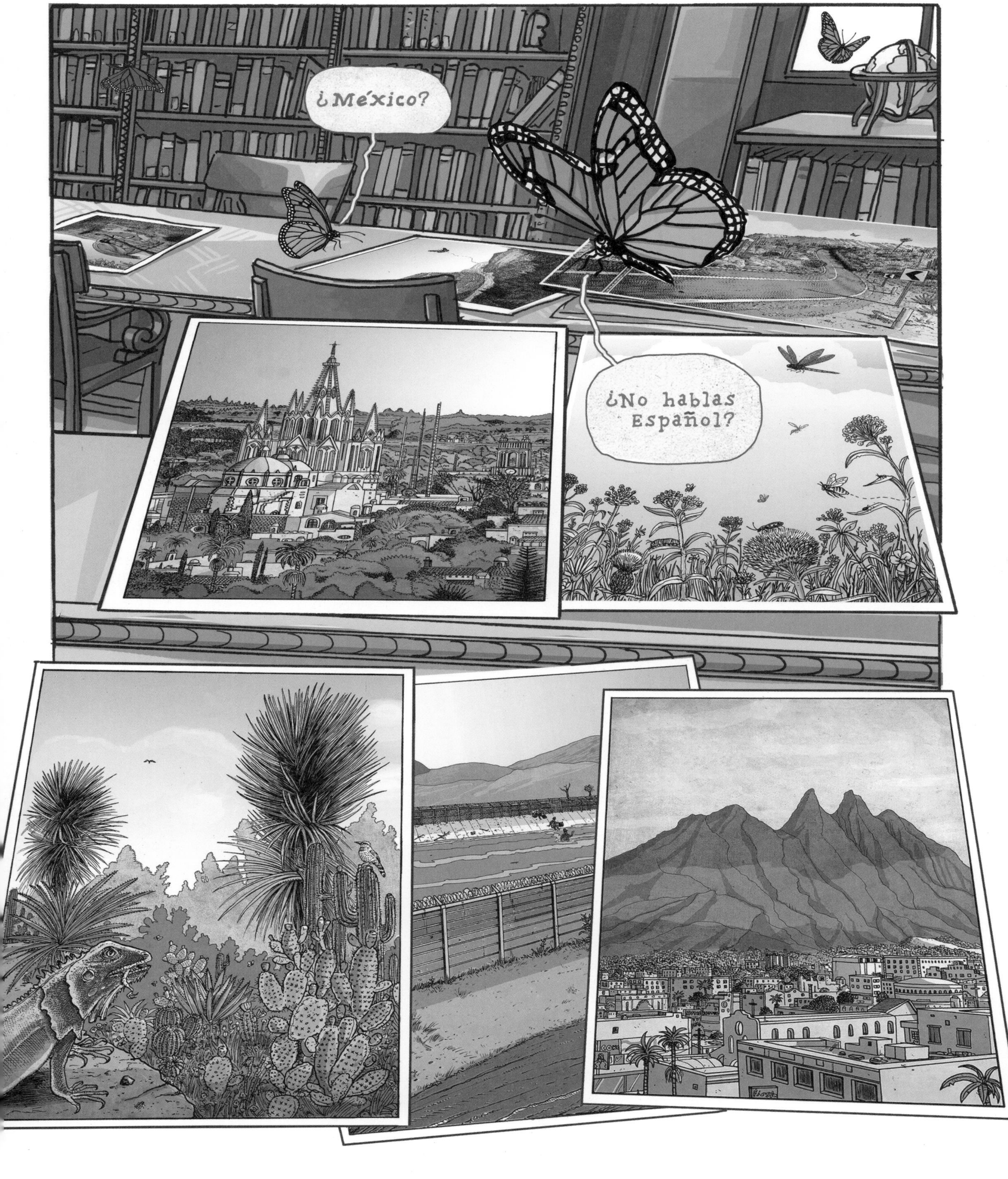
¿México?
¿No hablas Español?

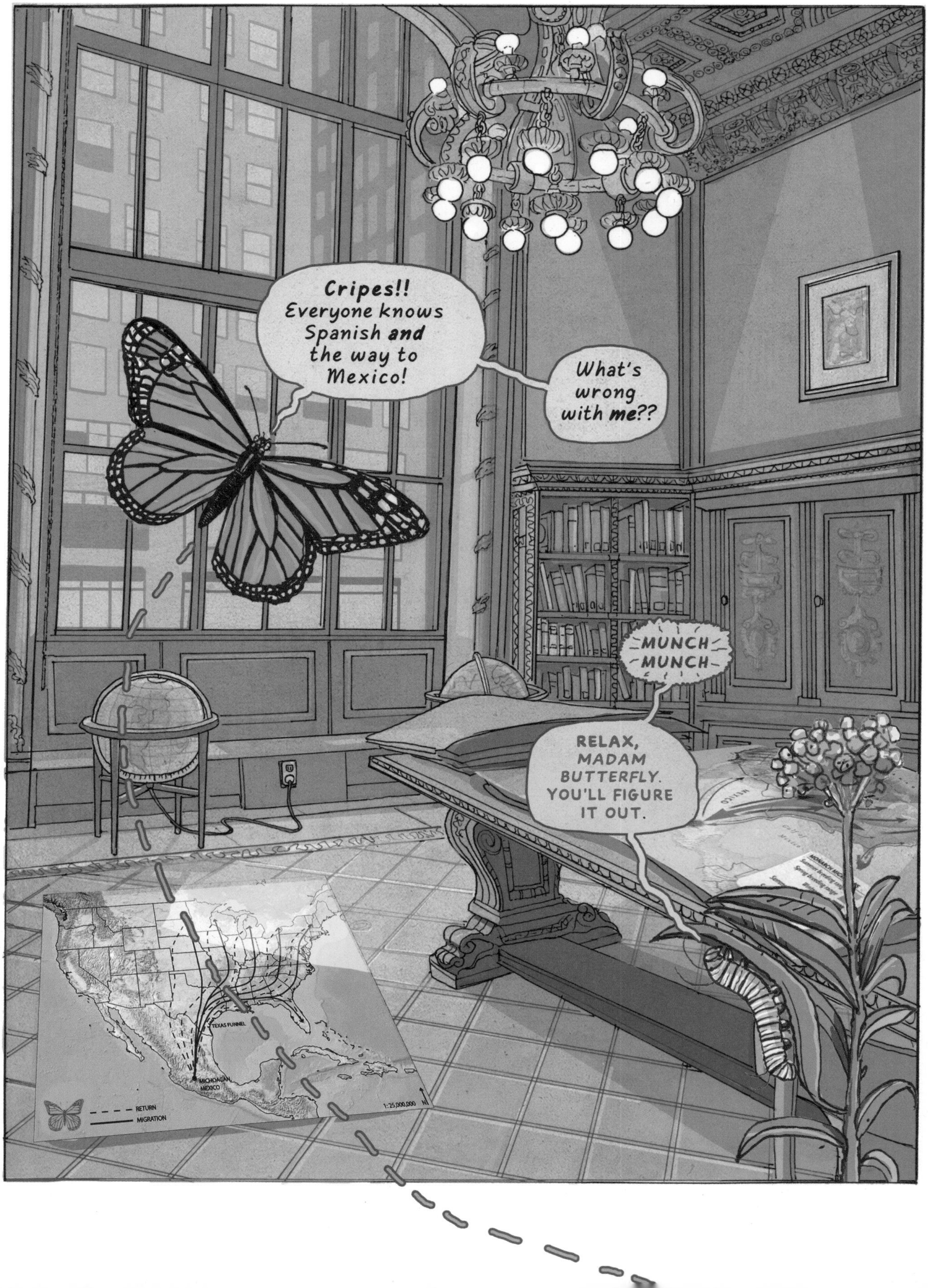
Cripes!! Everyone knows Spanish and the way to Mexico!
What's wrong with me??
MUNCH MUNCH
RELAX, MADAM BUTTERFLY. YOU'LL FIGURE IT OUT.
TEXAS FUNNEL
MICHOACÁN MEXICO
RETURN
MIGRATION
1:25,000,000

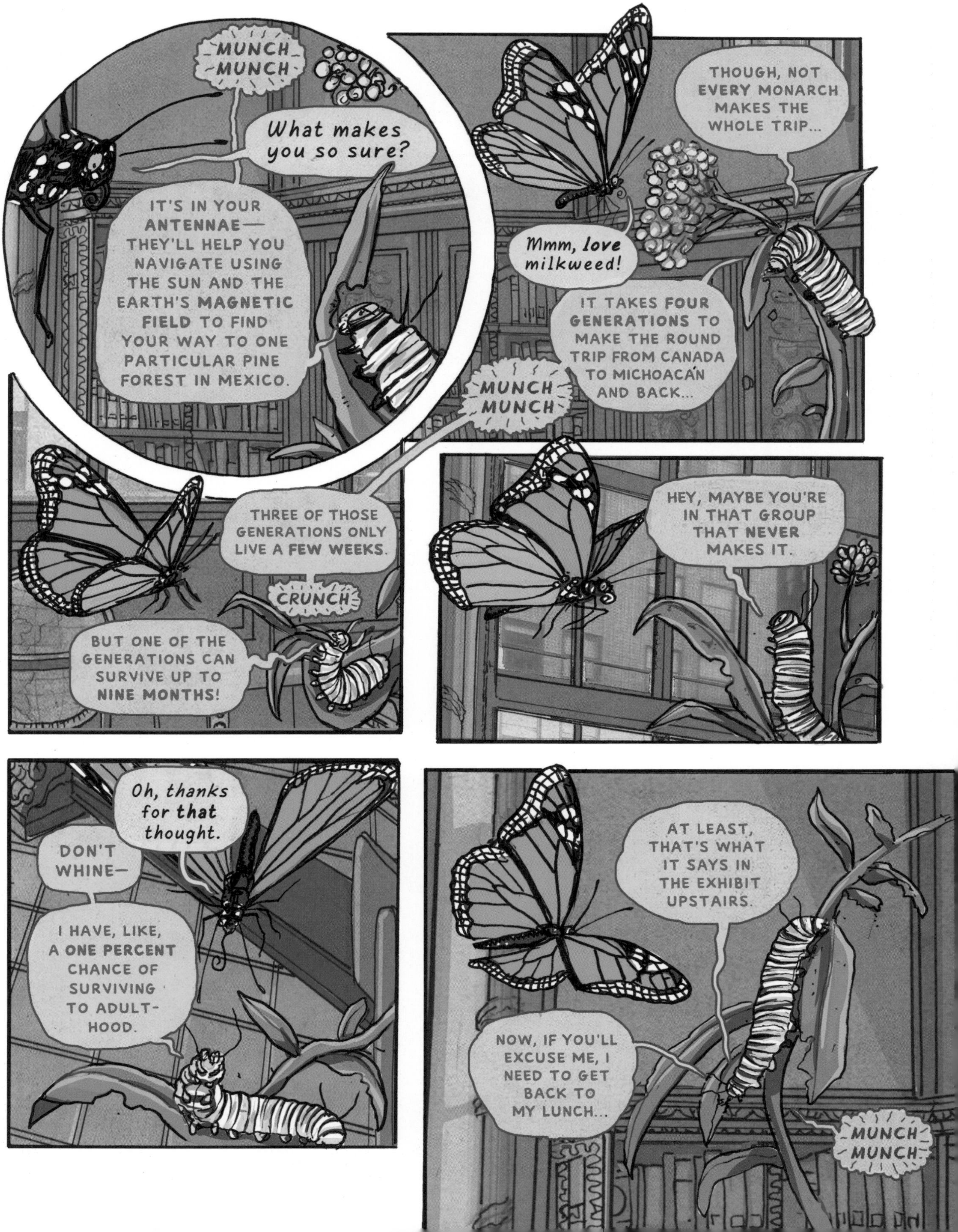
MUNCH MUNCH
What makes you so sure?
IT'S IN YOUR ANTENNAE— THEY'LL HELP YOU NAVIGATE USING THE SUN AND THE EARTH'S MAGNETIC FIELD TO FIND YOUR WAY TO ONE PARTICULAR PINE FOREST IN MEXICO.
THOUGH, NOT EVERY MONARCH MAKES THE WHOLE TRIP...
Mmm, love milkweed!
IT TAKES FOUR GENERATIONS TO MAKE THE ROUND TRIP FROM CANADA TO MICHOACÁN AND BACK...
MUNCH MUNCH
THREE OF THOSE GENERATIONS ONLY LIVE A FEW WEEKS.
CRUNCH
BUT ONE OF THE GENERATIONS CAN SURVIVE UP TO NINE MONTHS!
HEY, MAYBE YOU'RE IN THAT GROUP THAT NEVER MAKES IT.
Oh, thanks for that thought.
DON'T WHINE—
I HAVE, LIKE, A ONE PERCENT CHANCE OF SURVIVING TO ADULT-HOOD.
AT LEAST, THAT'S WHAT IT SAYS IN THE EXHIBIT UPSTAIRS.
NOW, IF YOU'LL EXCUSE ME, I NEED TO GET BACK TO MY LUNCH...
MUNCH MUNCH

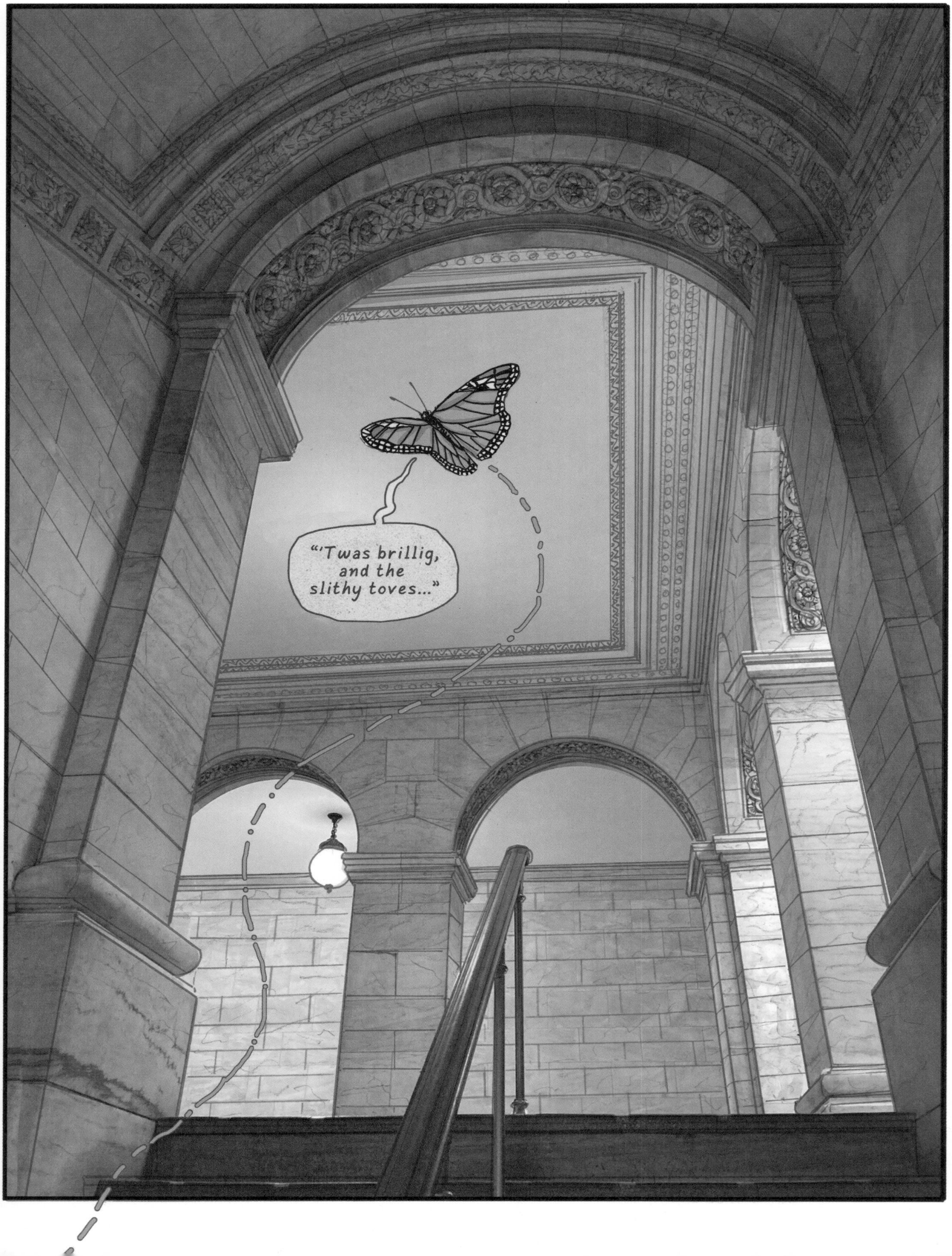
"'Twas brillig, and the slithy toves..."

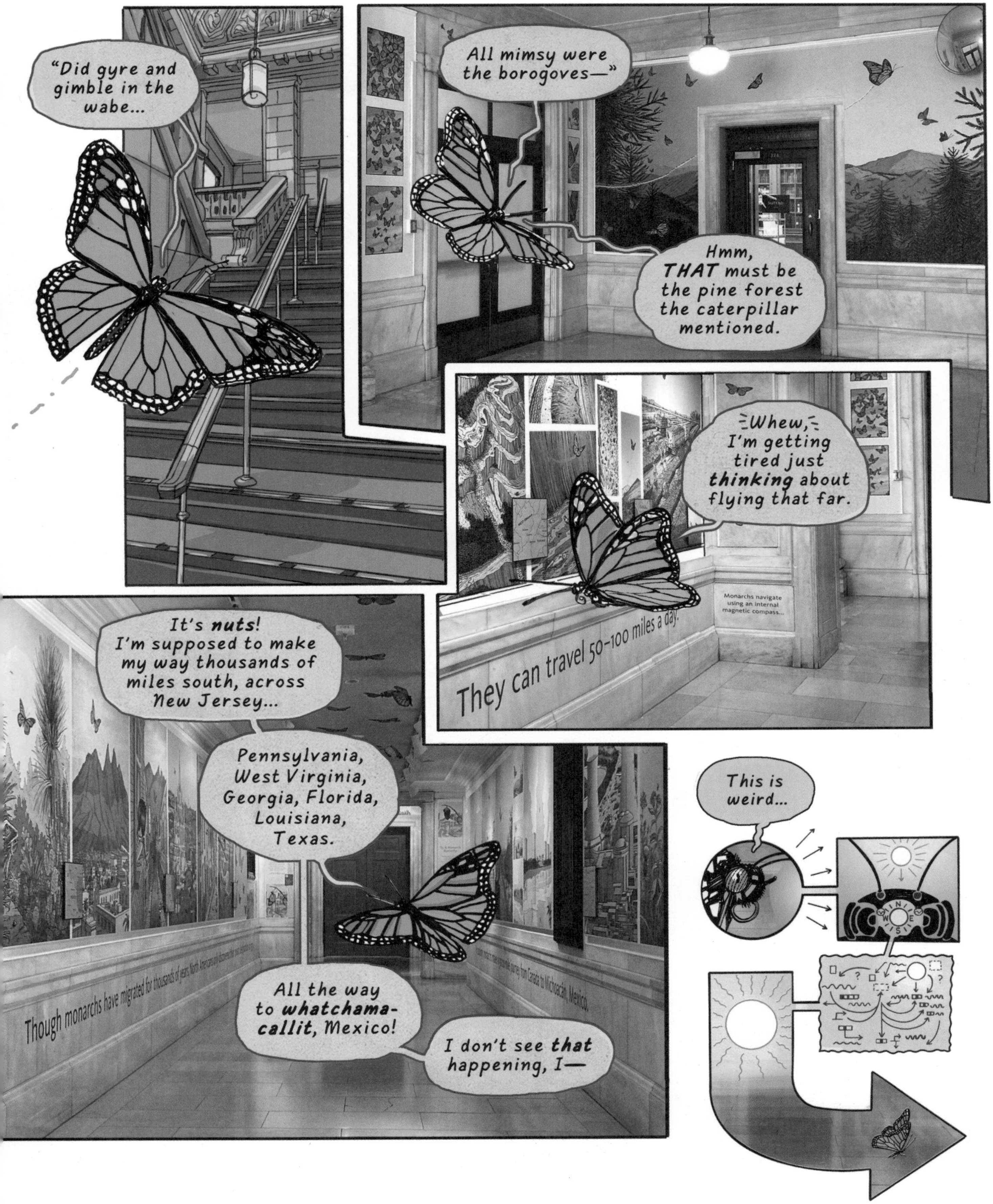
"Did gyre and gimble in the wabe...
All mimsy were the borogoves—"
Hmm, THAT must be the pine forest the caterpillar mentioned.
Whew, I'm getting tired just thinking about flying that far.
Monarchs navigate using an internal magnetic compass...
They can travel 50–100 miles a day.
It's nuts! I'm supposed to make my way thousands of miles south, across New Jersey...
Pennsylvania, West Virginia, Georgia, Florida, Louisiana, Texas.
All the way to whatchama-callit, Mexico!
I don't see that happening, I—
Though monarchs have migrated for thousands of years,
This is weird...

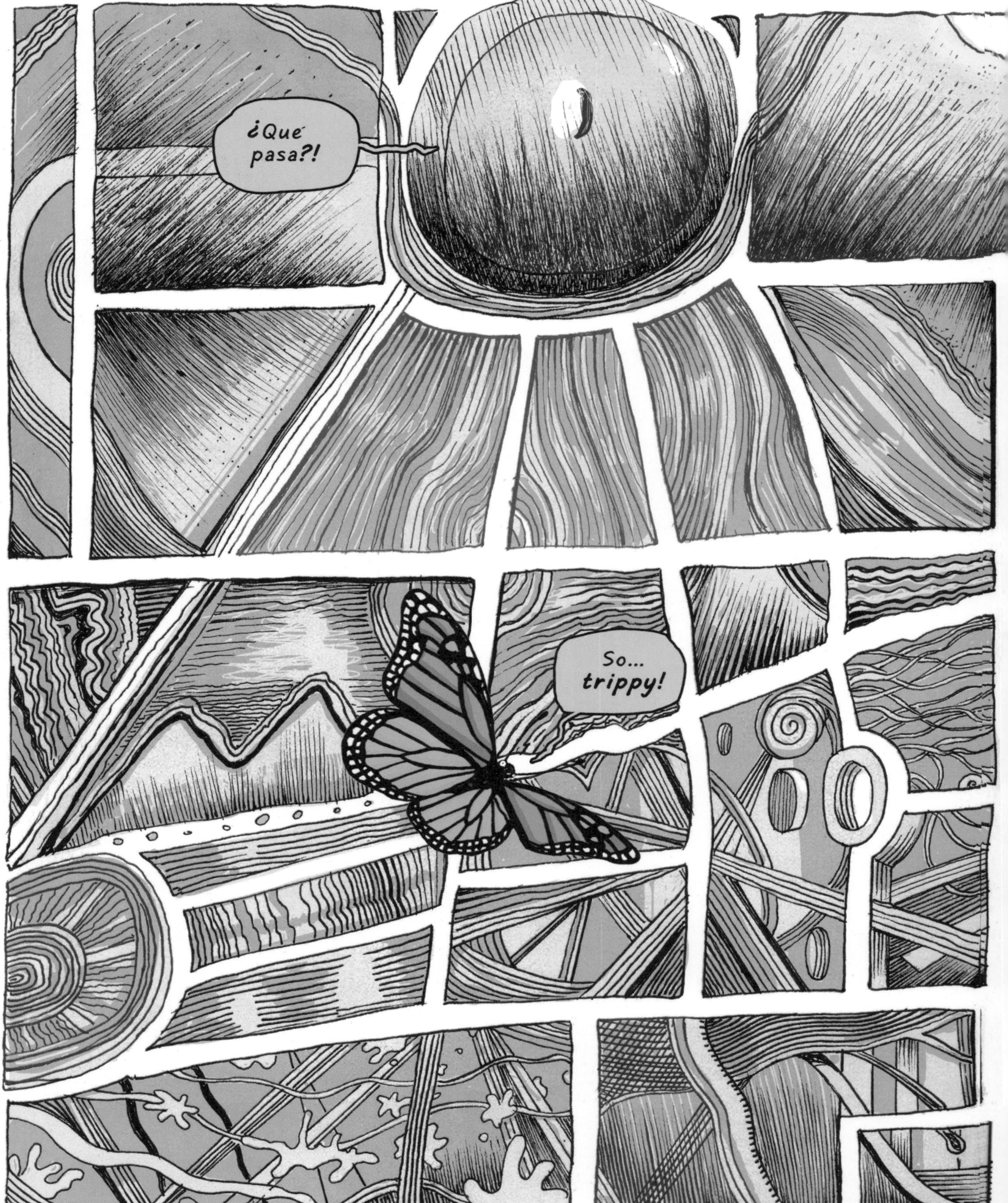
¿Qué pasa?!
So... trippy!

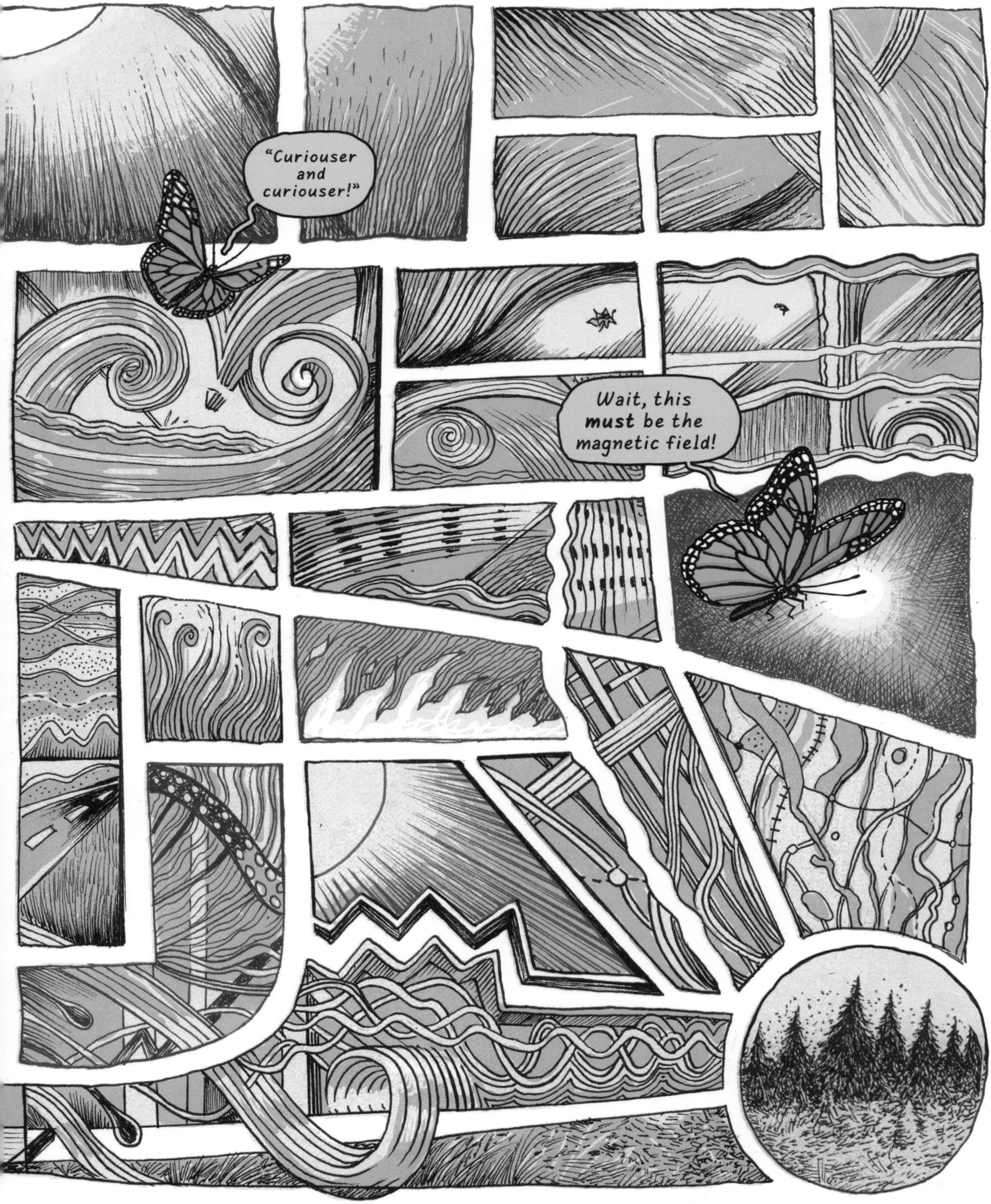
"Curiouser and curiouser!"
Wait, this **must** be the magnetic field!

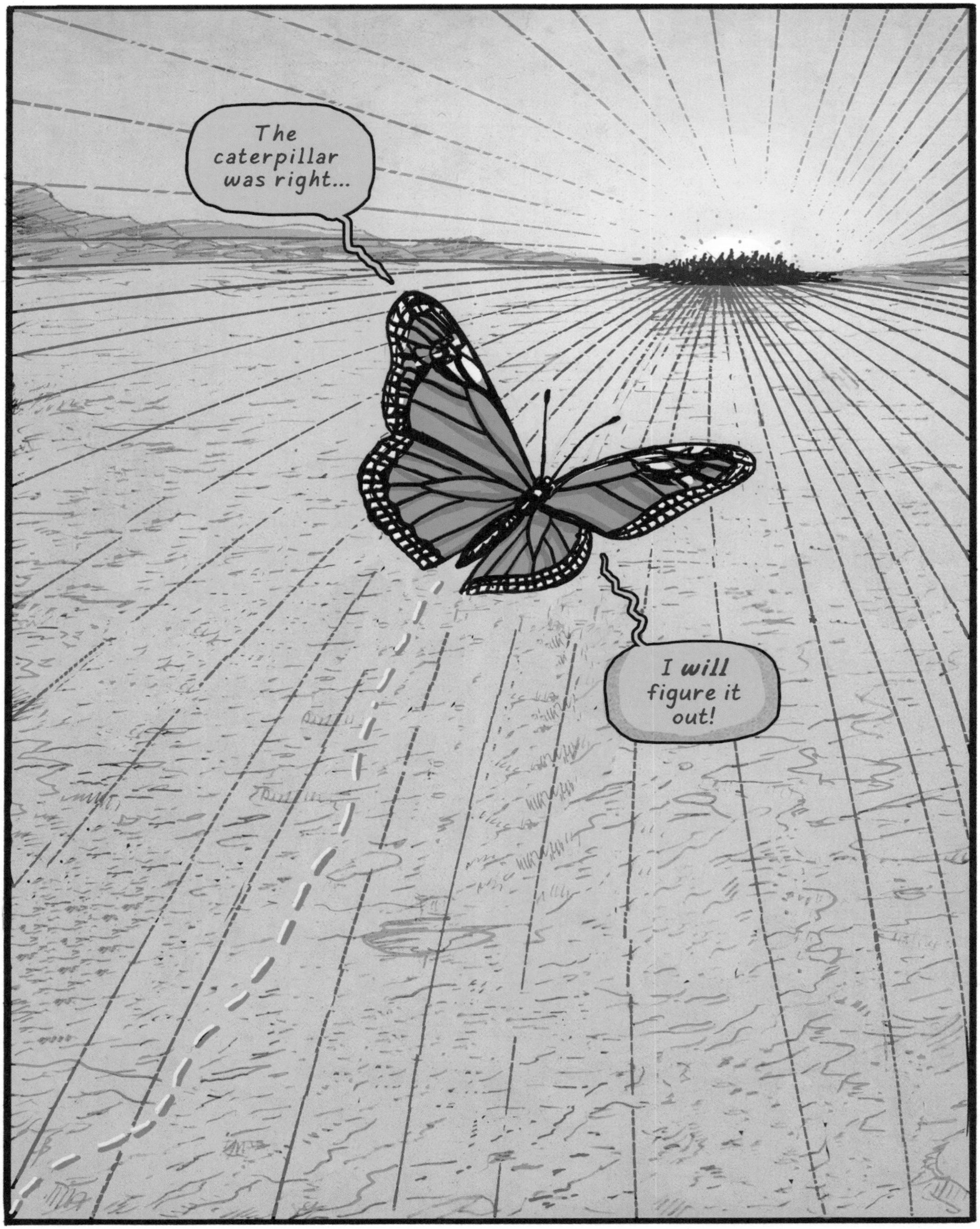
The caterpillar was right...
I will figure it out!

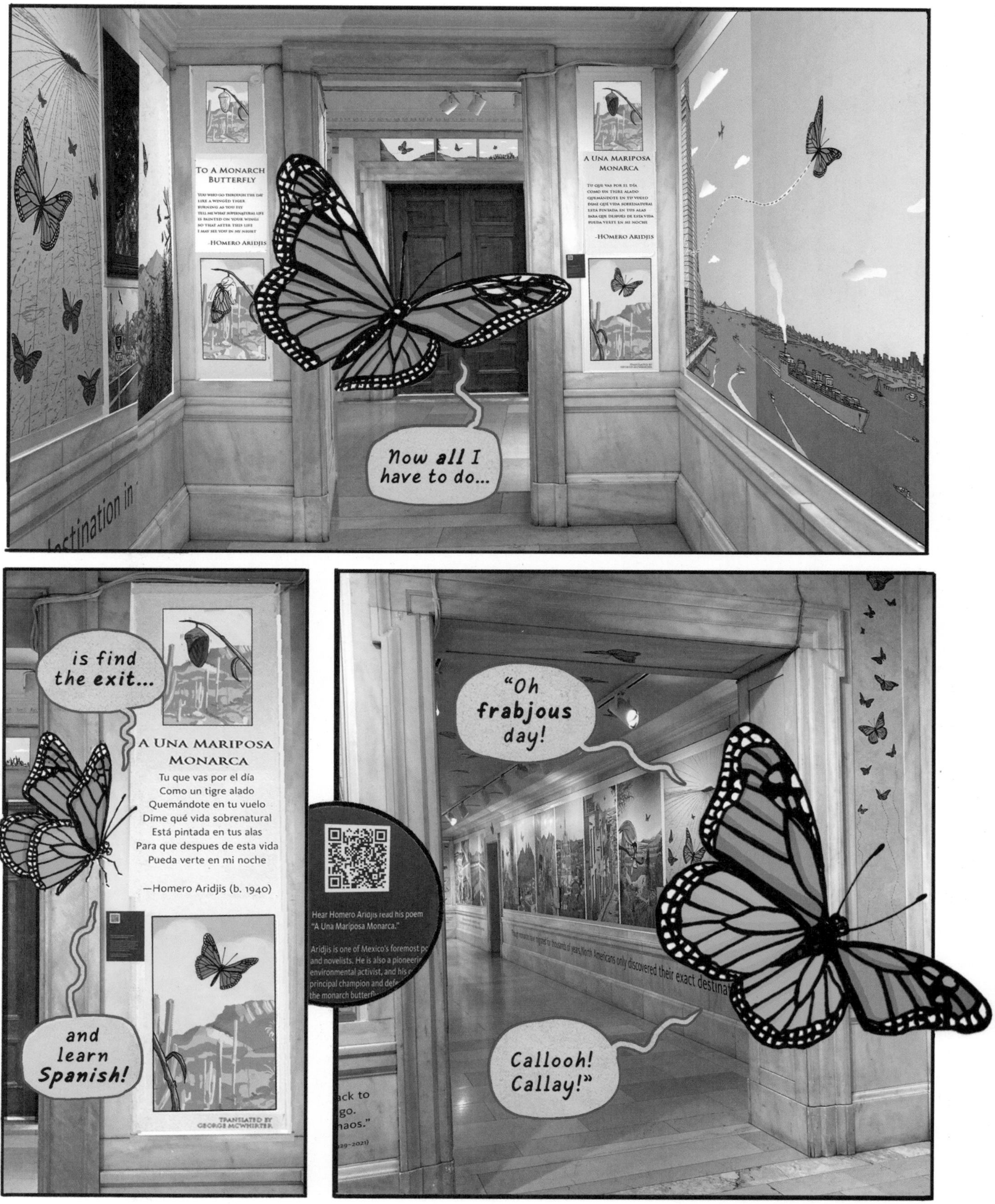
TO A MONARCH BUTTERFLY
-HOMERO ARIDJIS
A UNA MARIPOSA MONARCA
-HOMERO ARIDJIS
Now all I have to do...
is find the exit...
A Una Mariposa Monarca
Tu que vas por el día
Como un tigre alado
Quemándote en tu vuelo
Dime qué vida sobrenatural
Está pintada en tus alas
Para que despues de esta vida
Pueda verte en mi noche
—Homero Aridjis (b. 1940)
TRANSLATED BY GEORGE MCWHIRTER
and learn Spanish!
Hear Homero Aridjis read his poem "A Una Mariposa Monarca."
"Oh frabjous day!
Callooh! Callay!"

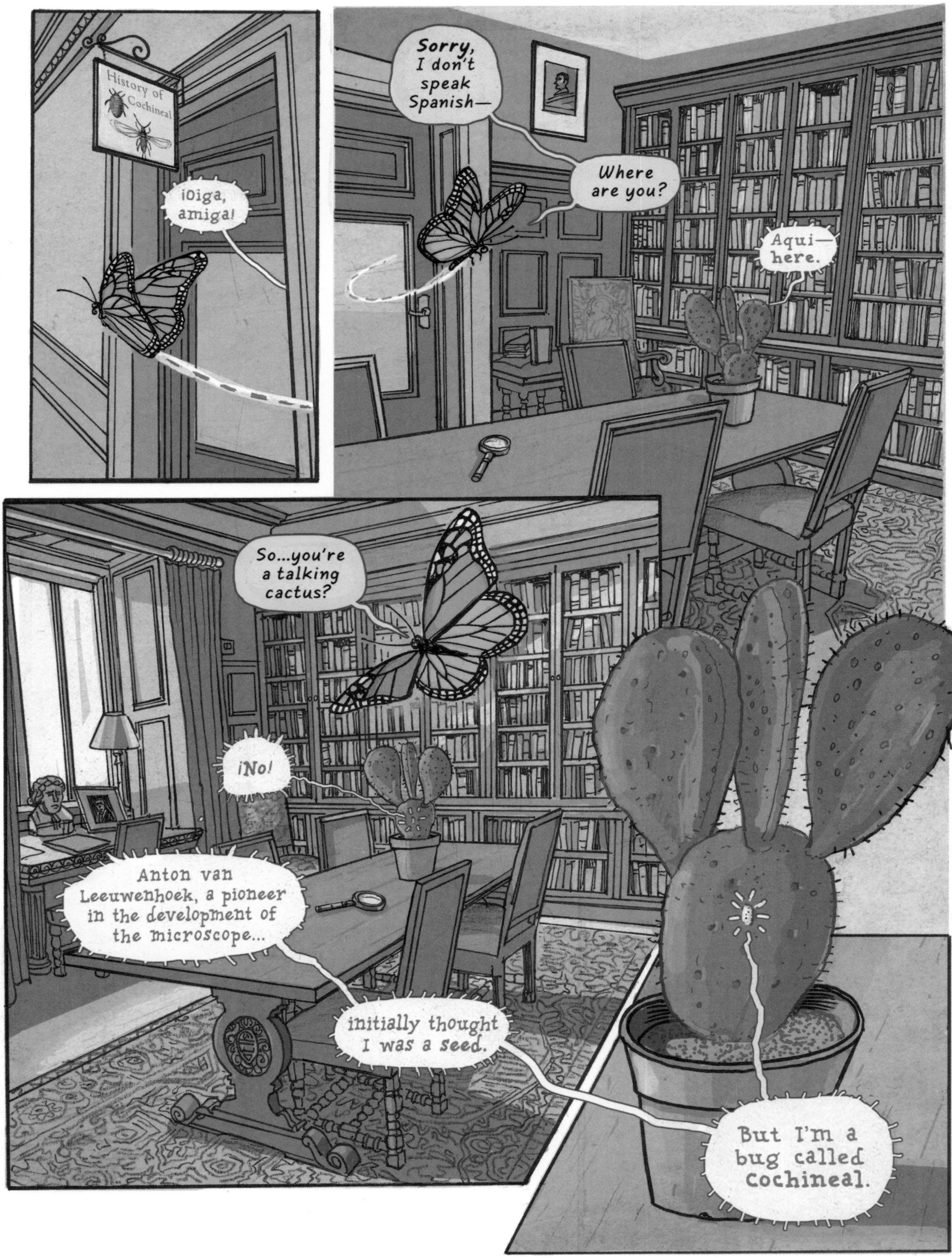

History of Cochineal
¡Oiga, amiga!
Sorry, I don't speak Spanish—
Where are you?
Aqui— here.
So...you're a talking cactus?
¡No!
Anton van Leeuwenhoek, a pioneer in the development of the microscope...
initially thought I was a seed.
But I'm a bug called cochineal.

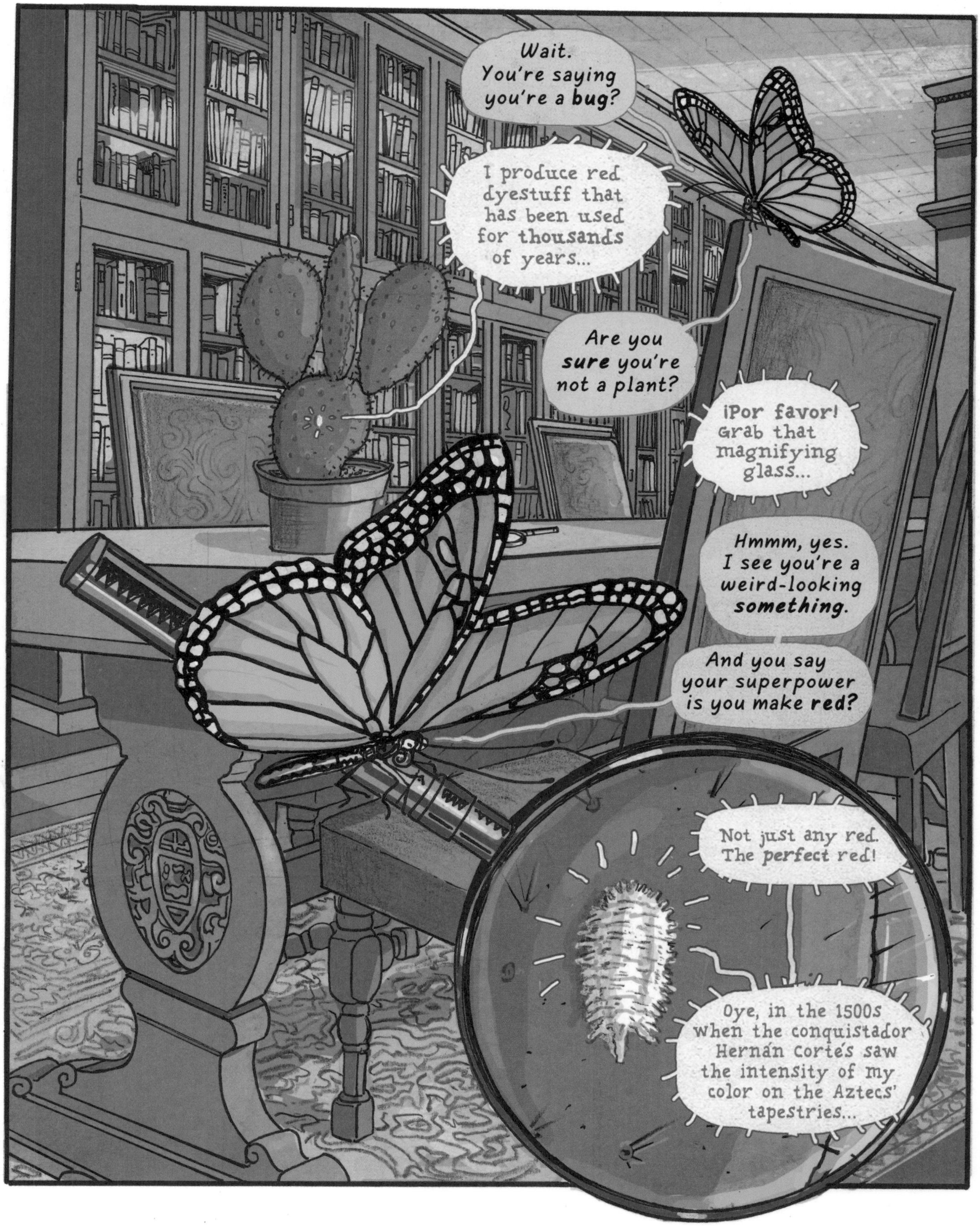
Wait. You're saying you're a **bug**?
I produce red dyestuff that has been used for thousands of years...
Are you ***sure*** you're not a plant?
¡Por favor! Grab that magnifying glass...
Hmmm, yes. I see you're a weird-looking ***something***.
And you say your superpower is you make **red**?
Not just any red. The *perfect* red!
Oye, in the 1500s when the conquistador Hernán Cortés saw the intensity of my color on the Aztecs' tapestries...

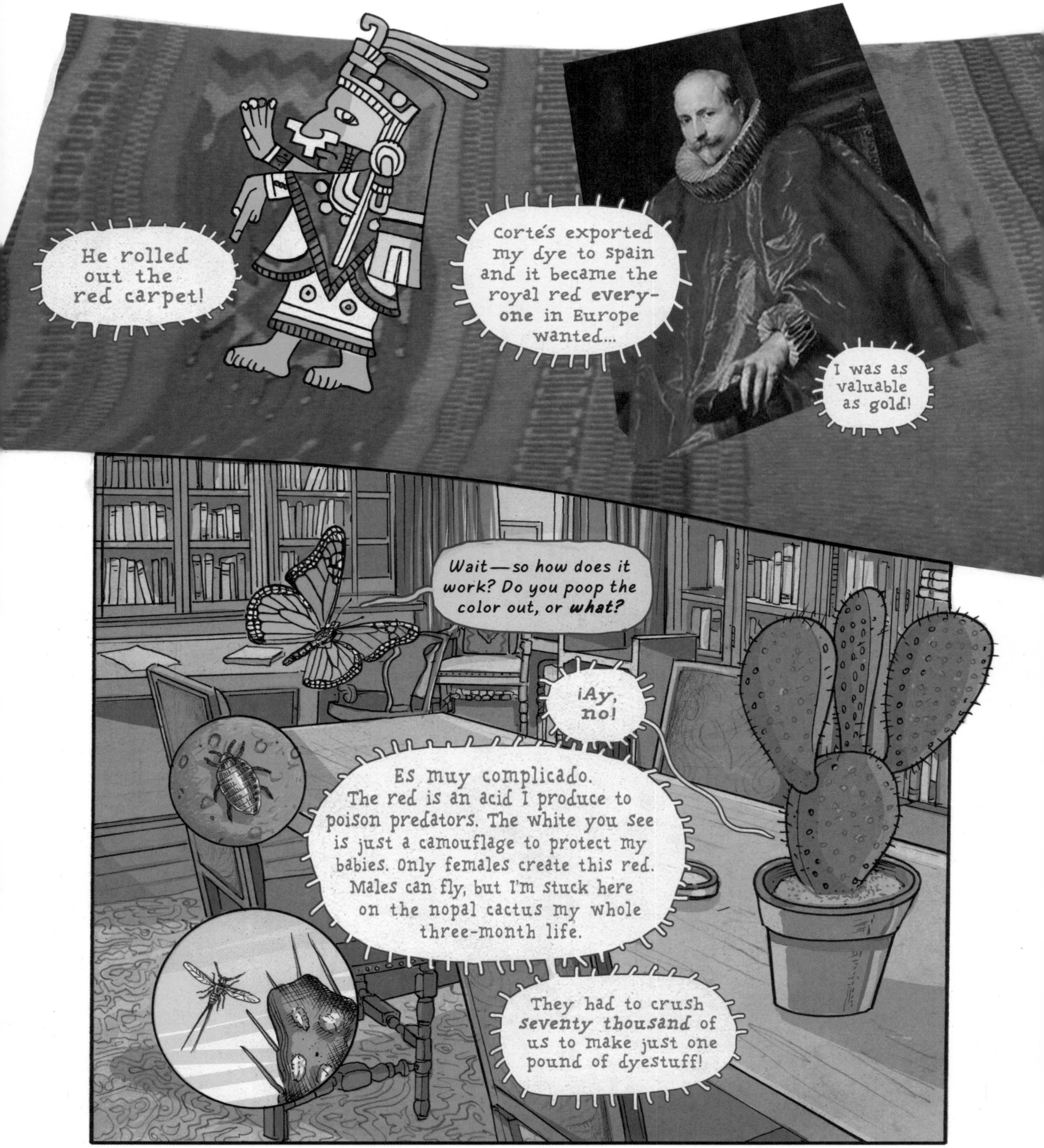
He rolled out the red carpet!
Cortés exported my dye to Spain and it became the royal red every-one in Europe wanted...
I was as valuable as gold!
Wait—so how does it work? Do you poop the color out, or ***what?***
¡Ay, no!
Es muy complicado. The red is an acid I produce to poison predators. The white you see is just a camouflage to protect my babies. Only females create this red. Males can fly, but I'm stuck here on the nopal cactus my whole three-month life.
They had to crush *seventy thousand* of us to make just one pound of dyestuff!

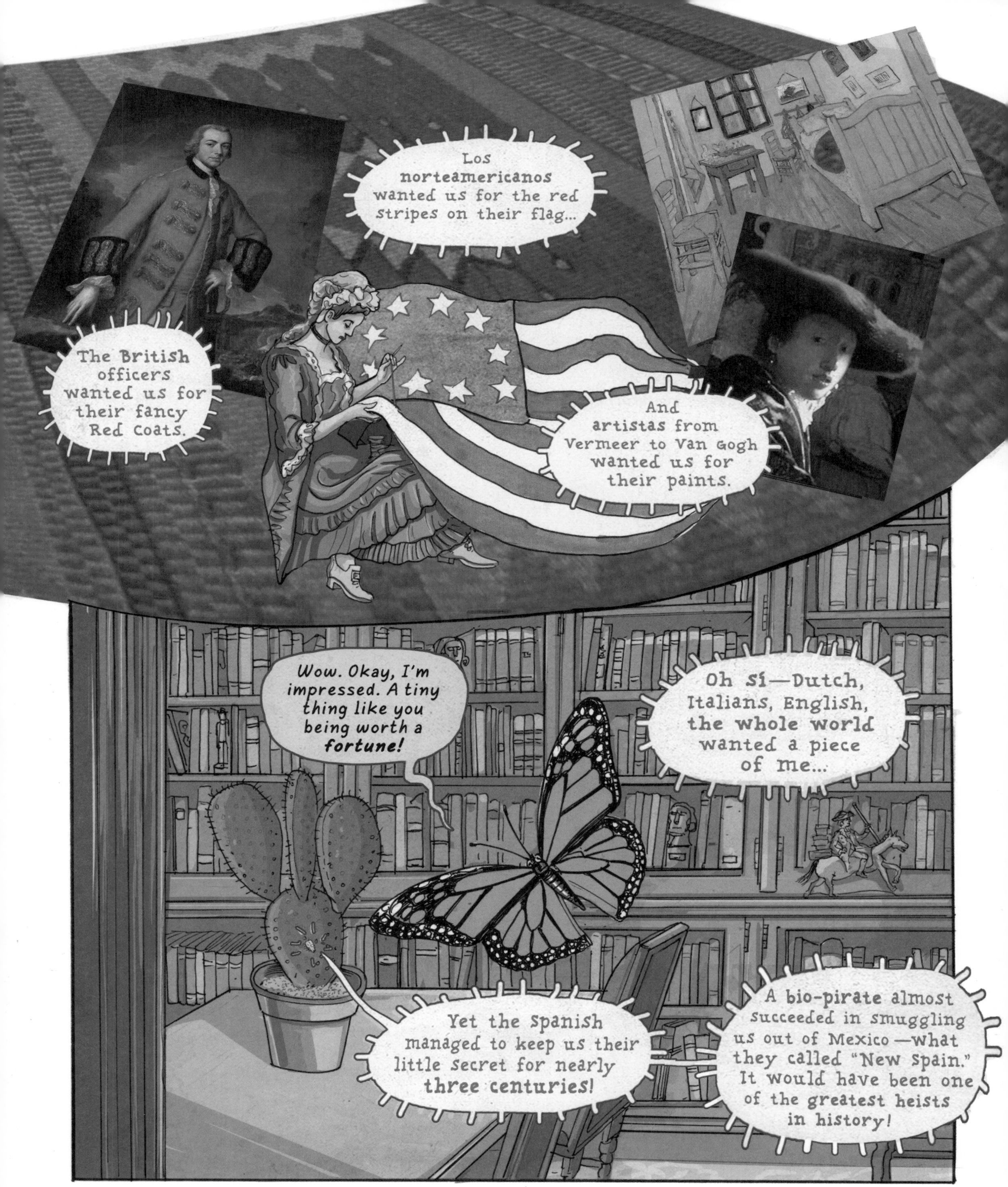

Los norteamericanos wanted us for the red stripes on their flag...
The British officers wanted us for their fancy Red Coats.
And artistas from Vermeer to Van Gogh wanted us for their paints.
Wow. Okay, I'm impressed. A tiny thing like you being worth a fortune!
Oh sí—Dutch, Italians, English, the whole world wanted a piece of me...
Yet the Spanish managed to keep us their little secret for nearly three centuries!
A bio-pirate almost succeeded in smuggling us out of Mexico—what they called "New Spain." It would have been one of the greatest heists in history!

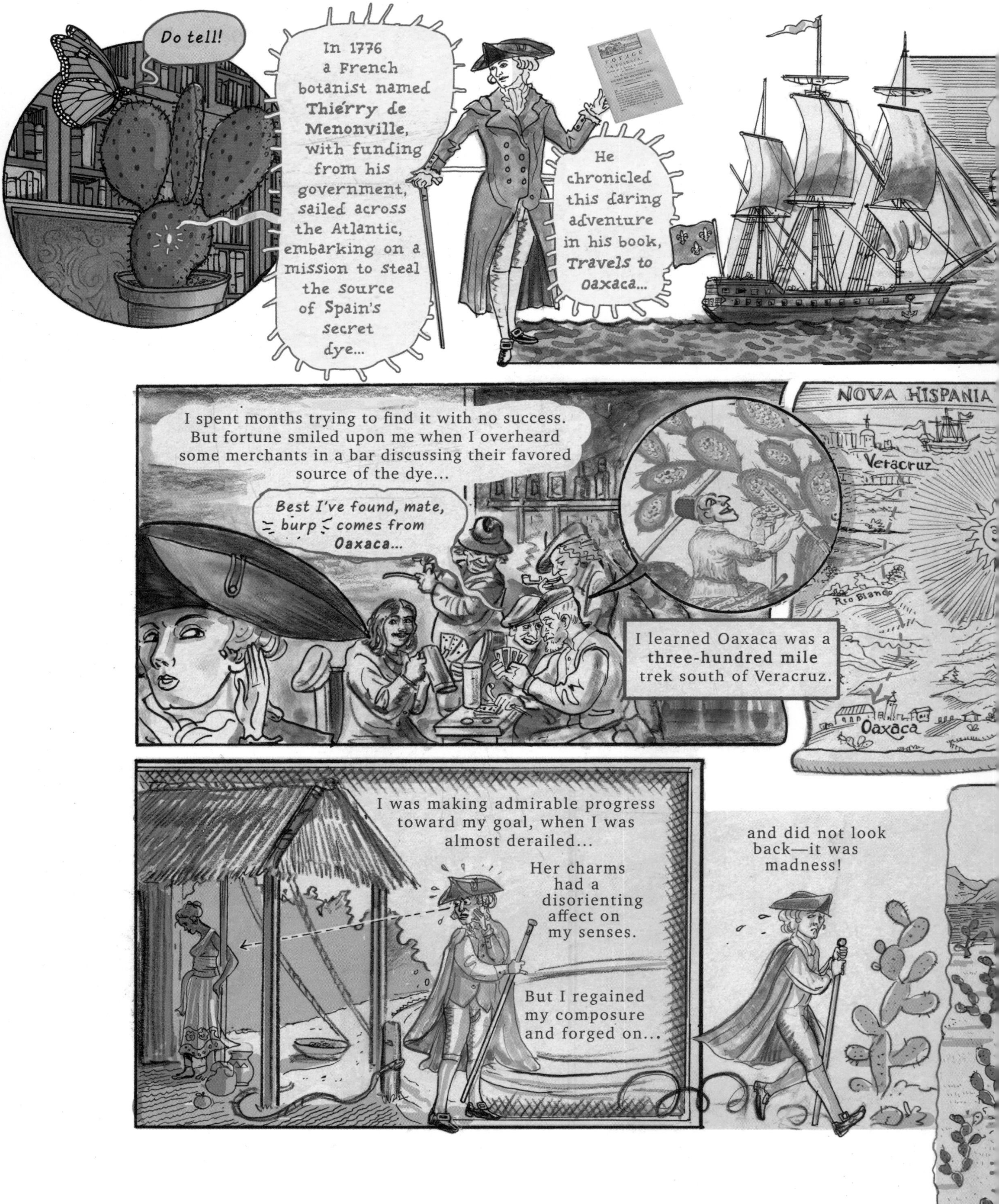

Do tell!
In 1776 a French botanist named **Thiérry de Menonville,** with funding from his government, sailed across the Atlantic, embarking on a mission to steal the source of Spain's secret dye...
He chronicled this daring adventure in his book, ***Travels to Oaxaca...***
I spent months trying to find it with no success. But fortune smiled upon me when I overheard some merchants in a bar discussing their favored source of the dye...
Best I've found, mate, burp comes from Oaxaca...
NOVA HISPANIA
Veracruz
Rio Blanco
Oaxaca
I learned Oaxaca was a **three-hundred mile** trek south of Veracruz.
I was making admirable progress toward my goal, when I was almost derailed...
Her charms had a disorienting affect on my senses.
But I regained my composure and forged on...
and did not look back—it was madness!

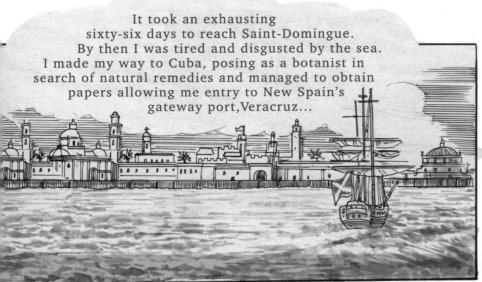
It took an exhausting sixty-six days to reach Saint-Domingue. By then I was tired and disgusted by the sea. I made my way to Cuba, posing as a botanist in search of natural remedies and managed to obtain papers allowing me entry to New Spain's gateway port, Veracruz...

Continuing my ruse, I wandered Veracruz always with a plant and magnifying glass in hand. Remarkably, I in fact discovered a jalapa root which alleviated constipation. This rendered me famous throughout the city and increased the believability of my charade as I searched for the origin of that priceless red!

I requested a permit to travel to the interior, but the suspicious Viceroy not only refused me, he confiscated my passport and demanded that I depart Veracruz on the next available ship— leaving just three weeks hence!

Rather than return to France a failure, I decided to sneak out of the city and attempt the six-hundred-mile round trip!

It was an arduous journey through forests and rain with little food or shelter.

I had to avoid roads, for I knew that if I were caught by authorities, it would mean certain death!

With no time to spare I at last arrived in Oaxaca and found the treasure I had been seeking.

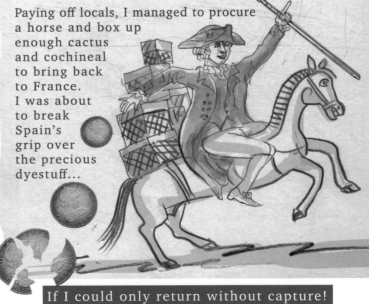
Paying off locals, I managed to procure a horse and box up enough cactus and cochineal to bring back to France. I was about to break Spain's grip over the precious dyestuff...

If I could only return without capture!

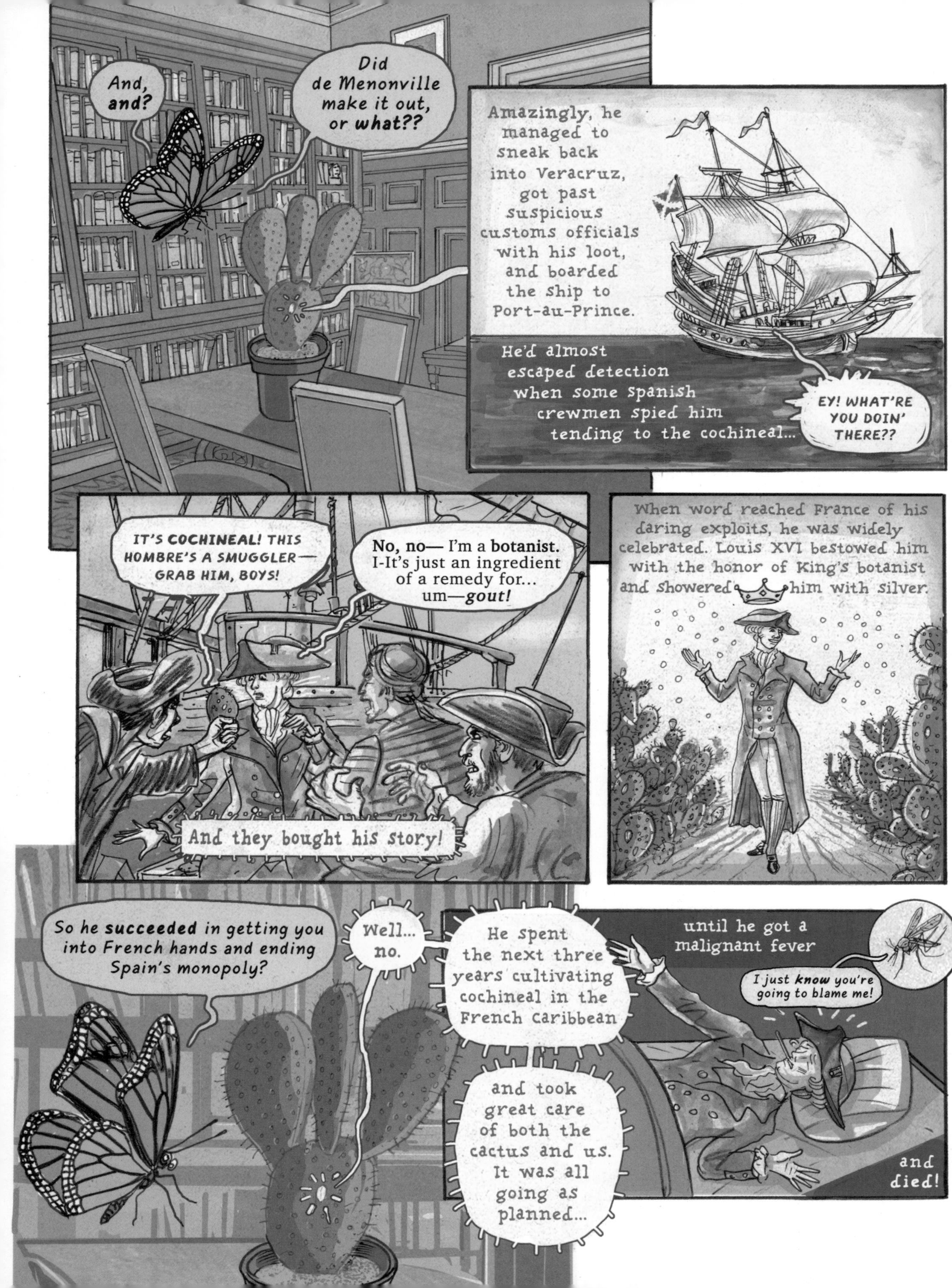
And, and?
Did de Menonville make it out, or what??
Amazingly, he managed to sneak back into Veracruz, got past suspicious customs officials with his loot, and boarded the ship to Port-au-Prince.
He'd almost escaped detection when some Spanish crewmen spied him tending to the cochineal...
EY! WHAT'RE YOU DOIN' THERE??
IT'S COCHINEAL! THIS HOMBRE'S A SMUGGLER— GRAB HIM, BOYS!
No, no— I'm a botanist. I-It's just an ingredient of a remedy for... um—gout!
And they bought his story!
When word reached France of his daring exploits, he was widely celebrated. Louis XVI bestowed him with the honor of King's botanist and showered him with silver.
So he succeeded in getting you into French hands and ending Spain's monopoly?
Well... no.
He spent the next three years cultivating cochineal in the French Caribbean
and took great care of both the cactus and us. It was all going as planned...
until he got a malignant fever
I just know you're going to blame me!
and died!

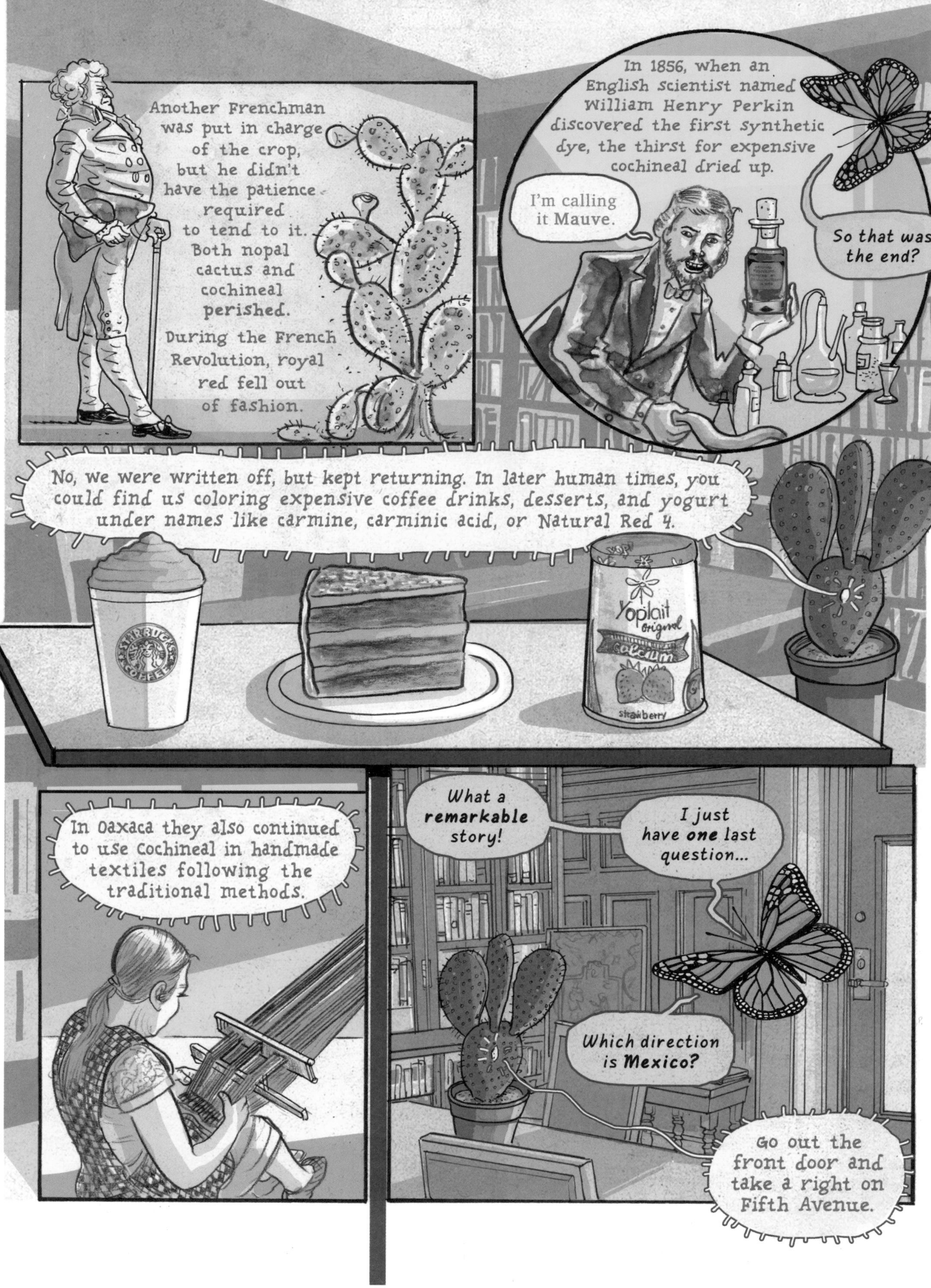
Another Frenchman was put in charge of the crop, but he didn't have the patience required to tend to it. Both nopal cactus and cochineal perished.
During the French Revolution, royal red fell out of fashion.
In 1856, when an English scientist named William Henry Perkin discovered the first synthetic dye, the thirst for expensive cochineal dried up.
I'm calling it Mauve.
So that was the end?
No, we were written off, but kept returning. In later human times, you could find us coloring expensive coffee drinks, desserts, and yogurt under names like carmine, carminic acid, or Natural Red 4.
STARBUCKS COFFEE
Yoplait Original
Calcium
strawberry
In Oaxaca they also continued to use cochineal in handmade textiles following the traditional methods.
What a remarkable story!
I just have one last question...
Which direction is Mexico?
Go out the front door and take a right on Fifth Avenue.

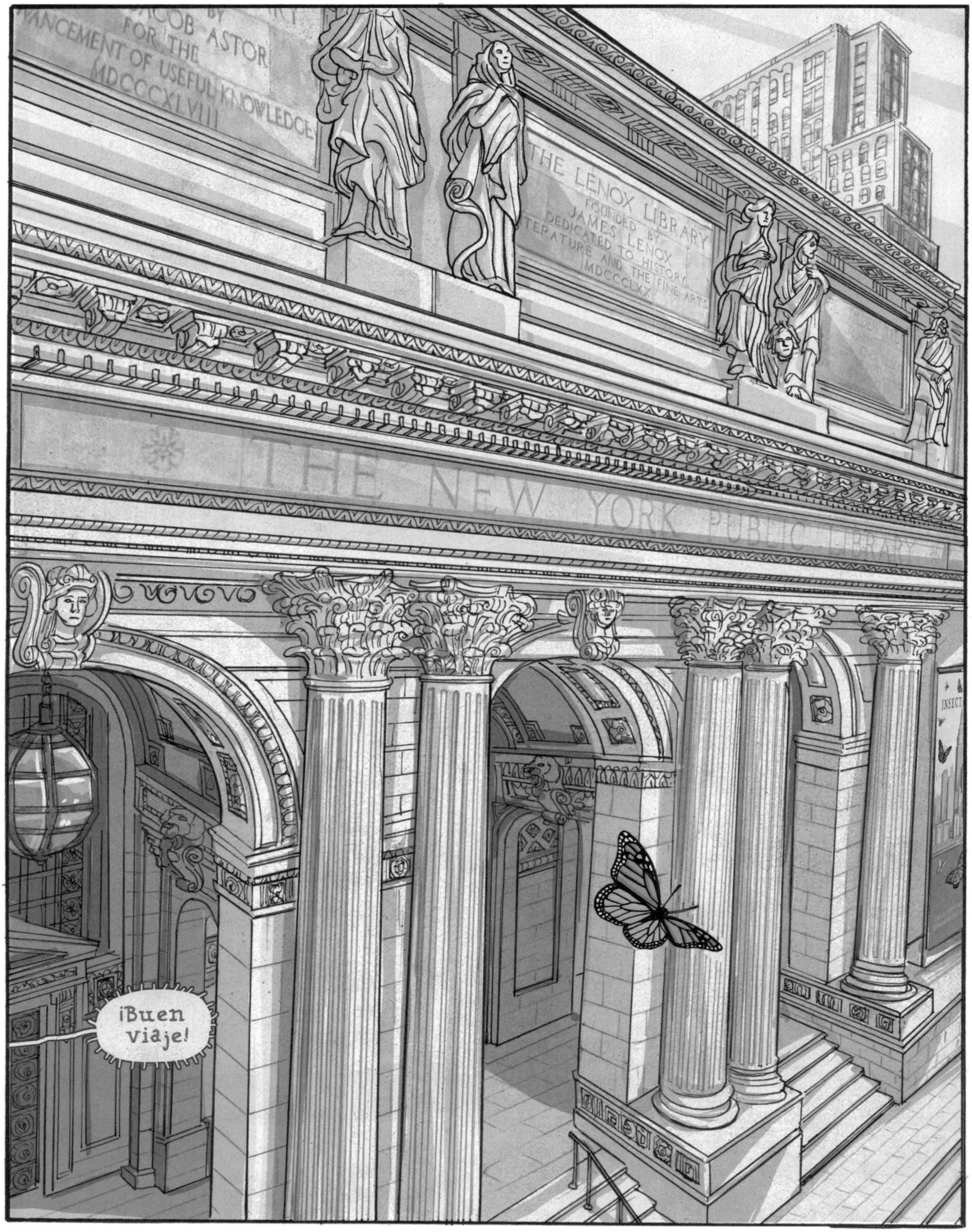
THE NEW YORK PUBLIC LIBRARY
THE LENOX LIBRARY
FOUNDED BY
JAMES LENOX
DEDICATED TO HISTORY
MDCCCLXX
¡Buen viaje!

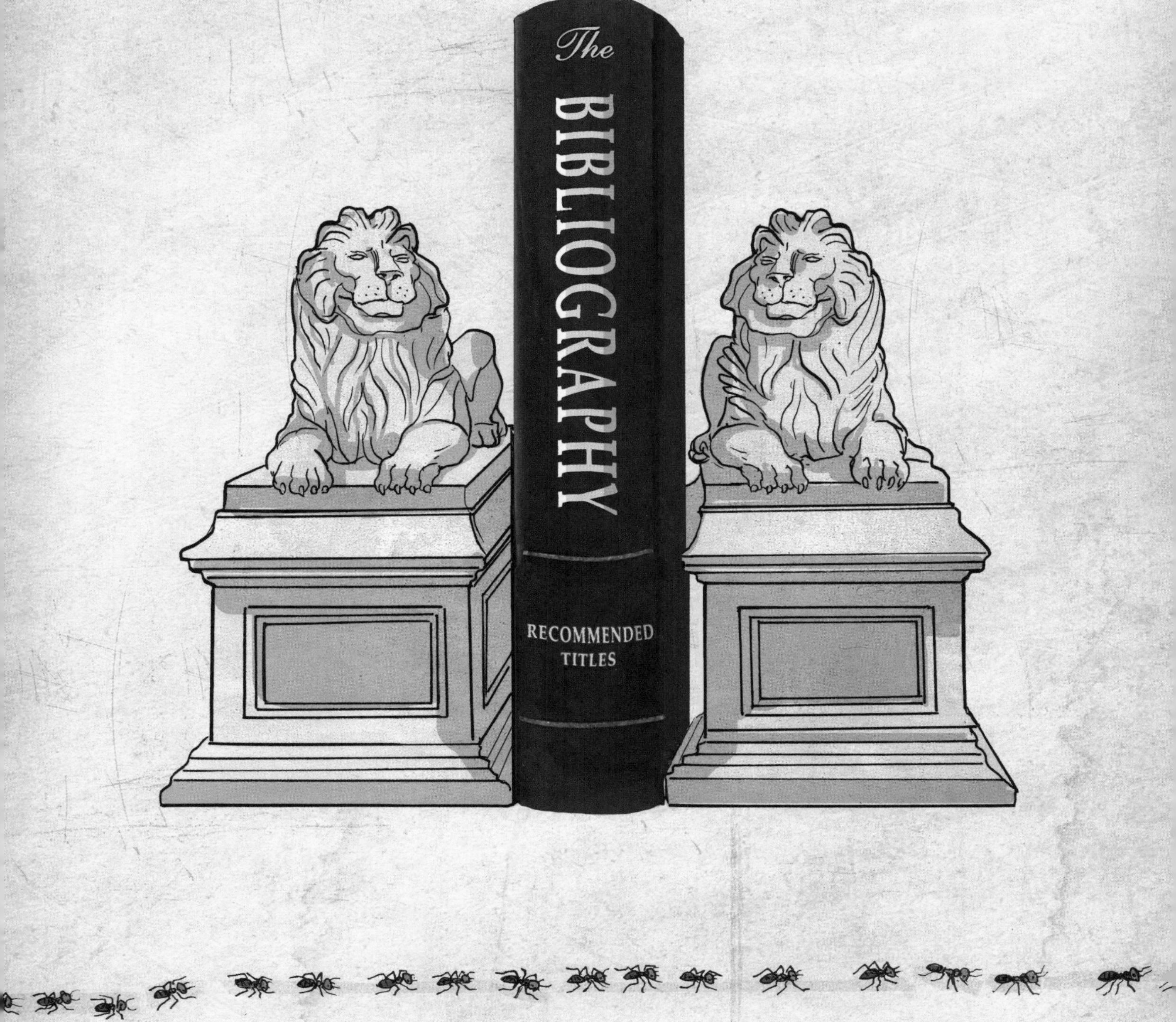
The
BIBLIOGRAPHY
RECOMMENDED
TITLES

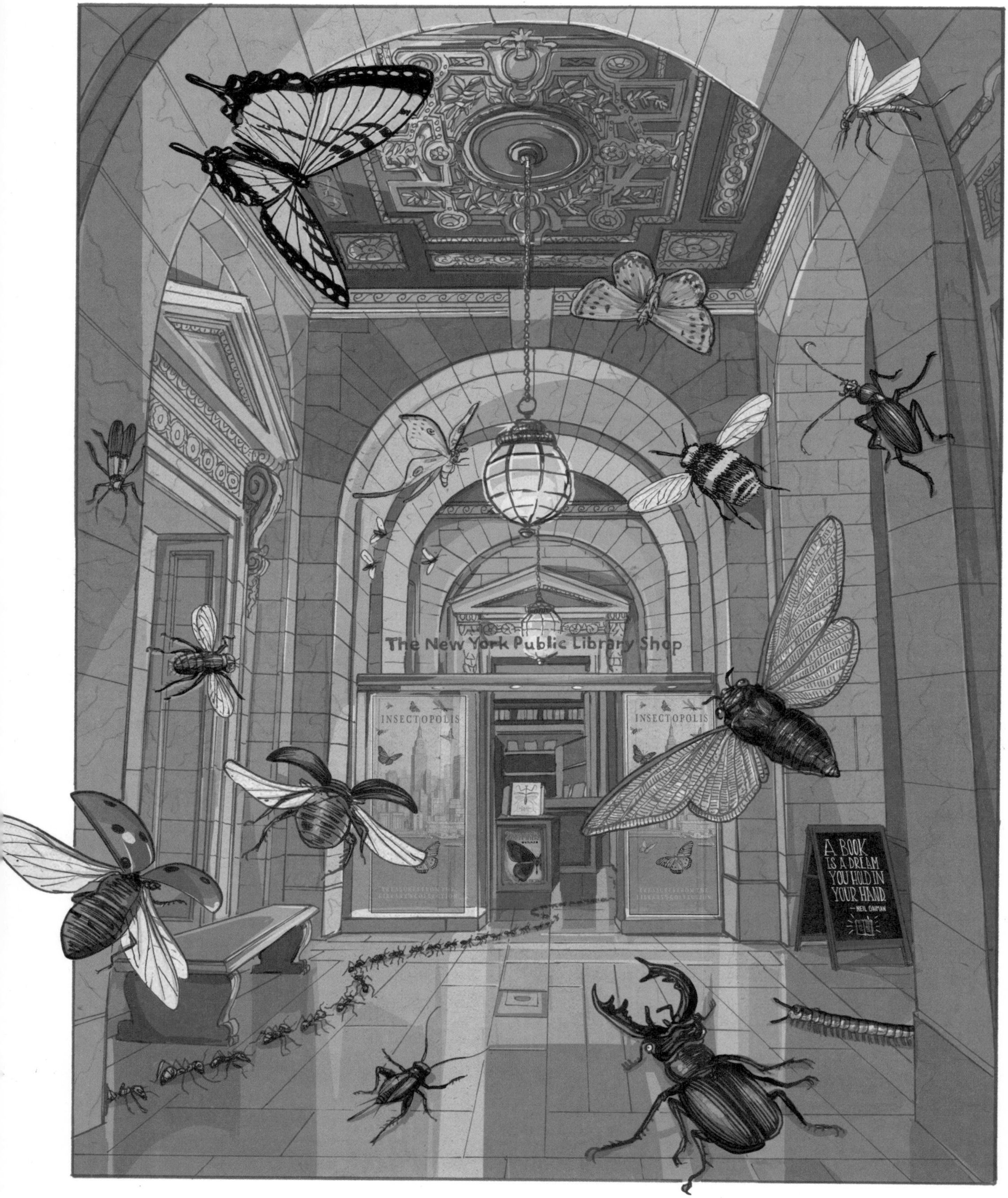
The New York Public Library Shop
INSECTOPOLIS
INSECTOPOLIS
A BOOK IS A DREAM YOU HOLD IN YOUR HAND.
—NEIL GAIMAN

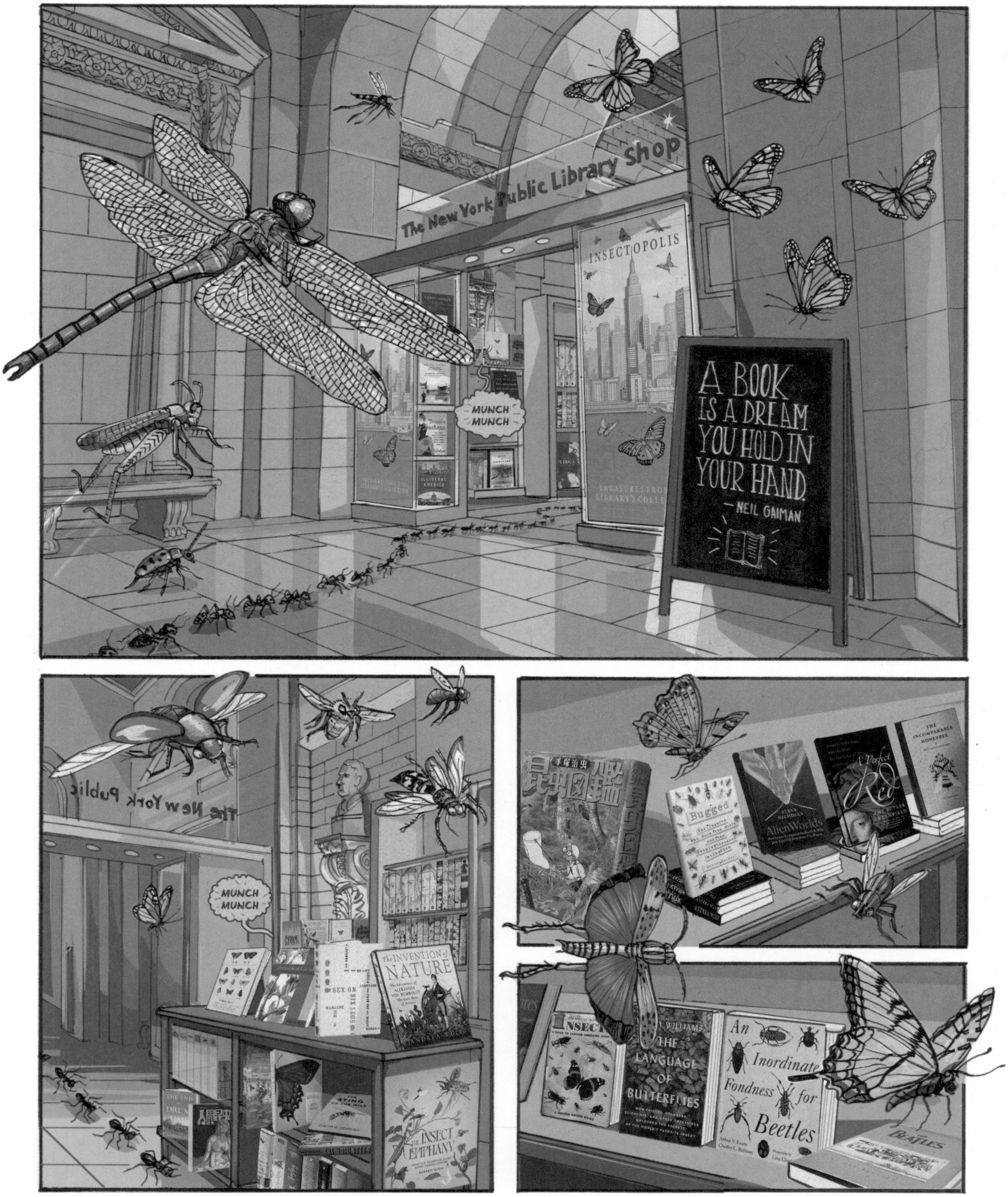

The New York Public Library Shop
INSECTOPOLIS
MUNCH MUNCH
A BOOK IS A DREAM YOU HOLD IN YOUR HAND. —NEIL GAIMAN
MUNCH MUNCH
The INVENTION of NATURE
SEX ON SIX LEGS
THE INSECT EPIPHANY
Bugged
AlienWorlds
A Perfect Red
THE INCOMPARABLE HONEYBEE
INSECTS
THE LANGUAGE OF BUTTERFLIES
An Inordinate Fondness for Beetles

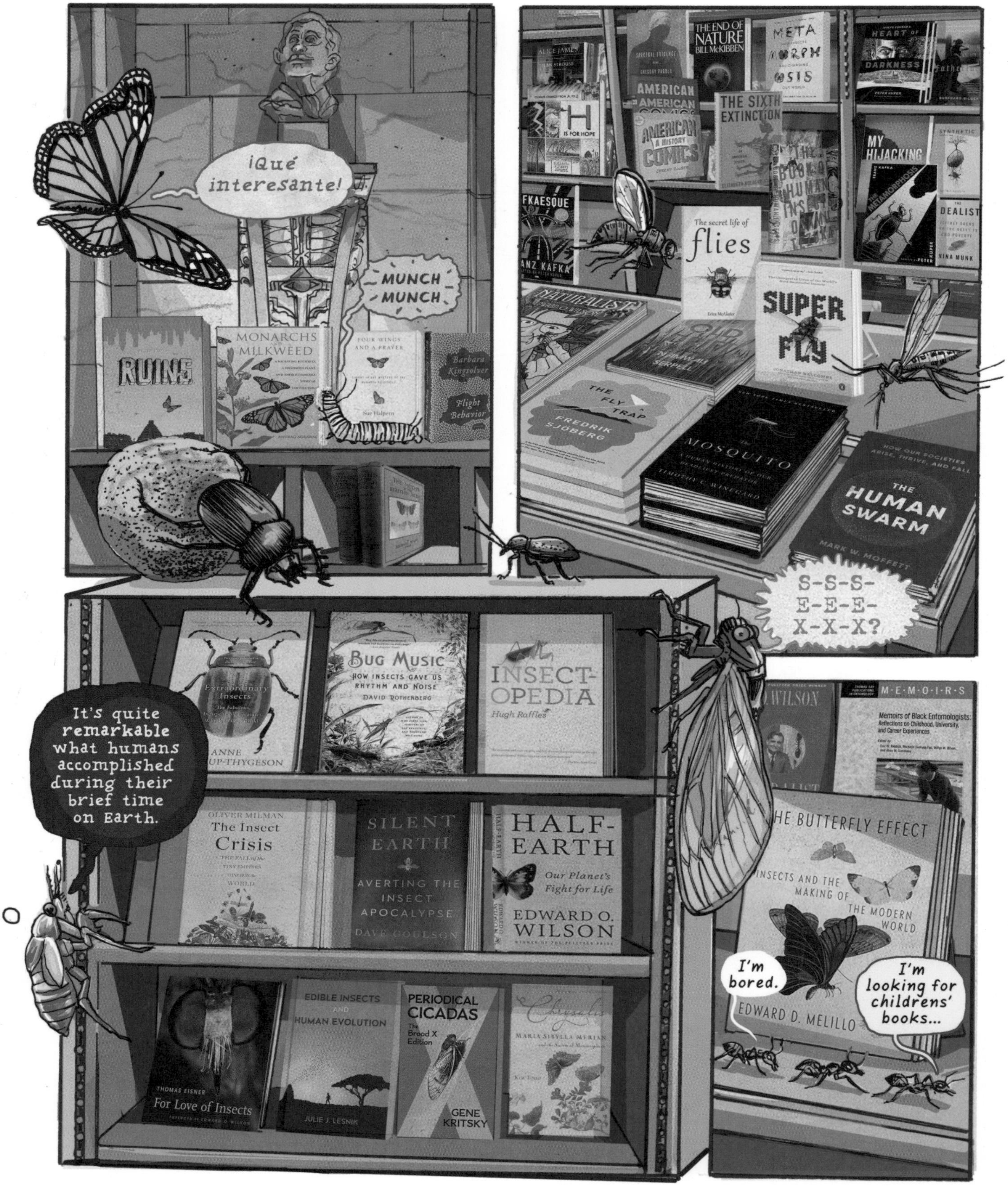
¡Qué interesante!
MUNCH MUNCH
S-S-S-E-E-E-X-X-X?
It's quite remarkable what humans accomplished during their brief time on Earth.
I'm bored.
I'm looking for childrens' books...

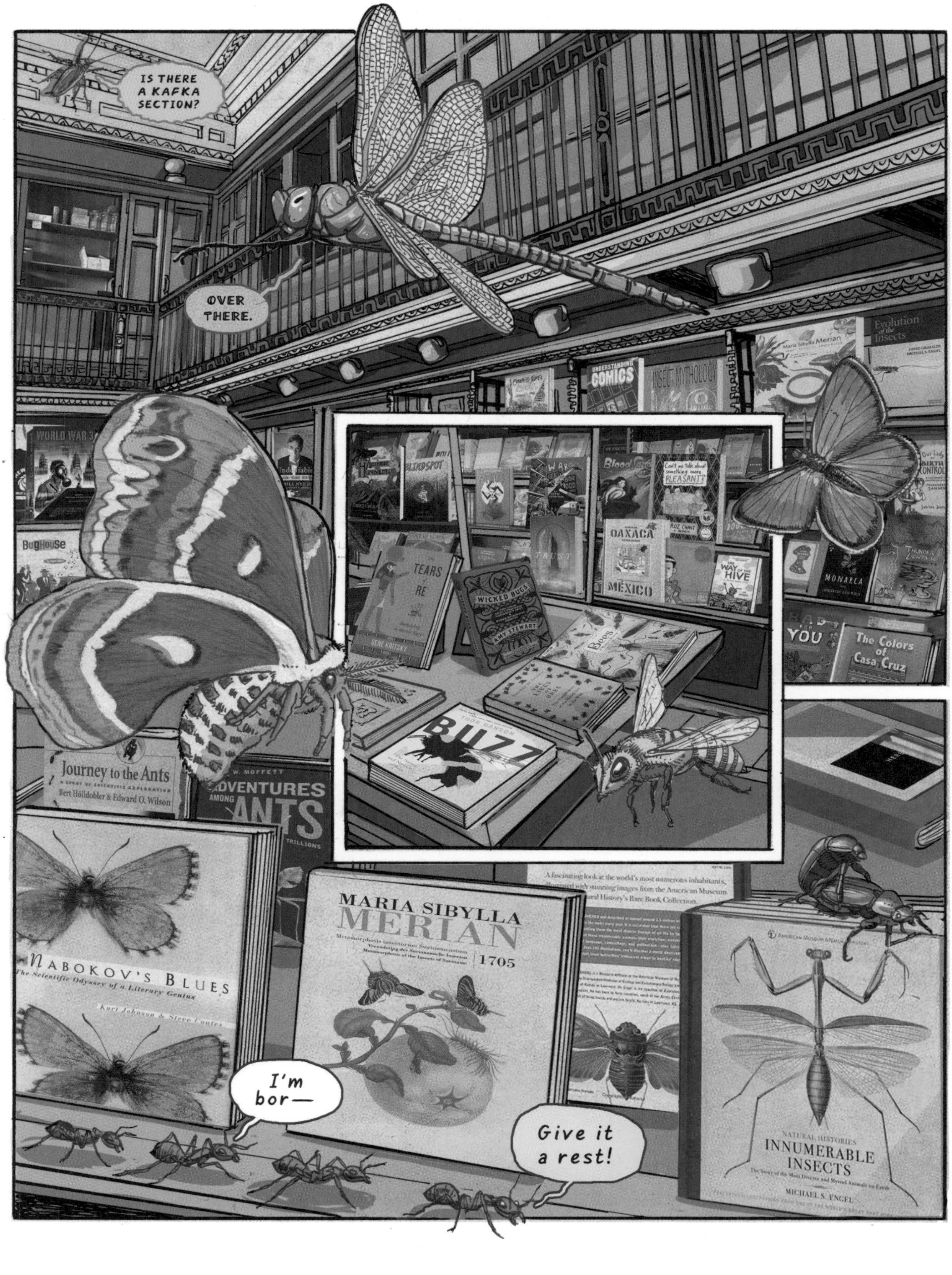
IS THERE A KAFKA SECTION?
OVER THERE.
I'm bor—
Give it a rest!

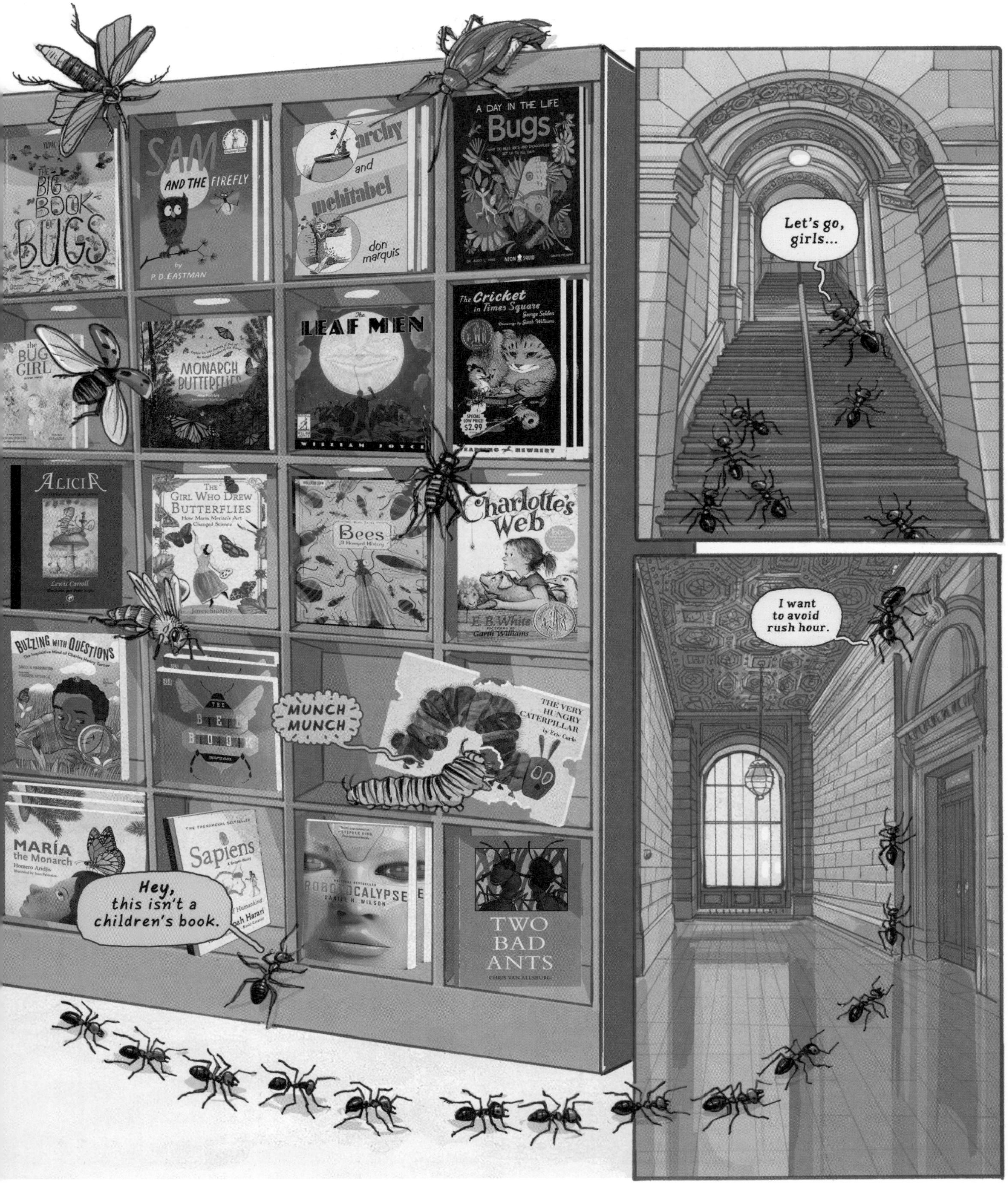
MUNCH MUNCH
Hey, this isn't a children's book.
Let's go, girls...
I want to avoid rush hour.

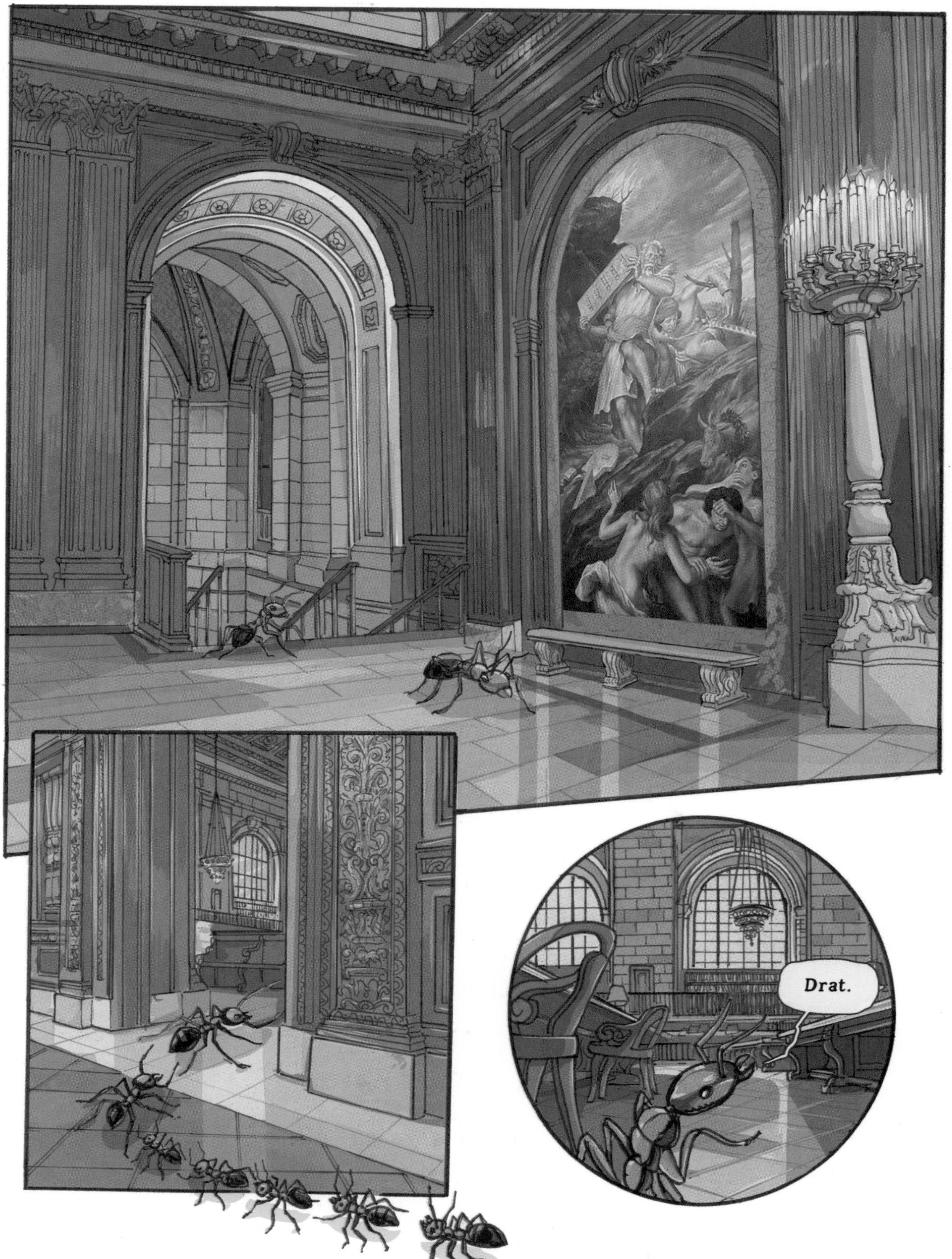
Drat.

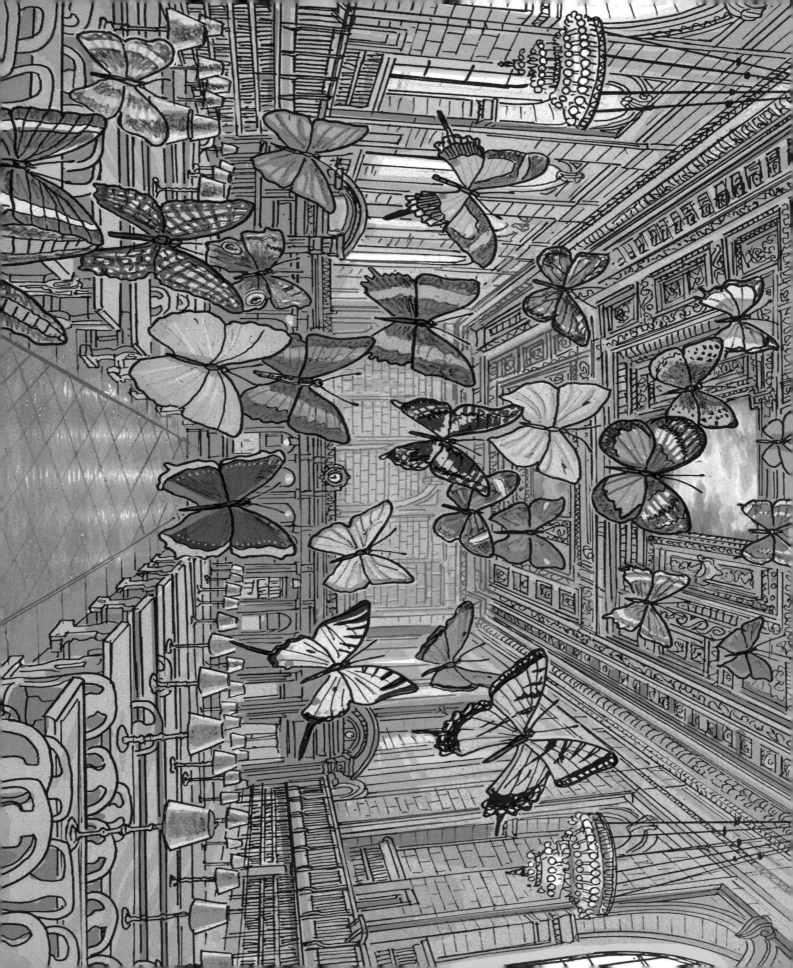

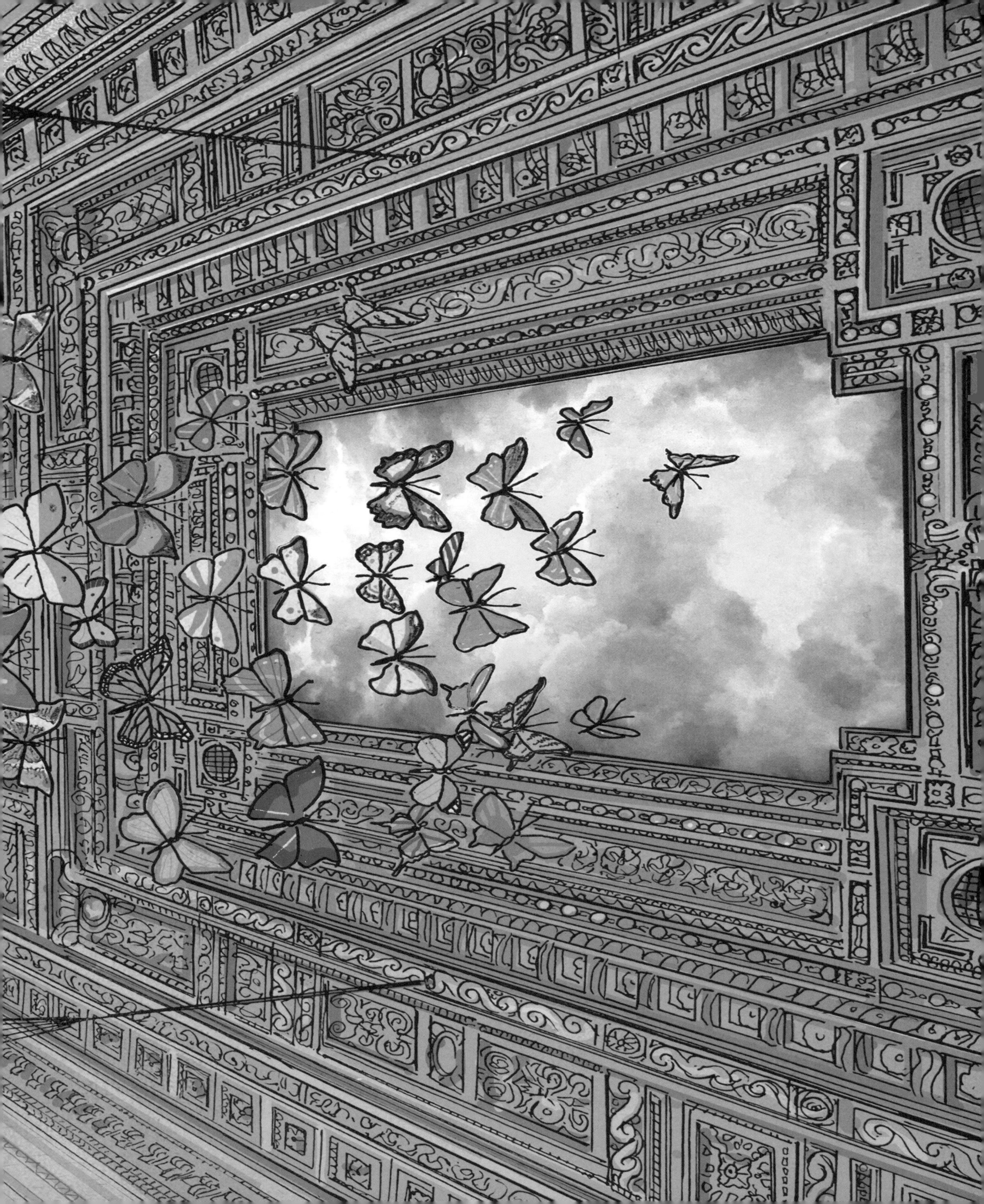

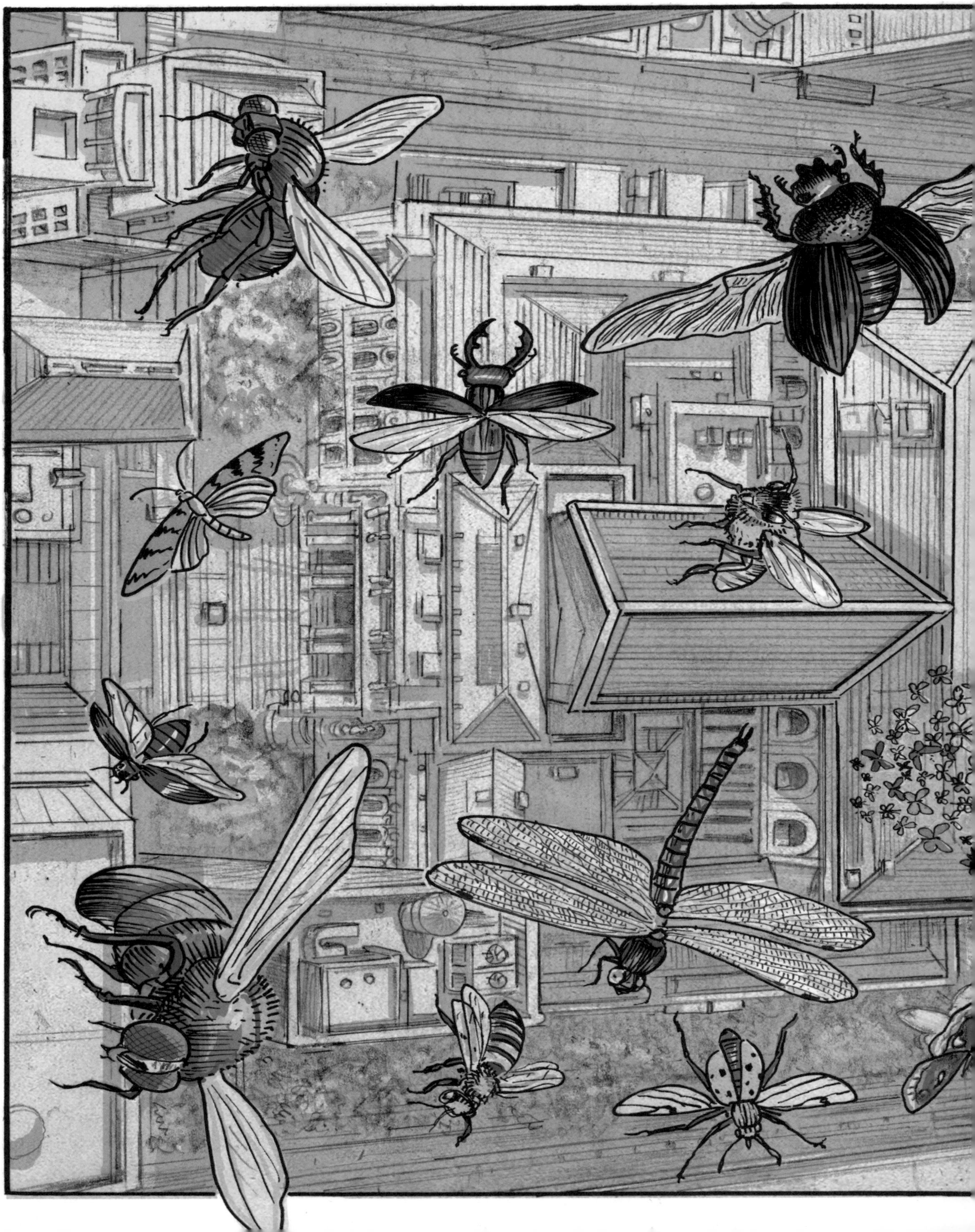

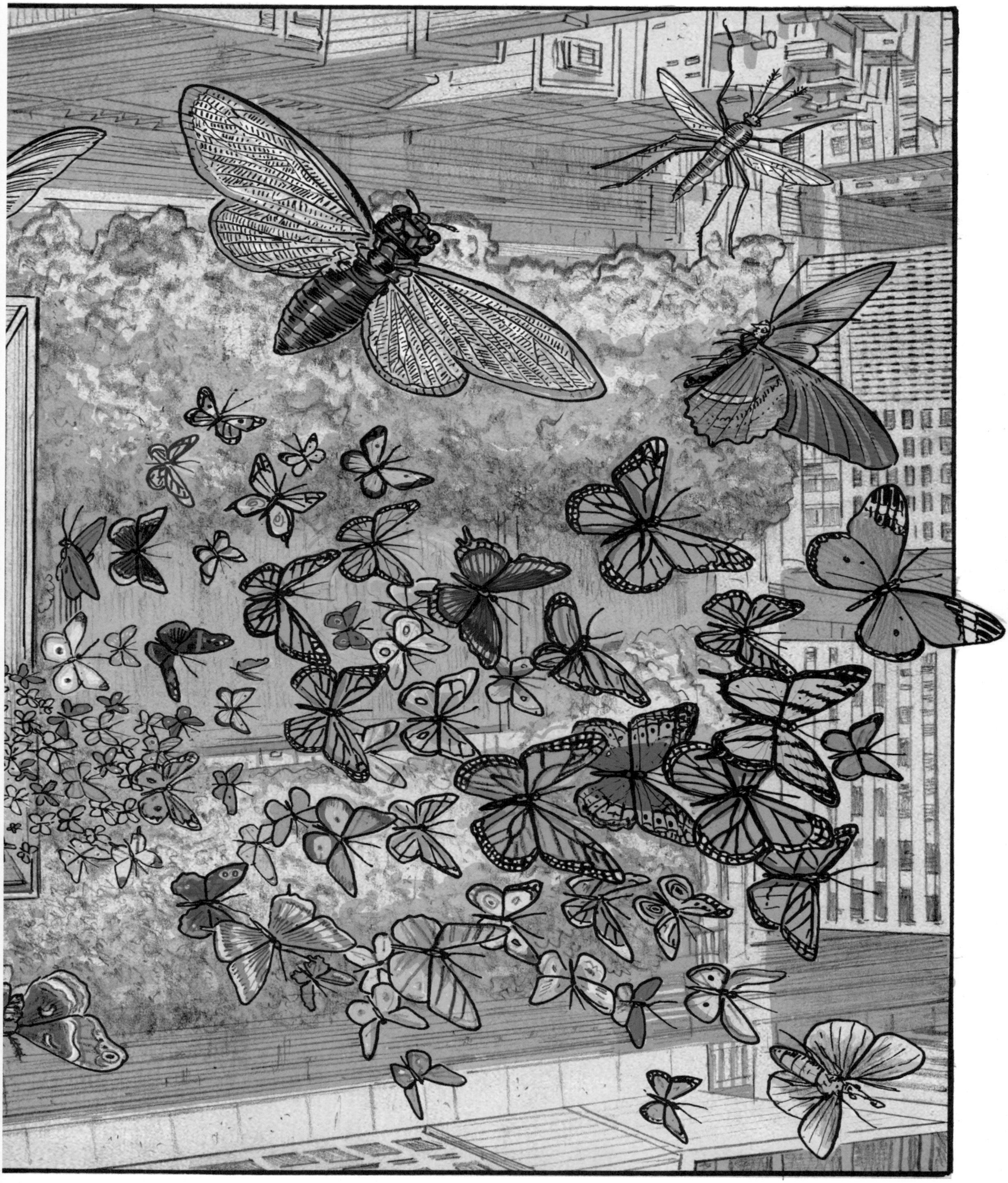

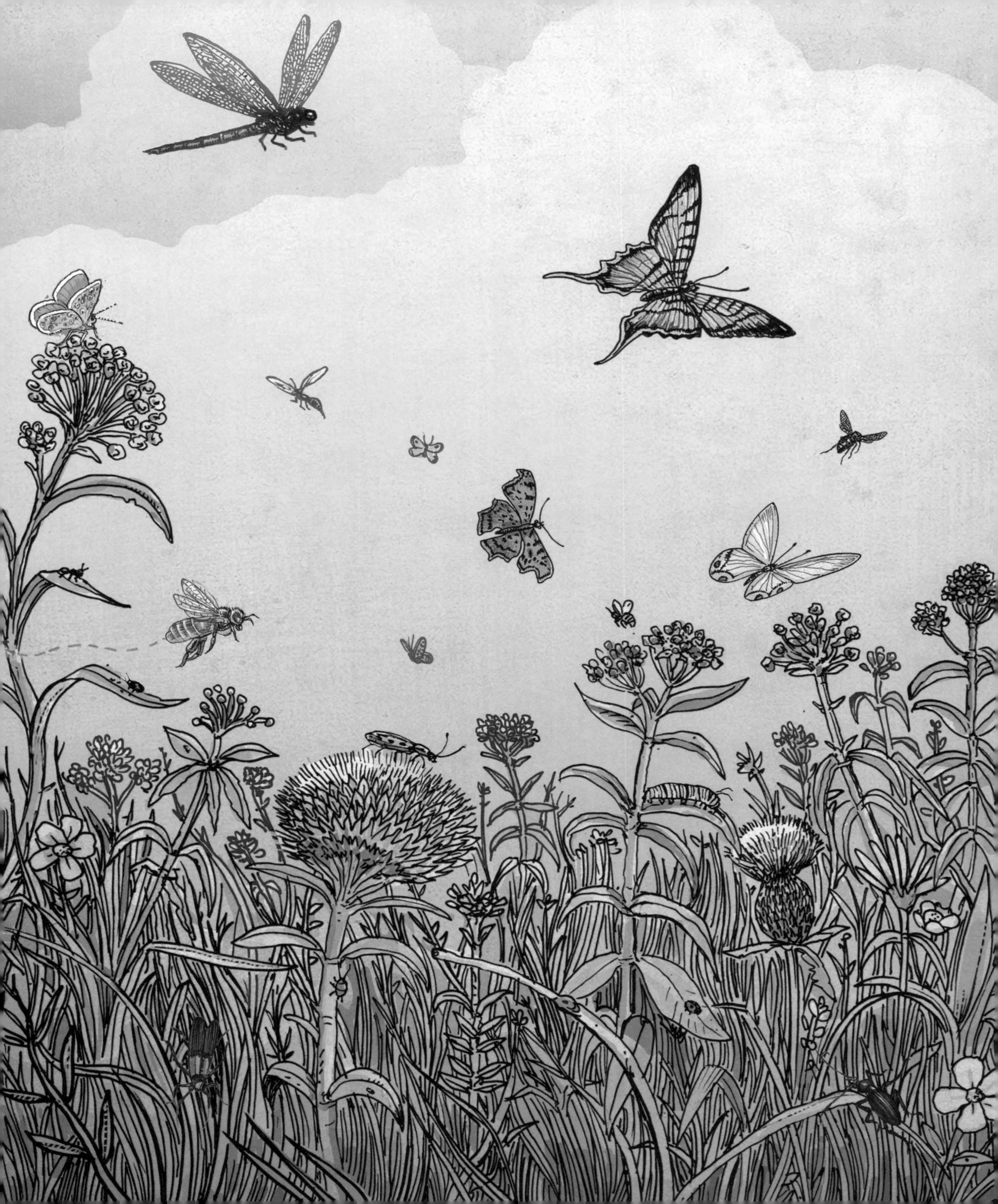

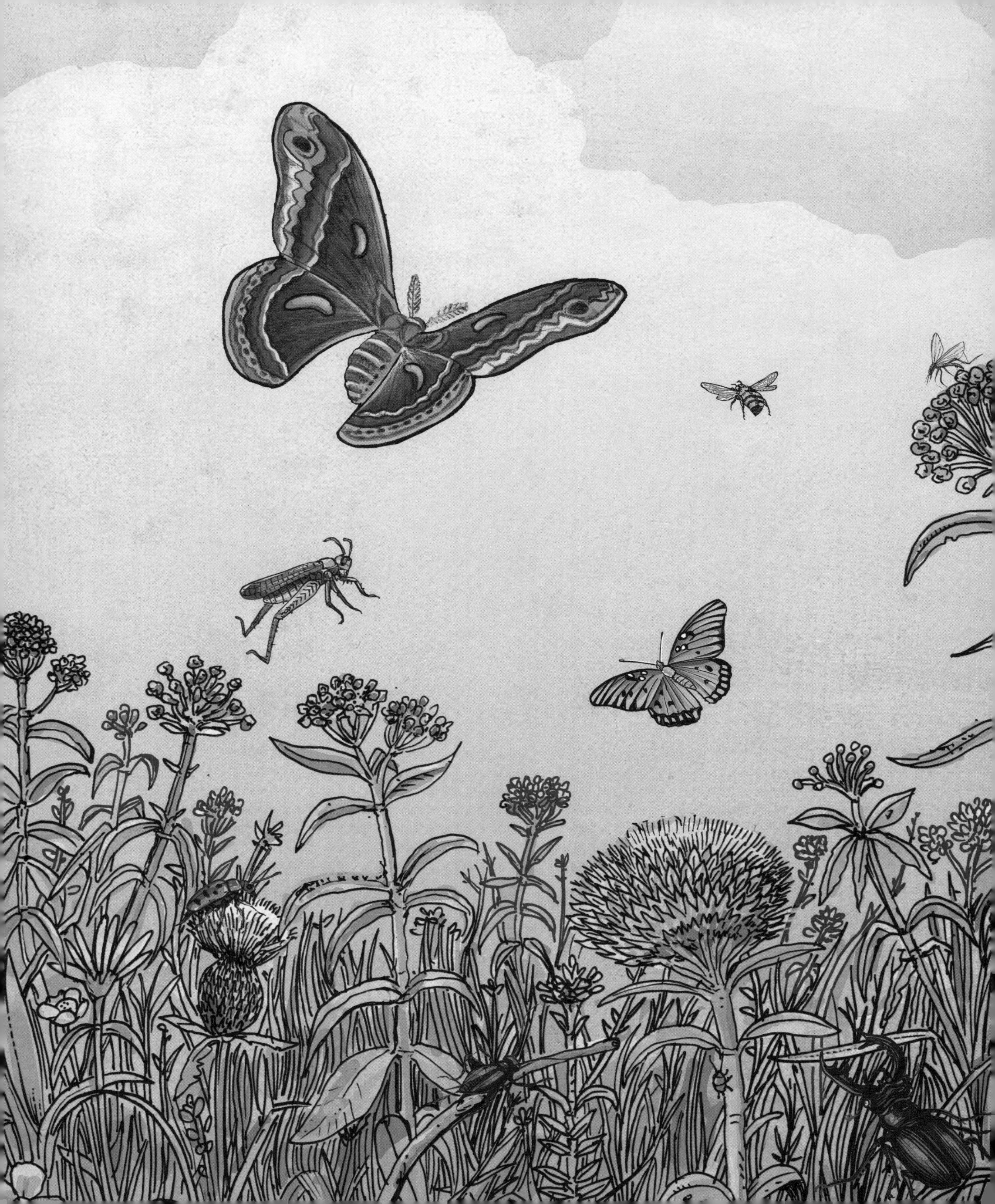

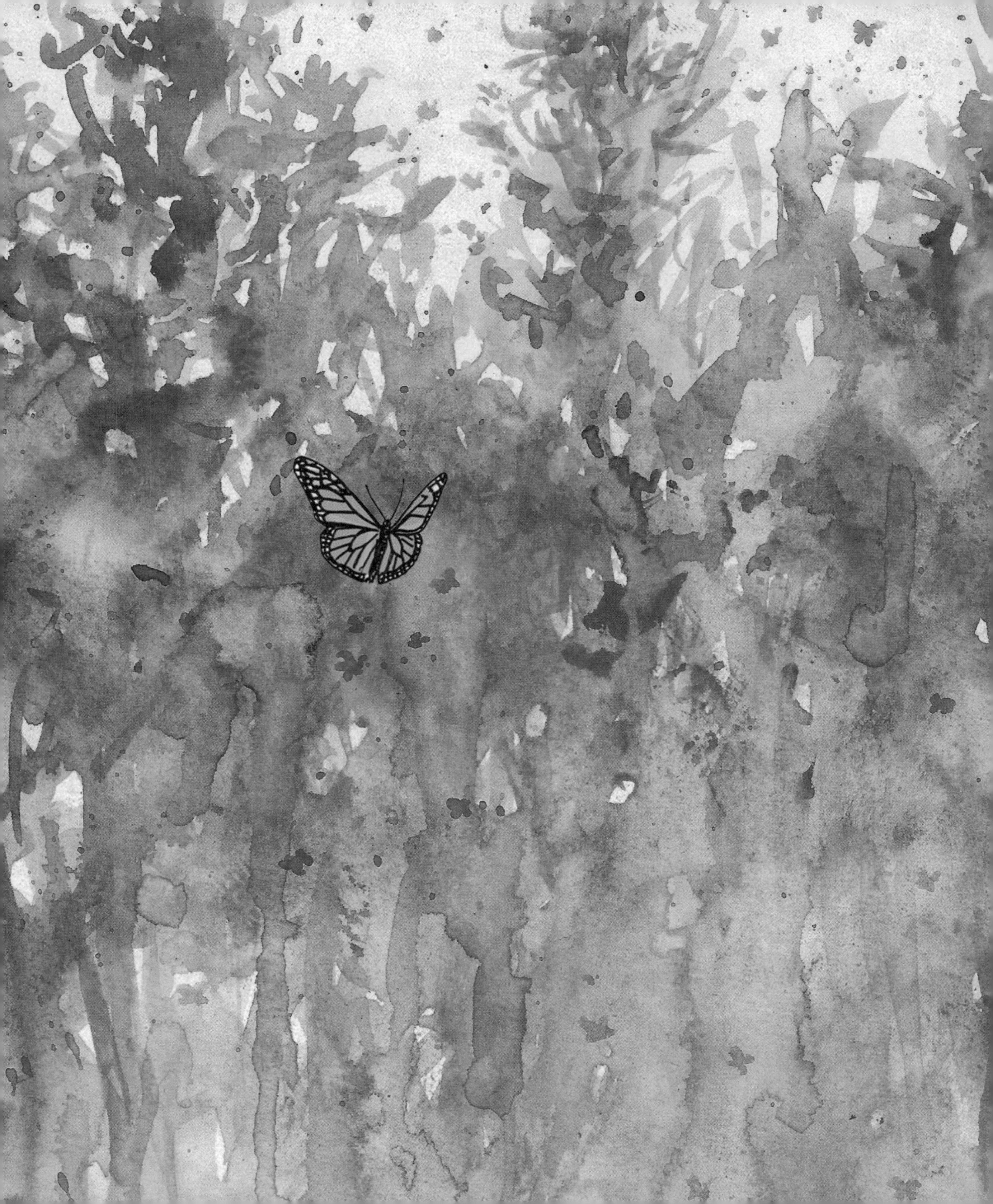

Time is rhythm: the insect rhythm
of a warm humid night, brain ripple,
breathing, the drum in my temple—
these are our faithful timekeepers...

—Vladimir Nabokov
(1899–1977)

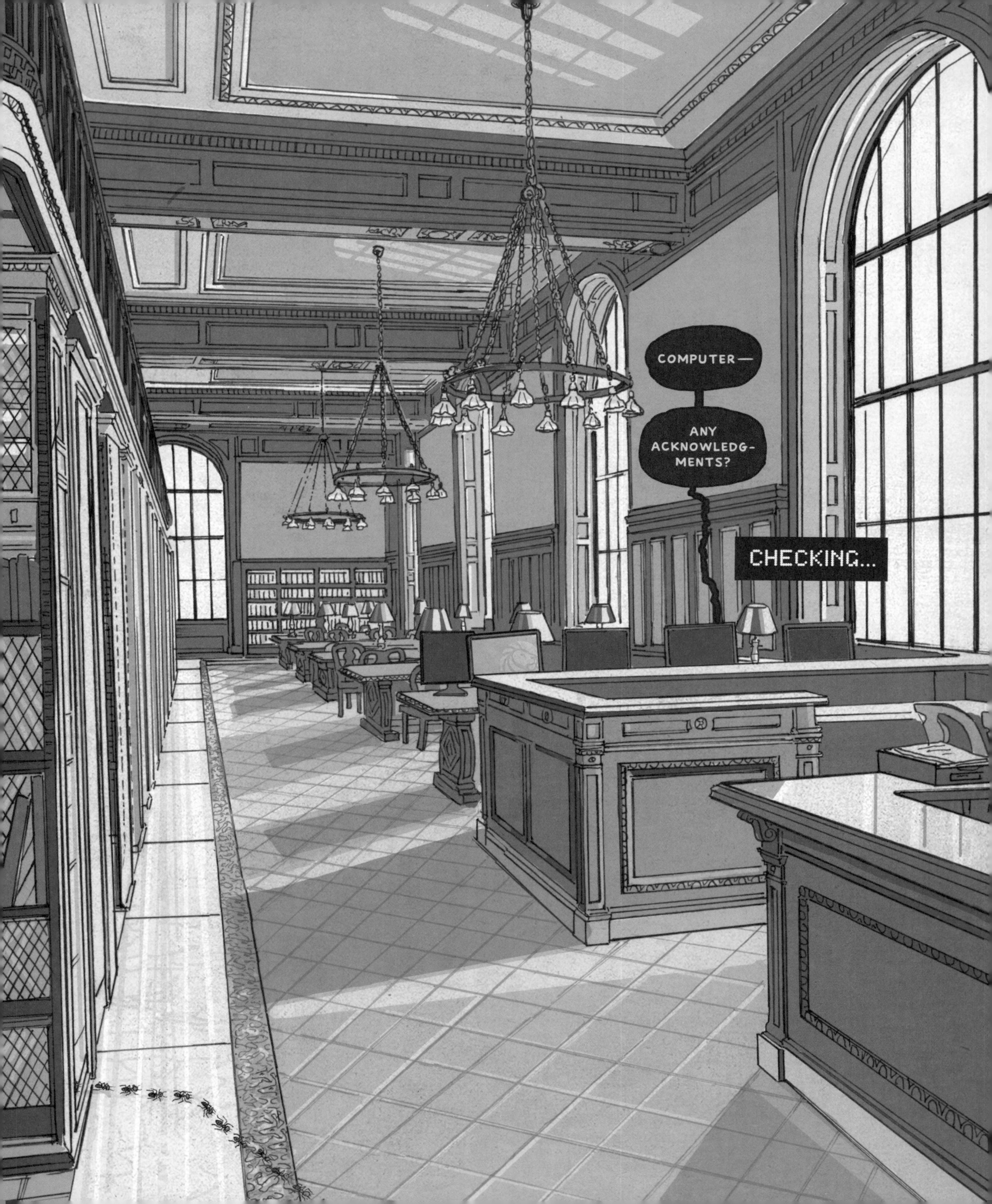
COMPUTER—
ANY ACKNOWLEDG-MENTS?
CHECKING...

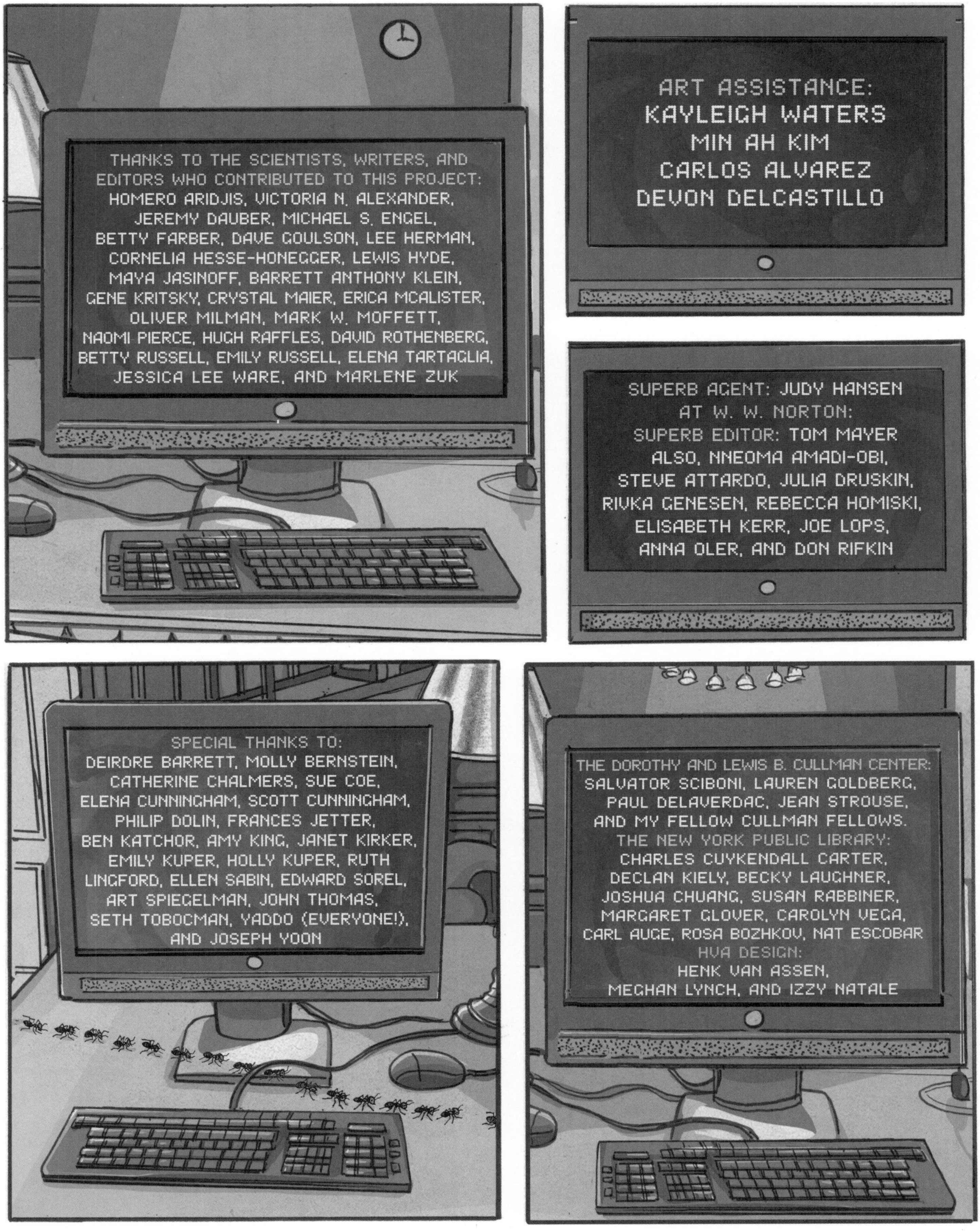
THANKS TO THE SCIENTISTS, WRITERS, AND EDITORS WHO CONTRIBUTED TO THIS PROJECT:
HOMERO ARIDJIS, VICTORIA N. ALEXANDER, JEREMY DAUBER, MICHAEL S. ENGEL, BETTY FARBER, DAVE GOULSON, LEE HERMAN, CORNELIA HESSE-HONEGGER, LEWIS HYDE, MAYA JASINOFF, BARRETT ANTHONY KLEIN, GENE KRITSKY, CRYSTAL MAIER, ERICA MCALISTER, OLIVER MILMAN, MARK W. MOFFETT, NAOMI PIERCE, HUGH RAFFLES, DAVID ROTHENBERG, BETTY RUSSELL, EMILY RUSSELL, ELENA TARTAGLIA, JESSICA LEE WARE, AND MARLENE ZUK
ART ASSISTANCE:
KAYLEIGH WATERS
MIN AH KIM
CARLOS ALVAREZ
DEVON DELCASTILLO
SUPERB AGENT: JUDY HANSEN
AT W. W. NORTON:
SUPERB EDITOR: TOM MAYER
ALSO, NNEOMA AMADI-OBI, STEVE ATTARDO, JULIA DRUSKIN, RIVKA GENESEN, REBECCA HOMISKI, ELISABETH KERR, JOE LOPS, ANNA OLER, AND DON RIFKIN
SPECIAL THANKS TO:
DEIRDRE BARRETT, MOLLY BERNSTEIN, CATHERINE CHALMERS, SUE COE, ELENA CUNNINGHAM, SCOTT CUNNINGHAM, PHILIP DOLIN, FRANCES JETTER, BEN KATCHOR, AMY KING, JANET KIRKER, EMILY KUPER, HOLLY KUPER, RUTH LINGFORD, ELLEN SABIN, EDWARD SOREL, ART SPIEGELMAN, JOHN THOMAS, SETH TOBOCMAN, YADDO (EVERYONE!), AND JOSEPH YOON
THE DOROTHY AND LEWIS B. CULLMAN CENTER:
SALVATOR SCIBONI, LAUREN GOLDBERG, PAUL DELAVERDAC, JEAN STROUSE, AND MY FELLOW CULLMAN FELLOWS.
THE NEW YORK PUBLIC LIBRARY:
CHARLES CUYKENDALL CARTER, DECLAN KIELY, BECKY LAUGHNER, JOSHUA CHUANG, SUSAN RABBINER, MARGARET GLOVER, CAROLYN VEGA, CARL AUGE, ROSA BOZHKOV, NAT ESCOBAR
HVA DESIGN:
HENK VAN ASSEN, MEGHAN LYNCH, AND IZZY NATALE

Peter Kuper is a regular contributor to
The New Yorker, *Nation*, and *Charlie Hebdo.*
He has authored over two dozen books including *Sticks and Stones,*
The System, Diario de Oaxaca, Drawn to New York, and *Ruins.*
He has also created adaptations of Upton Sinclair's *The Jungle*,
Joseph Conrad's *Heart of Darkness,* and many of
Franz Kafka's works, including *The Metamorphosis*
and fourteen short stories collected in *Kafkaesque.*
He has written and illustrated "Spy vs. Spy" for *Mad* magazine since 1997
and is the cofounder with Seth Tobocman of *World War 3 Illustrated,*
a graphics magazine that has been giving a forum
to political artists since 1979.
Kuper has received the 2024 RFK Human Rights Award for cartooning,
an Eisner Award in 2016, and was the 2020–21 Jean Strouse Cullman Fellow
at the New York Public Library.
He has lectured widely and exhibited his work around the world.
He teaches cartooning at Harvard University.

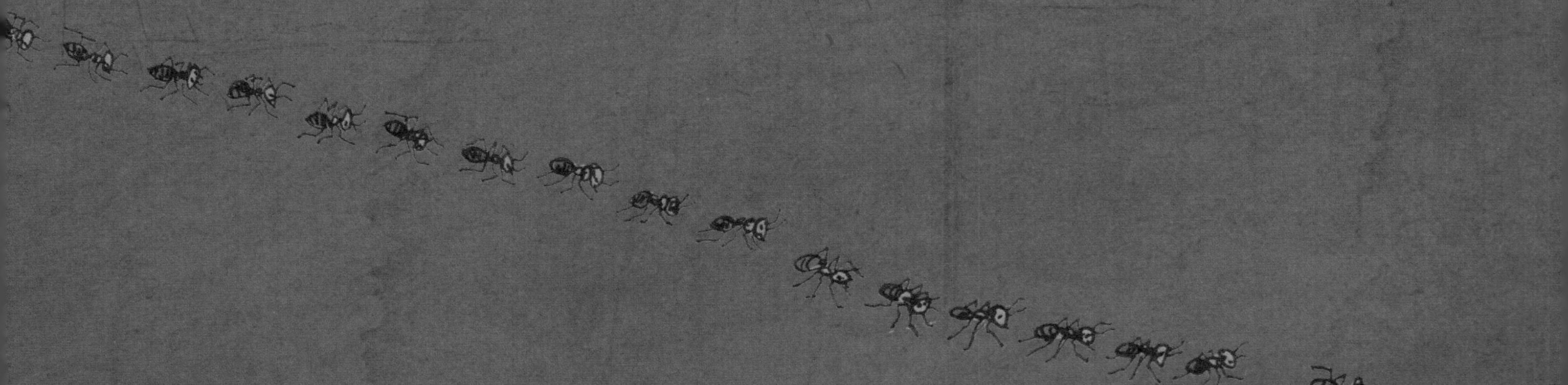

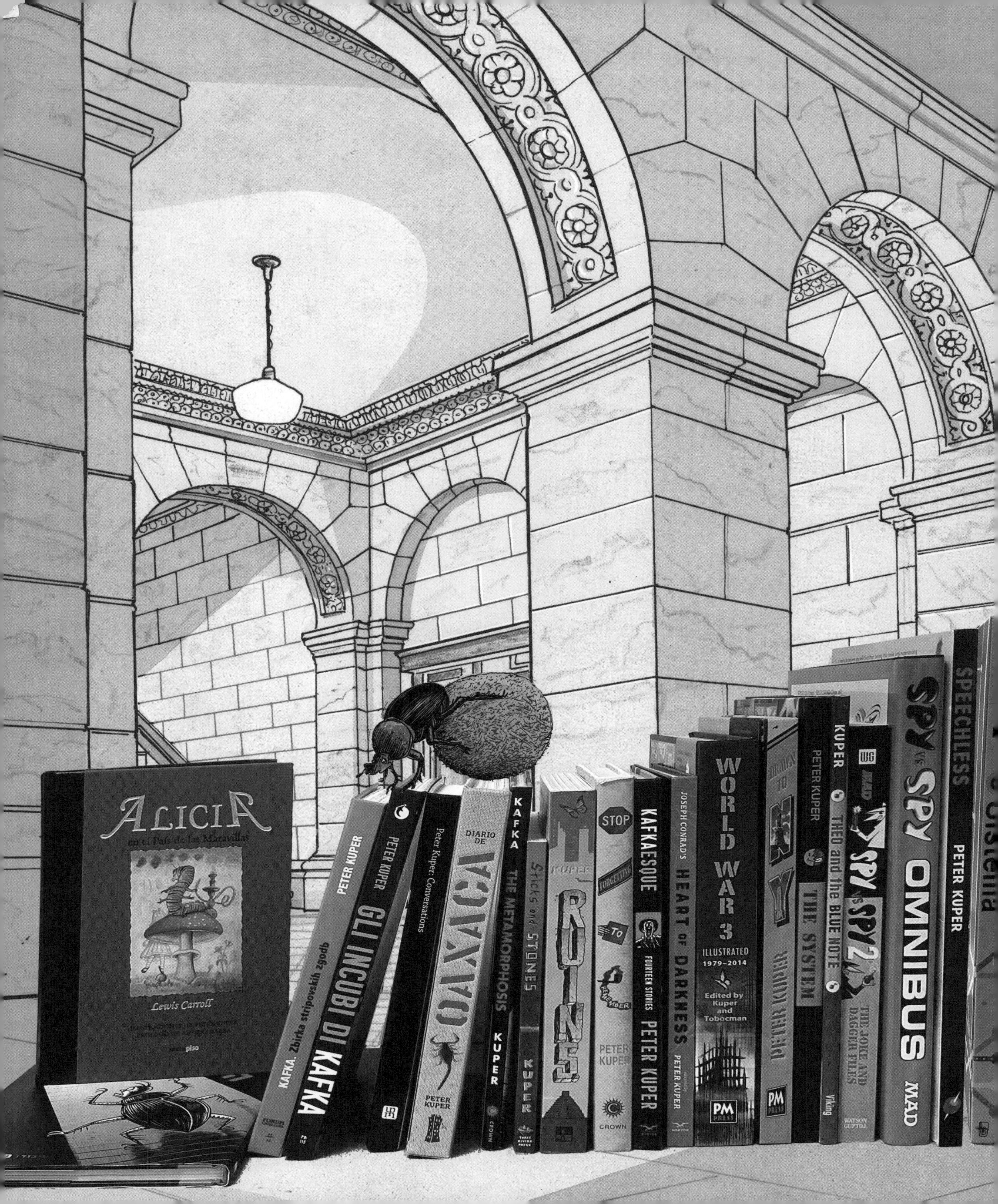
ALICIA
en el País de las Maravillas
Lewis Carroll
PETER KUPER
KAFKA. Zbirka stripovskih zgodb
PETER KUPER
GLI INCUBI DI KAFKA
Peter Kuper: Conversations
DIARIO DE OAXACA
PETER KUPER
KAFKA
THE METAMORPHOSIS
KUPER
CROWN
Sticks and Stones
KUPER
KUPER
RUINS
STOP
FORGETTING TO REMEMBER
PETER KUPER
CROWN
KAFKAESQUE
FOURTEEN STORIES
PETER KUPER
JOSEPH CONRAD'S
HEART OF DARKNESS
PETER KUPER
NORTON
WORLD WAR 3
ILLUSTRATED
1979–2014
Edited by Kuper and Tobocman
PM PRESS
PM PRESS
PETER KUPER
THE SYSTEM
KUPER
THEO and the BLUE NOTE
Viking
SPY vs SPY 2
THE JOKE AND DAGGER FILES
WATSON GUPTILL
SPY vs SPY
OMNIBUS
MAD
SPEECHLESS
PETER KUPER

Copyright © 2025 by Peter Kuper

This book's QR code-linked audio features (produced, co-directed and co-written by Charles Cuykendall Carter; co-written and co-directed by Peter Kuper) originally appeared in Peter Kuper's 2022 installation at The New York Public Library, INterSECTS. They appear here courtesy of NYPL.
Lions image and logo © The New York Public Library. Printed with permission.
Cornelia Hesse-Honegger: Insects affected by low-level radiation left to right: Scorpion Fly from near Swissnuclear power plant, Leibstadt, Switzerland 1988. Ladybird beetle from West of Richland, Hanford Area, USA 1989. Soft Bug, Miridae, from near Swiss nuclear power plant Gösgen, Switzerland 1988. © 2024 Artists rights Society (ARS), New York / ProLitteris, Zurich.
QR code soundtracks by David Rothenberg from his album Bug Music © 2013 Published by Mysterious Mountain Music (BMI). All rights reserved. Used by permission.

There are six additional QR codes to be found in this book.

All rights reserved
Printed in Malaysia
First Edition

For information about permission to reproduce selections from this book, write to Permissions, W. W. Norton & Company, Inc., 500 Fifth Avenue, New York, NY 10110

For information about special discounts for bulk purchases, please contact W. W. Norton Special Sales at specialsales@wwnorton.com or 800-233-4830

Manufacturing by Imago Group
Book design by Peter Kuper
Production manager: Anna Oler

ISBN 978-1-324-03571-8

W. W. Norton & Company, Inc., 500 Fifth Avenue, New York, NY 10110
www.wwnorton.com

W. W. Norton & Company Ltd., 15 Carlisle Street, London W1D 3BS

1 2 3 4 5 6 7 8 9 0

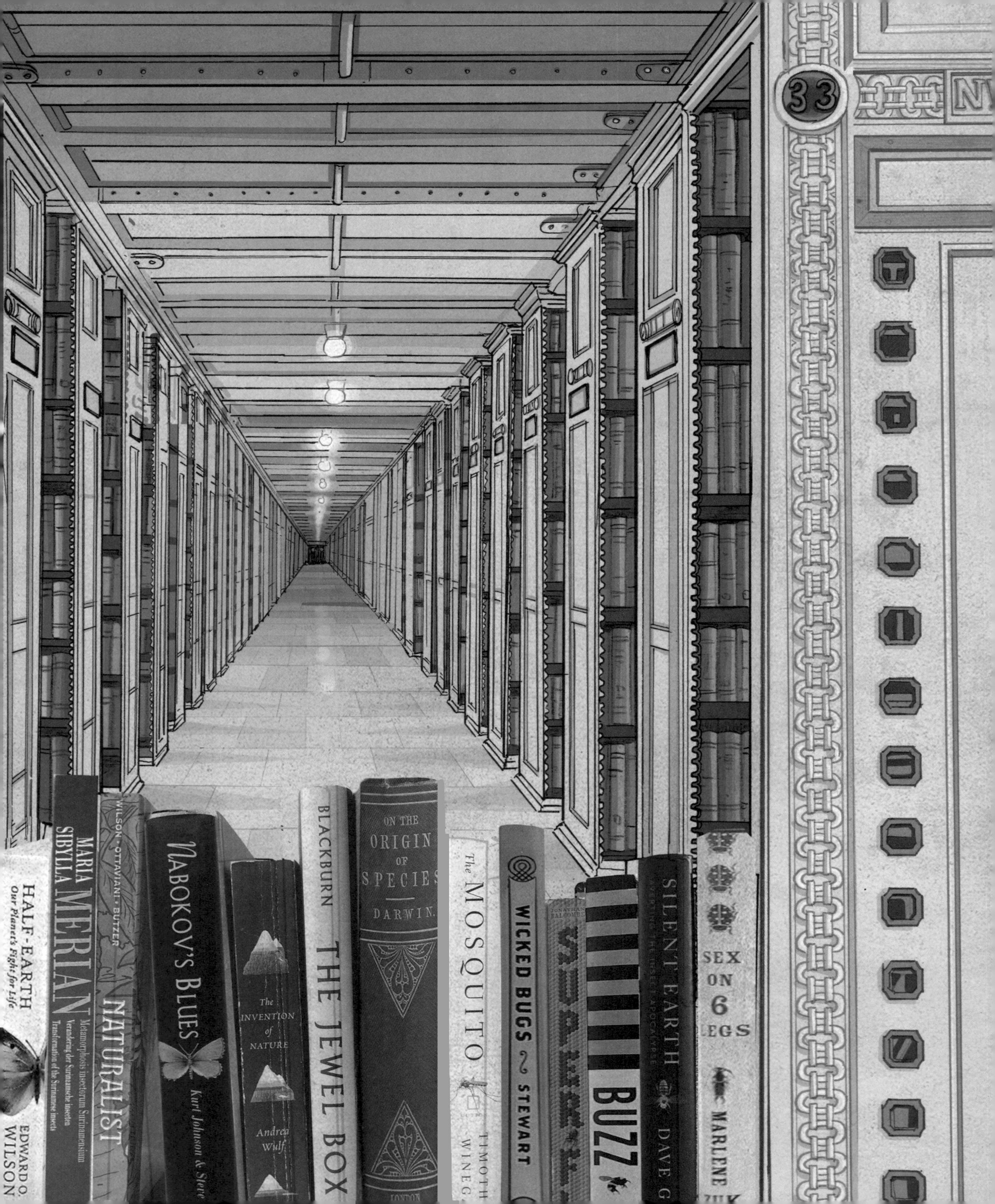

33
NW
HALF-EARTH
Our Planet's Fight for Life
EDWARD O. WILSON
MARIA SIBYLLA MERIAN
Metamorphosis insectorum Surinamensium
Verandering der Surinaamsche Insecten
Transformation of the Surinamese insects
WILSON · OTTAVIANI · BUTZER
NATURALIST
NABOKOV'S BLUES
Kurt Johnson & Stev
The INVENTION of NATURE
Andrea Wulf
BLACKBURN
THE JEWEL BOX
ON THE ORIGIN OF SPECIES
DARWIN.
LONDON
The MOSQUITO
TIMOTH
WINEG
WICKED BUGS
STEWART
BUZZ
SILENT EARTH
DAVE G
SEX ON 6 LEGS
MARLENE